Integrated Satellite-Terrestrial Network Fundamentals for Mobile Communications

For a listing of recent titles in the *Artech Mobile Communications Library*,
turn to the back of this book.

Integrated Satellite-Terrestrial Network Fundamentals for Mobile Communications

Lixia Xiao, Pei Xiao, and Tao Jiang

ARTECH
HOUSE

BOSTON | LONDON
artechhouse.com

Library of Congress Cataloging-in-Publication Data
A catalog record of this book is available from the U.S. Library of Congress.

British Library Cataloguing in Publication Data
A catalog record for this book is available from the British Library.

ISBN 978-1-68569-009-0

Cover design by Joi Garron

685 Canton Street
Norwood, MA 02062

10 9 8 7 6 5 4 3 2 1

Contents

Preface *xi*

CHAPTER 1

Concept of Satellite-Terrestrial Integrated Communication 1
1.1 Terrestrial Mobile Communication 1
1.2 Satellite Mobile Communication 4
1.3 Typical Satellite Communication Systems 5
1.3.1 High-Orbit Narrowband System 7
1.3.2 High-Orbit Broadband System 8
1.3.3 Low-Orbit Narrowband Systems 8
1.3.4 Low-Orbit Broadband Systems 9
1.4 Satellite-Terrestrial Integrated Communication 10
1.4.1 Competition with Terrestrial Communication 11
1.4.2 Complement to Terrestrial Communication 11
1.4.3 Convergence with Terrestrial Communication 12
1.4.4 The Vision of Integrated Communication 12
References 14

CHAPTER 2

The Evolution for Satellite-Terrestrial Integrated Communication 15
2.1 Demand for Integrated Communication 15
2.2 Typical Application Scenarios 17
2.3 Integration Models 19
2.3.1 Service Models 20
2.3.2 Networking Models 21
2.3.3 Terminal Development Models 23
2.4 Evolution of International Standards 23
2.4.1 Release-15 for NR NTN 25
2.4.2 Release-16 for New Radio NTN 25
2.4.3 Release-17 for NR NTN 26
2.4.4 Release-18 for NR NTN 27
2.4.5 Other Initiatives 28
2.5 Possible Challenges 28
References 30

CHAPTER 3

Constellation Design for Satellite-Terrestrial Integrated Communication 33
3.1 Overview of Satellite Constellations 33
3.1.1 Definition of Satellite Constellation 33
3.1.2 Development of Satellite Constellations 34
3.2 Classification of Satellite Constellations 36
3.2.1 Walker Constellation 36
3.2.2 Star Constellation 37
3.2.3 Flower Constellation 38
3.2.4 Classical Satellite Constellation Design Solution 40
3.3 Satellite Constellation Design 43
3.3.1 Configuration Design 43
3.3.2 Coverage Design 46
3.3.3 Design Factor Analysis 47
References 48

CHAPTER 4

Intersatellite Free-Space Optical Communication 51
4.1 Fundamentals 51
4.2 Key Techniques 54
4.2.1 Link Construction 54
4.2.2 Signal Modulation Technique 56
4.2.3 Laser Antenna Technology 59
4.2.4 Microwave Antenna Technology 61
4.3 Current Status and Possible Challenges 62
4.3.1 Current Status 62
4.3.2 Possible Challenges 65
References 67

CHAPTER 5

Channel Models for Satellite-Terrestrial Integrated Communication 69
5.1 Wireless Channel Fundamentals 69
5.2 Satellite-Terrestrial Channel Characteristics 70
5.2.1 Free-Space Loss 71
5.2.2 Ionospheric Scintillation 71
5.2.3 Shadow Fading and Clutter Loss 72
5.2.4 Rain Fading 72
5.2.5 Multipath Fading 72
5.2.6 Doppler Effect 73
5.2.7 Atmospheric Absorption 74
5.2.8 Building Penetration Loss 74
5.3 Classical Satellite-Terrestrial Channel Models 75
5.3.1 The C.Loo Model 76
5.3.2 Corazza Model 77

5.3.3 Lutz Model 78
5.3.4 TDL Satellite Mobile Channel Models 79
5.4 Evolution of Satellite-Terrestrial Channel Standards 80
References 81

CHAPTER 6

Channel Coding for Satellite-Terrestrial Integrated Communication 83
6.1 Classical Channel Coding 83
6.1.1 Linear Block Code 84
6.1.2 Convolutional Code 89
6.2 Channel Coding for Terrestrial Communication 91
6.2.1 Cellular Mobile Communication 91
6.2.2 Wireless Local Area Network Communication 94
6.3 Channel Coding for Satellite Communication 95
6.3.1 Deep-Space Communication 95
6.3.2 Near-Space Communication 98
6.4 Integrated Satellite-Terrestrial Channel Coding 100
6.4.1 The Current Research Status 101
6.4.2 Possible Challenges 103
6.4.3 Possible Channel Coding Schemes 104
References 108

CHAPTER 7

Signal Modulation for Satellite-Terrestrial Integrated Communication 111
7.1 Classic Modulation Waveforms 111
7.1.1 Baseband Modulation 111
7.1.2 OFDM-Based MC Modulation 115
7.1.3 DFT-s-OFDM-Based SC Modulation 117
7.2 Modulation Standard for Cellular Mobile Communication 119
7.2.1 Modulation for 1G Communication 119
7.2.2 Modulation for 2G Communication 119
7.2.3 Modulation for 3G Communication 120
7.2.4 Modulation for 4G Communication 121
7.2.5 Modulation for 5G Communication 121
7.3 Modulation Standard for Satellite Communication 122
7.3.1 Modulation for DVB-S Communication 123
7.3.2 Modulation for DVB-S2 Communication 123
7.3.3 Modulation for DVB-S2X Communication 124
7.3.4 Modulation for DVB-SH Communication 125
7.3.5 Modulation for ATSC Communication 126
7.3.6 Modulation for ISDB-S Communication 126
7.4 Potential Modulation for Integrated Communication 127
7.4.1 Irregular Baseband Modulation 128
7.4.2 FBMC Multicarrier Modulation 128
7.4.3 UFMC Multicarrier Modulation 133
7.4.4 GFDM Multicarrier Modulation 136

7.4.5 OTFS Multicarrier Modulation 139
7.4.6 OCDM Multicarrier Modulation 141
7.4.7 AFDM Multicarrier Modulation 143
7.4.8 Performance Analysis 146
7.5 Design Guidelines 148
7.5.1 Irregular Constellation Configuration Design 148
7.5.2 Integrated Coding and Modulation 148
7.5.3 Versatile Carrier Waveform Design 148
7.5.4 AI-Aided Adaptive Waveform 149
References 150

CHAPTER 8

Multiantenna Technique for Satellite-Terrestrial Integrated Communication 153
8.1 Antenna Technology Introduction 153
8.1.1 Satellite Antenna Classification 153
8.1.2 Beamforming Technique 158
8.2 Satellite-Terrestrial User Link Antenna Technology 162
8.2.1 Single Satellite Beamforming 162
8.2.2 Multisatellite Beamforming 164
8.2.3 Characteristics of User Terminal Antennas 169
8.3 Satellite-Terrestrial Feeder Link Antenna Technology 169
References 170

CHAPTER 9

Multiple Access for Satellite-Terrestrial Integrated Communication 173
9.1 Classic OMA Schemes 173
9.1.1 FDMA 173
9.1.2 TDMA 174
9.1.3 CDMA 175
9.1.4 OFDMA 177
9.1.5 SC-FDMA 178
9.2 Classic NOMA Schemes 180
9.2.1 PD-NOMA 180
9.2.2 MUSA 181
9.2.3 SCMA 182
9.2.4 PDMA 184
9.3 MA for Terrestrial Cellular Communication 185
9.4 MA for Satellite Communication 186
9.4.1 MF-TDMA 186
9.4.2 Hybrid TDMA/CDMA 189
9.5 Potential MA for Integrated Communication 190
9.5.1 Rate Splitting Multiple Access 190
9.5.2 Interleave Division Multiple Access 193
9.5.3 Lattice Partition Multiple Access 194
References 195

CHAPTER 10
Resource Management for Satellite-Terrestrial Integrated Communication 197
10.1 Overview of Multidimensional Resources 197
10.1.1 Spectrum Resources 197
10.1.2 Power Resources 200
10.1.3 Time Slot Resources 203
10.2 Resource Management Technology 205
10.2.1 Frequency Reuse Technology 205
10.2.2 Beam Hopping Technology 208
10.3 Intersatellite Resource Management 211
10.3.1 Limited On-Satellite Power 212
10.3.2 Dynamic Time Slot Allocation 212
10.3.3 Channel Availability in Short Time Slots 213
10.3.4 Cross-Orbit Multilayer Cooperative Transmission 213
10.3.5 Intersatellite Transmission Based on OPA 214
10.4 Interference Management 214
10.4.1 Natural Interference 214
10.4.2 Space Interference 220
10.5 Interference Management Technology 230
10.5.1 Adaptive Antenna Anti-Interference Technology 230
10.5.2 On-Satellite Processing Technology 231
10.5.3 Spread Spectrum Technology 232
10.5.4 Adaptive Modulation and Coding Technology 234
10.5.5 Digital Predistortion Technology 235
References 236

CHAPTER 11
Mobility Management for Satellite-Terrestrial Integrated Communication 241
11.1 Overview of Mobility Management 242
11.2 Link Layer Management Technology 244
11.2.1 5G Handover Management 244
11.2.2 Beam Handover 245
11.2.3 Interstellar Handover 248
11.3 Network Layer Management Technology 250
11.3.1 MIPv6 Technology 250
11.3.2 PMIPv6 Technology 252
11.3.3 HiMIPv6 Technology 254
11.3.4 VMIPv6 Technology 254
11.4 Transport Layer Management Technology 255
11.4.1 SIGMA Technology 255
11.4.2 Predictive SIGMA Technology 257
11.5 Potential Mobility Management Technology 259

11.5.1 SDN-Based Mobility Management 259
11.5.2 Mobility Management Based on O-RAN 259
References 261

List of Acronyms and Abbreviations 263

About the Authors 269

Index 271

Preface

For the sake of meeting people's demand for information capacity, mobile communication technology has evolved from the first-generation mobile communication system (1G) that only supported voice calls to the fifth-generation mobile communication system (5G), which is capable of supporting enhanced mobile bandwidth, massive connections, and ultra-low latency reliable transmission. The development of mobile communication not only accelerates the evolution of the scientific and technological revolution and industrial changes, but also greatly promotes the prosperity and development of economic society. Innovative applications such as virtual reality and augmented reality, industrial internet, internet of vehicles, telemedicine, and smart cities based on 5G communication are bringing qualitative benefits to people's lives.

However, the terrestrial network only covers about 20% of the earth's surface land, which is less than 6% of the earth's surface area. To provide seamless coverage services across the globe, satellite networks are essential to supplement the terrestrial network to jointly build a full-space three-dimensional communication network covering land, sea, air, and space. With the continuous evolution of the 3rd Generation Partnership Project (3GPP), 5G nonterrestrial network (NTN), and 6G space-terrestrial integrated technology, satellite and terrestrial networks are gradually forming a trend of deep integration. With the help of the advanced communication technology of the terrestrial network and the satellite industry, this integrated satellite-terrestrial network has entered a new and rapid development stage, realizing the ubiquitous connection of the land, sea, and air three-dimensional space for remote areas, deserts, oceans, aviation, and other areas, and providing individuals, enterprises, and governments with new services such as direct connection of mobile phones, wide-area IoT, and emergency rescue support.

Unlike terrestrial mobile communication systems, an integrated satellite-terrestrial network faces more complex challenges, mainly including:

1. Huge differences of channel environment: Satellite communication has the characteristics of long propagation distance, high latency, and high mobility, which seriously affects the performance;
2. Limited capabilities of the satellite network element platform: The satellite load is limited, and resources such as communication, computing, and storage are severely limited, which restricts the capacity of the satellite communication system;
3. High-speed dynamic changes in network topology including routing and resource allocation. Only by solving these challenges can we fully realize the huge potential of the satellite-terrestrial integrated converged network,

> bring revolutionary changes to the field of communications, and provide more extensive, reliable, and efficient communication services. To address these issues, this book focuses on the theory and technology of integrated satellite-terrestrial network, which provides a comprehensive discussion on the research status, key technologies, open issues, and future development direction from the physical layer to the network layer.

This book consists of 11 chapters.

Chapter 1 briefly describes the history of mobile communications, including the development of terrestrial communication and satellite communication.

Chapter 2 introduces the evolution of integrated satellite-terrestrial communication including 5G NR over satellites.

Chapter 3 introduces satellite constellation, including its definitions, types, and design methods.

Chapter 4 introduces intersatellite free-space optics satellite communication including its fundamentals, key techniques, and challenge issues.

Chapter 5 introduces the channel characteristics of integrated satellite-terrestrial communication.

Chapter 6 introduces the channel coding for integrated communication including classical channel coding, design challenges, and possible solutions.

Chapter 7 introduces the signal modulation for integrated communication including classical modulation waveforms and design guidelines.

Chapter 8 introduces the multiple antenna technology for integrated communication including user link and feed link.

Chapter 9 introduces the multiple access technology for integrated communication including classic access techniques and potential access techniques.

Chapter 10 introduces the resources and interference management for integrated communication.

Chapter 11 introduces the mobility management for integrated communication, including location management and route management.

This book has benefited from the support and contributions of many experts who have rich experience and knowledge in the field of satellite-terrestrial converged communications. Through the publication of this book, we hope to provide a valuable reference for students, researchers, and industry professionals to help them better understand the concepts, key technology, and opportunities of the integrated satellite-terrestrial network.

CHAPTER 1

Concept of Satellite-Terrestrial Integrated Communication

Italian inventor Guglielmo Marconi successfully conducted his long-distance electromagnetic wave communication experiment in 1896 [1], humanity ushering in a new era of information transmission at the speed limit of the universe—the speed of light—marking the beginning of mobile communication. With advancements in microelectronics technology as well as large-scale integrated circuits, the development of mobile communications is gradually maturing and moving toward global commercialization.

This chapter provides an overview of the development of terrestrial mobile communications, satellite mobile communications, and satellite-terrestrial integrated communications.

1.1 Terrestrial Mobile Communication

Since the 1980s, mobile communication has undergone a significant evolution approximately every decade, progressing from the first-generation (1G) technology, which only supported voice calls, to fifth-generation (5G) technology, which enables high bandwidth, extensive connectivity, low latency, and highly reliable transmission. The upcoming sixth-generation (6G) mobile communication is envisioned as the intelligent interconnection of all devices and the digital twin [2, 3]. Figure 1.1 illustrates the historical development of mobile communication. The advancement of mobile communication not only accelerates scientific and technological revolution and industrial transformation but also significantly contributes to economic and social prosperity [4–6].

1G emerged in the 1980s and primarily utilized analog signal-based cellular mobile communication mechanisms and frequency division multiple access (FDMA) technology for voice services. Specifically, electromagnetic waves are frequency modulated to carry voice signals over wireless channels; receivers employ carrier technology to demodulate the signals. During the 1G era, there was a lack of uniform communication standards across countries, hindering global roaming and impeding the advancement of wireless communication technology. Notable representations of 1G include Bell Labs' Advanced Mobile Phone Service, the United Kingdom's All-Access Communication System, Nordic four-country Nordic Mobile telephone System, and Japan's High-Capacity Mobile System. Additionally, due to its reliance on analog signal transmission, 1G is limited in its ability to transmit only voice signals with low spectral efficiency, data rate, capacity, coverage area, and susceptibility to interference.

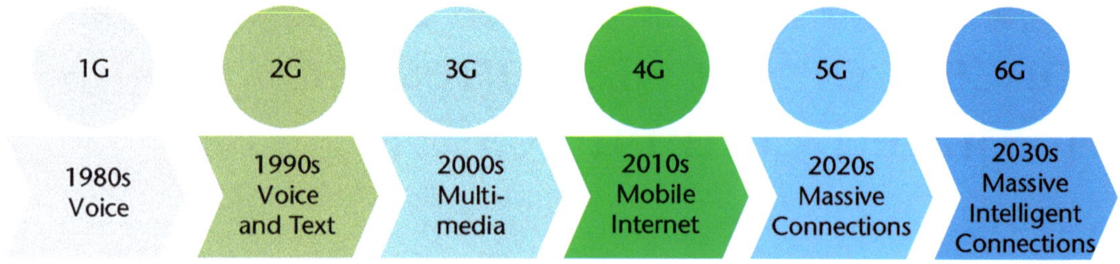

Figure 1.1 Evolution of mobile communications.

In order to address the technical challenges of 1G analog communication and meet the growing demand for communication, the emergence of second-generation mobile communication (2G) systems has been crucial. The 2G system adopts digital modulation and advanced multiple access technology, effectively improving communication metrics such as spectrum efficiency and data transmission rate while providing voice and low-rate data services. Representative standards of 2G include IS-95 in the United States and Global System for Mobile (GSM) in Europe. However, 2G communication is limited to supporting voice and text message transmission, unable to meet the increasing demand for multimedia services such as pictures and videos.

With the rising need for mobile multimedia services, third-generation mobile communication (3G) systems are gradually evolving. The 3G system continues to utilize digital modulation and code division multiple access (CDMA) access technology while expanding into a new electromagnetic spectrum with a bandwidth of 5 MHz and a data rate ranging from 384 kbps to 2 Mbps. Not only capable of providing voice and multimedia services, but also offering a wide range of other services, typical standards of 3G include broadband code division multiple access (BCDMA) in Europe, CDMA2000 in the United States, and time division synchronous code division multiple access (TD-SCDMA) in China. Furthermore, with intelligent terminals advancing rapidly alongside this development phase, individuals can efficiently handle images and music streaming video content among other media services, marking humanity's official entry into the era of mobile multimedia.

The fourth-generation mobile communication (4G), further developed on top of its predecessor and adopting more advanced technologies and protocols while broadening its spectrum even further is now boasting an expanded bandwidth reaching up to 100 MHz along with peak upstream and downstream rates hitting speeds between 500 Mbps and 1 Gbps, respectively. The 4G system adopts a new type of multicarrier modulation, namely orthogonal frequency division multiplexing (OFDM) technology, which converts serial data streams into high-speed parallel carriers not only significantly improving spectral efficiency, but also effectively combating multipath fading and enhancing transmission reliability.

The deployment of 4G networks has significantly facilitated the rapid expansion of the internet. However, with the exponential growth of smart devices, new services such as augmented reality, virtual reality, and intelligent industrial IoT are continuously emerging, placing higher demands on communication speed and latency. This trend further propels the advancement of 5G mobile communications. In response to diverse communication needs, 5G delineates three primary application scenarios: enhanced mobile bandwidth, ultra-high reliability and low-latency communication, and massive machine-type connectivity. Notably, the peak uplink and downlink rates in the enhanced mobile bandwidth scenario can reach up to 10 and 20 Gbps, respectively. Technologically speaking, 5G adopts OFDM multicarrier modulation with varying parameters along with massive multiple-input multiple output (MIMO) technology. Furthermore, it expands spectrum resources to encompass both low-frequency and high-frequency millimeter-wave communications.

With the widespread commercialization of 5G, the global industry has initiated research and exploration into 6G. As we approach 2030 and beyond, human society is poised to enter an era of intelligence, characterized by balanced and high-end social services, scientific and precise social governance, as well as green and energy-saving social development. Progressing from mobile internet to the Internet of Everything, and ultimately to the intelligent connectivity of all things, 6G will facilitate a transition from serving individuals and their interactions with objects to supporting efficient connections among intelligent entities. This will be achieved through the intelligent interconnection of people, machines, and objects in a synergistic symbiosis that meets the demands for high-quality economic and societal development. It will serve to enable intelligent production and living while promoting the construction of an inclusive and intelligent human society.

In the future, 6G communications will be capable of:

- Integrating advanced computing, big data, artificial intelligence, and block chain and other information technologies to achieve deep integration of communications with sensing, computing, and control;
- Holographic interaction, common-sense interconnection, digital twins, and intelligent interaction, which will make full use of emerging technologies such as brain-computer interaction, artificial intelligence (AI), and molecular communication to shape new forms of life with high efficiency in learning, convenient shopping, collaborative office work, and healthy living;
- Applying emerging information technology for existing agricultural production and industrial production in-depth empowerment, which can add impetus to the healthy development of production, and thus promote rapid development of a digital economy;
- Using a mobile communication network to build a smart society and integrating land-based, air-based, space-based, and sea-based networks to achieve ubiquitous coverage, which will greatly extend the coverage of public services, narrow the digital divide in different areas, which in turn will effectively enhance the level of social governance refinement.

The ultimate goal of 6G is to realize efficient and intelligent interconnection among individuals in the physical world by constructing a ubiquitous real-time digital world that accurately reflects and predicts the real state of the physical world. This advancement can help humanity enter a new era of intelligent interconnection between human beings, machines, and things through deep fusion of virtual reality ultimately realizing the vision of intelligent connection of everything.

1.2 Satellite Mobile Communication

Satellite communication systems utilize artificial satellites in orbit as relay stations to transmit radio signals for information exchange between terrestrial and space-based users [7, 8]. As depicted in Figure 1.2, the satellite communication system comprises a space segment and a ground segment. The space segment primarily encompasses the orbiting satellite and the ground station responsible for its control. The ground station is chiefly responsible for tracking, telemetry, remote control, and providing essential management and control functions to ensure the proper functioning of satellites in orbit. Ground systems for satellite communication typically include gateway stations and user stations. Gateway stations are equipped with high-capacity and large-aperture antennas to connect with the ground network and satellite feed beams, while user stations communicate with gateway stations via communication satellites. Typically, user stations transmit minimal traffic to gateway stations, while backhaul data traffic is more substantial. The communication link between the communication satellite and user terminal is referred to as the user link; furthermore, the link from user to satellite may be referred to as the user uplink, while that from satellite to user is known as the user downlink. Additionally, feeder links refer to communication links between satellites and gateway stations; intersatellite links pertain to links used for intersatellite communications.

The gateway station serves as the interface between satellite and terrestrial networks, comprising a radio frequency subsystem and a baseband subsystem. The baseband subsystem encompasses a satellite modem, access service network, web accelerator, network routing, and security system. Meanwhile, the radio frequency (RF) subsystem of the gateway station consists of an antenna, RF components, and intermediate frequency (IF) components. Typically, with a diameter ranging from several meters to over 10 meters, the antenna is accompanied by filters, low-noise amplifiers, and power amplifiers within the RF components, while

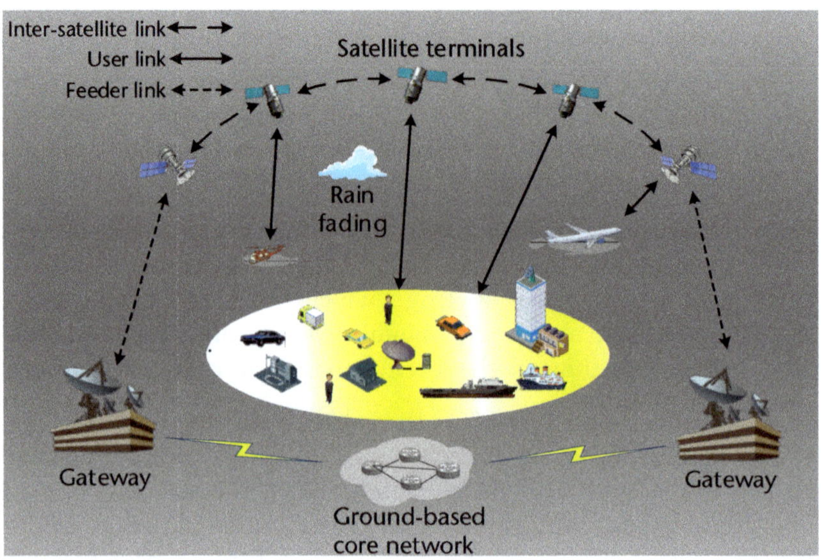

Figure 1.2 Basic satellite communications system.

IF components mainly consist of up- and downconversion devices. The RF subsystem facilitates the transformation of RF signals from satellites to IF before transmitting them to the baseband subsystem, similarly handling upconversion of IF signals from the baseband subsystem to RF for transmission back to satellites after amplification. Within the baseband subsystem lies a modem responsible for managing data traffic between user terminals and gateway routers/servers as well as controlling power/frequency in both forward/reverse links along with satellite network bandwidth. Additionally, tasked with authenticating/authorizing user access and implementing quality-of-service management is the access service network; web accelerators are utilized for enhancing throughput/end-user performance in Hypertext Transfer Protocol/Transmission Control Protocol (HTTP/TCP)-based applications. Lastly, ensuring traffic security/user quality-of-service falls under the responsibility of the network routing/security system.

The subscriber station comprises three primary components: the antenna, the outdoor unit, and the indoor unit. The outdoor unit is composed of a feeder, receiving equipment (including a low-noise amplifier and downconverter), and transmitting equipment (comprising a high-power amplifier and upconverter). An IF connection cable links the outdoor unit to the indoor unit. The indoor unit includes baseband receiving equipment (IF downconverters and demodulators) and baseband transmitting equipment (IF upconverters and modulators), among other components. Typically, the user station's antenna measures around 1 meter in length and utilizes a solid-state amplifier.

The intersatellite link is a wireless link established between satellites. Through the intersatellite link, data communication, information exchange, and relative position measurement can be achieved. The intersatellite link connects multiple satellites together, making each satellite a node in the space communication network. This forms a space communication network with satellites as interchange nodes, reducing the dependence of satellite-to-satellite communication on ground communication networks and effectively improving the transmission efficiency of the space communication network. Different types of intersatellite links can be categorized according to orbital types (space domain) and frequency types (frequency domain). Intersatellite links can be categorized into intraorbit intersatellite links and interorbit intersatellite links according to different orbital types. According to different frequency types, intersatellite links can be categorized into millimeter-wave/microwave band intersatellite links and laser intersatellite links. Millimeter-wave/microwave intersatellite links use radio waves as the carrier, while laser intersatellite links use lasers as the carrier. Compared with millimeter-wave/microwave intersatellite links, laser intersatellite links have an ultralarge bandwidth and can achieve higher communication rates and smaller range measurement errors. On the other hand, millimeter-wave/microwave links have a wider beam and are easier to be pointed, tracked, and captured by the receiver.

1.3 Typical Satellite Communication Systems

Satellite communication systems are an extension of terrestrial communication systems, utilizing various technologies such as computers, microelectronics, onboard processing, and space-satellite multibeams to establish global communication networks and achieve worldwide connectivity.

As well, satellite communication spectrum resources encompass various electromagnetic wave frequency bands utilized in satellite communications, each impacting functionality differently. These bands include:

- L-band (1–2 GHz), primarily used for mobile communications, satellite phones, and remote sensing, it supports services like voice calls, text messaging, and earth observation data transmission.
- S-band (2–4 GHz), utilized for meteorology and satellite communications, including National Aeronautics and Space Administration (NASA's) deep space communications, it has limited bandwidth availability due to prioritization of ground services.
- C-band (4–8 GHz), used for satellite communications and broadcasting, particularly in tropical areas due to its rain resistance, it's the first band allocated for commercial satellite telecom.
- X-band (8–12 GHz), known for its anti-interference capabilities and efficiency in military applications, particularly for beyond-line-of-sight communication.
- Ku-band (12–18 GHz), popular for direct broadcasting and satellite communications, allowing high data rates and smaller antennas, but susceptible to rain fade.
- Ka-band (26.5–40 GHz), supports high-bandwidth communications and is used for internet access via satellites, although it faces greater rain attenuation challenges.

These frequency bands facilitate a range of applications from military communication to commercial broadcasting, each chosen based on specific operational needs and environmental conditions.

These systems can be categorized into narrowband and broadband satellite communication based on the services they provide. To be more specific, narrowband satellite communications offer low-speed voice data as well as Internet of Things (IoT) services, while broadband counterparts provide high-speed, low-latency internet services for mass connections. Additionally, as shown in Figure 1.3, satellite communication systems can be classified into low Earth orbit (LEO), medium Earth orbit (MEO), and geostationary orbit (GEO) satellites. Specifically,

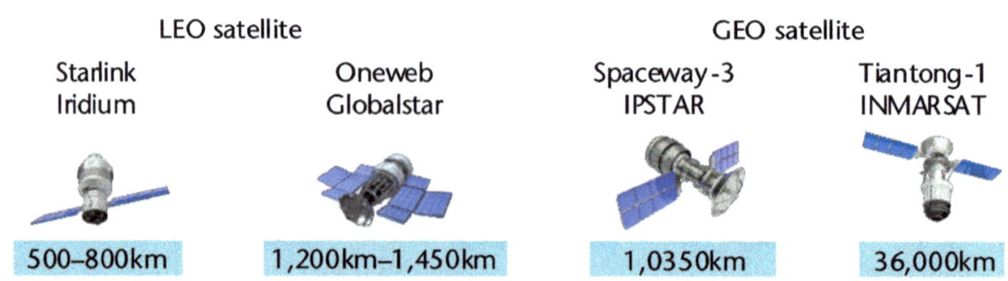

Figure 1.3 Typical satellite communication systems.

GEO satellites are positioned at an altitude of 35,786 km from the ground while LEO satellites range from approximately 200 to 2,000 km in altitude. A summary of typical existing satellite communication systems is presented in Table 1 of [8, 9].

1.3.1 High-Orbit Narrowband System

A well-known high-orbit narrowband satellite system was the International Maritime Satellite Organization (Inmarsat), which was established in 1982 to provide global maritime satellite communications services. Subsequent enhancements from 1985 to 1989 also extended the system's capabilities to include aeronautical and terrestrial communication services. The Inmarsat satellite system comprises ground stations, mobile stations, and space stations. The initial three generations of Inmarsat satellite systems primarily offered narrowband services. During the late 1970s, Inmarsat leased satellites such as Marecs from Europe and Marisat from the United States, which formed the first generation of the Inmarsat system providing global maritime satellite communication services for ocean vessels. The second generation of Inmarsat was deployed in the 1990s. For early first- and second-generation Inmarsat systems, effective communication could only be achieved between ships and shore stations; ship-to-ship communication required connection through shore stations to establish two-hop communication. With the third-generation Inmarsat system, vessels can communicate directly with each other and support direct communication with portable telephone terminals. Communication between the satellite and vessel is conducted using L-band frequency band, while a dual frequency band of C-band and L-band is utilized for communication between the satellite and shore station C-band for voice signals and L-band for telegrams.

Presently, Inmarsat is currently transitioning toward broadband, with the goal of providing broadband internet and satellite communication services for various platforms such as vehicles, airborne terminals, and ships. The fourth generation of Inmarsat (Inmarsat-4, I-4) offers L-band voice services, IoT services, global broadband regional network services, maritime broadband connectivity services, and aviation broadband connectivity services. Inmarsat-4 represents the world's first global satellite-based 3G network; its initial three satellites were launched in 2005 and 2008, respectively. Additionally, the system capacity was expanded with the launch of the Alphasat satellite in 2013. Each individual satellite of Inmarsat-4 is positioned in a high Earth orbit at an altitude of 35,786 km and can generate 19 wide beams along with more than 200 narrow spot beams to support both previous narrowband services and new broadband services.

Inmarsat has initiated the deployment of its sixth-generation maritime satellites (Inmarsat-6, I-6) featuring both L-band and K-band frequencies. Designed to deliver enhanced communication network coverage and capacity for mobile terminals, governments, and IoT customers worldwide, Inmarsat successfully launched its inaugural I-6 satellite, I-6 F1, in 2021. The company plans to launch three Inmarsat-8 satellites into geostationary orbit by 2040 to ensure that I-8, along with the I-6 satellites, can provide resilient high-capacity satellite internet services. Each of these new satellites is five times smaller than conventional GEO satellites with a volume of just 1.5 cubic meters. Furthermore, I-8 will serve as a radio-navigation transponder for satellite-based augmentation system services aimed at enabling precise tracking for safe navigation of aircraft and maritime safety while also unlocking new capabilities for machine-to-machine tracking in industries such as agriculture and transportation.

1.3.2 High-Orbit Broadband System

1.3.2.1 IPSTAR

In 2005, the Thai company Thaicom launched IPSTAR, heralding the advent of broadband satellite services in the Asia Pacific region. IPSTAR holds the distinction of being the heaviest commercial geostationary orbit satellite ever constructed, boasting a launch mass of nearly 6,500 kilograms. The IPSTAR satellite is a high-throughput broadband communications satellite, providing a total two-way link capacity of approximately 50.5 Gbps. It enables high-speed, two-way broadband communications across the Asia-Pacific region through multiple high-speed narrow beams. Compared to conventional Ku-band satellites, the IPSTAR system maximizes transmission frequencies and increases bandwidth by a factor of 20, thereby enhancing operational efficiency and meeting the growing demand for high-speed broadband internet access and data. Covering 22 countries and regions in Asia and the Pacific, IPSTAR provides data and internet services to rural areas within its coverage area.

1.3.2.2 Spaceway-3

On August 14, 2007, a Boeing 702 launched Spaceway-3 with a weight of approximately 6,075 kg. Hughes utilizes Spaceway-3 for the provision of broadband internet services. At that time, Hughes Networks was the world's leading manufacturer of very small aperture terminal (VSAT) equipment. With the successful launch of the Spaceway-3 satellite, Hughes Networks pioneered the Ka-band satellite broadband service.

The Spaceway-3 satellite boasts a capacity of approximately 10 Gbit/s, exceeding that of other satellites in the same generation by more than fivefold. In addition to delivering substantial communications capacity, it also facilitates single-hop communications between VSATs. Spaceway-3 leverages advanced antenna technology to establish a dynamic spot beam, offering a more adaptable approach to managing satellite communications capacity and enabling on-demand broadband services. Primarily catering to businesses, government agencies, and individuals in North America, Alaska, Hawaii, and parts of Latin America, Spaceway-3 delivers broadband services.

1.3.3 Low-Orbit Narrowband Systems

1.3.3.1 Iridium

In 1990, Motorola of the United States proposed the Iridium low-orbit satellite communication system with the aim of utilizing 66 low-orbit satellites to deliver global satellite phone, paging, and low-speed data services to users. The initial iteration of the Iridium communications system was deployed in 1997, made commercially available in 1998, and achieved global coverage by 2002. This first-generation system utilizes a constellation of 66 satellites operating at an altitude of 780 km with an inclination of 86.4 degrees. Each orbital plane consists of 11 operational satellites and one to two backup satellites at a slightly lower altitude of 648 km. The Iridium satellites employ Ka band for communication with gateway stations and L band for end-user interaction, with each satellite weighing approximately 680 kg. Featuring 48 spot beams and a communication rate of 2400 bit/s, each satellite is capable of supporting up to 1,100 concurrent calls.

In 2007, Iridium initiated the Iridium Next Generation (Iridium-NEXT) program. The first launch of 10 Iridium-NEXT satellites took place in January 2017, with the final launch occurring in January 2019, completing the deployment of 75 Iridium-NEXT satellites. All Iridium services were transferred to the Iridium NEXT satellite system in February 2019, completely replacing the first-generation Iridium system. The Iridium-NEXT satellite communications system comprises 66 LEO satellites, nine in-orbit backup satellites, and six ground-based backup satellites. Each satellite is equipped with 48 spot beams, providing a coverage diameter of approximately 400 km for each spot beam and up to a full coverage area diameter of 4,500 km per satellite. The spot beams can be overlapped to effectively reduce drop rates. Through the configuration of phased-array antennas with 48-beam transceivers, the addition of Ka frequency band for user link, and software-defined reproducible processing payloads, Iridium-NEXT achieves higher service rates, greater transmission capacity, and enhanced functionality.

1.3.3.2 Globalstar

In June 1991, the Globalstar Project was established as a joint venture between Laura Corporation and Qualcomm Incorporated, both based in the United States. This collaboration led to the founding of Globalstar Corporation. The company successfully deployed its inaugural satellite in February 1998, and by February 2000, with a total of 48 operational satellites and four backup satellites, it commenced full commercial services.

The Globalstar system can provide voice, fax, paging, and data transmission services to subscribers worldwide, with the exception of the Arctic and Antarctic regions. The system comprises 48 satellites orbiting at an altitude of 1,414 km across eight orbital planes, each containing six satellites. Each satellite is equipped with L-band and S-band active multibeam phased array antennas. The L-band antenna consists of 61 units for receiving signals from ground-based mobile terminals, while the S-band antenna consists of 91 units for transmitting signals to these terminals. System maintenance and upgrades are efficiently conducted through ground-based software deployment. The satellite employs a bent pipe architecture for signal relay back to Earth with minimal processing once received from the ground. During a call, Globalstar transmits the caller's signal via CDMA technology to the appropriate gateway satellite antenna before routing it through terrestrial communications systems. With no interplanetary links or on-planet processing required, the Globalstar system significantly reduces investment costs in order to provide cost-effective satellite communication services.

1.3.4 Low-Orbit Broadband Systems

1.3.4.1 Starlink

Starlink is a satellite broadband communications system led by SpaceX, aiming at providing global internet coverage through the construction of low-orbit satellites. Since 2019, SpaceX has been launching Starlink satellites. As of August 2023, a total of 4,903 satellites have been successfully deployed and are delivering high-speed satellite internet services to over one million locations worldwide. The initial phase of the Starlink constellation operates in Ku-band, Ka-band, and V-band, with an orbital altitude ranging from 540 to 570 km, respectively. Each satellite enhances its capacity through the use of four powerful phased array antennas and two parabolic antennas. Individual satellites cover an area of approximately 2.77 million square kilometers with a single-beam coverage radius extending up to 8 km. Due to the large number of satellites launched as part of the LEO constellation, SpaceX employs leading recoverable rocket technology and multistar systems to effectively reduce launch costs. Furthermore, each satellite features a compact design with flat panels that minimize size while utilizing a single solar array for simplified system design.

In June 2020, SpaceX applied to the United States for a license to use the Constellation II network in the E frequency band, which is planned to include 30,000 satellites. With the license of SpaceX's 2G network, it is expected to provide faster communication rates to more users and achieve full global coverage. This new license enables SpaceX to launch more greatly improved satellites with a throughput per satellite that is much higher than the 1G system. The Starlink II satellites launched in 2023 will be able to use T-Mobile's PCS frequency band (1.9 GHz for CDMA, GSM, and TDMA services in North America) to provide SMS and low-rate data services in the United States. The Starlink II satellites weigh more than 1250 kg and have a capacity of more than 200 Gbps, while supporting onboard processing and intersatellite link technology, which can reduce the dependence on ground gateways to achieve rapid landing of services.

1.3.4.2 OneWeb

In 2014, the OneWeb satellite constellation was proposed. It is a low-Earth orbit satellite broadband communication system, with an initial phase consisting of 648 low-Earth orbit satellites and a long-term plan of 6,372 satellites. OneWeb launched the construction of the satellite constellation in 2015. Due to issues such as spectrum licensing and lack of funding, satellite launches did not begin until 2019. Each satellite in the initial phase is expected to weigh approximately 150 kg, have an altitude of 1,200 km, an inclination of 87.9°, and be deployed in 12 orbital planes. Each satellite generates 16 elliptical Ku-band user beams and 2 Ka-band feed beams. The throughput of each satellite is expected to be 6 Gbit/s, and ground terminals use phased array antennas, allowing the entire system to provide downlink speeds of up to 50 Mbit/s to ground users. The first-generation OneWeb satellite constellation does not have intersatellite links and uses ground-based gateway stations for routing. OneWeb plans to provide services to regions north of 50 degrees, including the United Kingdom, Northern Europe, Greenland, and Alaska.

1.4 Satellite-Terrestrial Integrated Communication

The development of satellite internet began with Motorola's Iridium system in the 1980s, and has mainly gone through three development stages. From the perspective of service content, satellite internet has evolved from a constellation system mainly providing traditional low-speed voice, narrowband data, and IoT services to a broadband constellation system that can provide high-speed, low-latency, and massive internet data services. From the perspective of market positioning, it has gradually shifted from competition with ground communication systems to complementary and collaborative relationships. Figure 1.4 illustrates the evolution of satellite and terrestrial networks, which are described in detail as follows.

1.4.1 Competition with Terrestrial Communication

The first stage of satellite communication systems (1980–2000) mainly competed with terrestrial communication. During this period, the represented satellite constellations were the Iridium constellation, Globalstar, and the Orbit Communications (Orbcomm) system, which were separately proposed by Motorola, the United States' Laura, and Qualcomm, and the Orbit Communications companies. The service development of low orbit constellation in this stage was divided into two categories: voice and low-speed data and internet access. With the rapid development of terrestrial communication systems, they were far superior to satellite communications in terms of features such as communication quality and tariff prices, so that satellite systems at this stage declared bankruptcy around 2000.

1.4.2 Complement to Terrestrial Communication

The second stage of satellite communication systems (2000–2014) mainly were complementary to the terrestrial communications network. Early high-throughput satellites such as IPSTAR and Anik F2 had a system throughput between 2 and 50 Gbps and a data download rate of less than 5 Mbps, which could only meet 2G communication services. At that time, most of the world's satellite internet business was concentrated in the United States, mainly dial-up internet households, and the scale of the global satellite internet industry was less than US 1 billion. In 2005, WildBlue Corporation of the United States launched home broadband internet access services using AnikF2 and WildBlue-1 satellites, but limited by the total system capacity, it could only provide users with a downlink rate of 512 kbps.

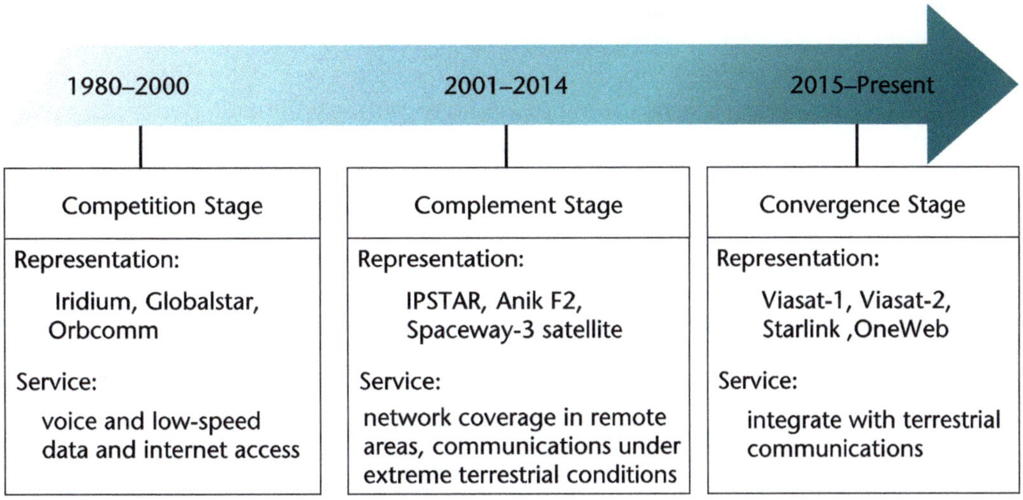

Figure 1.4 The evolution of satellite and terrestrial networks.

In addition, during this period, satellite communications were mainly used as a supplement to terrestrial communications networks, playing an irreplaceable and important role in network coverage in remote areas and in communications under extreme terrestrial conditions.

1.4.3 Convergence with Terrestrial Communication

In 2014, satellite communications began to integrate with terrestrial communications. With the high-throughput satellite capacity increasing to 100 Gbps and the maturity of key technologies such as packet technology and compression modulation, the download speed of user data can reach 50 Mbps. High-throughput satellites were gradually able to meet the broadband internet application requirements of high-definition video, multimedia, and so on, thus ushering in the era of satellite internet applications [9, 10]. The scale of the satellite internet industry has been steadily growing, with average annual revenue per user stable at 100, and the number of global users increasing by 1 million per year on average. For example, Viasat-1 and Viasat-2, launched by Viasat, have significantly improved the broadband service speed provided, gradually supporting small screens (12 Mbps/360 p), standard definition (25 Mbps/480 p), high definition (50 Mbps/720 p), and full high definition (100 Mbps/1080 p) from the perspective of video transmission quality [11].

Since 2018, as the market application of high-throughput satellites has gradually matured, the scale of the satellite internet industry has undergone a steplike development, accompanied by the rapid growth of user numbers and the rapid decline of resource prices. In the past three years, the average annual revenue per household for global satellite internet users has dropped to 96, and the number of global users has increased by 3 million per year on average. At the same time, the application development of 4G and 5G has led to exponential growth in the number of global internet users, the number of connected devices in the internet of things, and the amount of communication data. To meet the demand for large-scale, high-volume terminal access, the mobile management capabilities and terminal antenna technologies of high-throughput satellite-terrestrial systems have developed rapidly, and satellite internet has begun to provide application services for users with high-speed mobility, such as aviation and maritime users. The next-generation high-throughput satellites, represented by Viasat-3 and SES-26, adopt full-digital beamforming technology, which will further enhance the on-orbit reconfigurable software-defined capabilities and support full-dynamic capacity generation and utilization [12]. The single-satellite capacity of Viasat-3 launched in 2023 will even reach the level of 1 Tbps, enabling operators to provide users with higher up/download speeds and support the high-speed access applications of large civilian aircraft and cruise ships. Moreover, with the deployment of low-Earth-orbit satellite internet constellations such as Starlink and OneWeb, satellite internet will enter a new era of integration of low- and high-orbit systems [13].

1.4.4 The Vision of Integrated Communication

Figure 1.5 presents a 6G-oriented satellite-terrestrial integrated communication, mainly composed of ground networks, near-space networks, and space-based

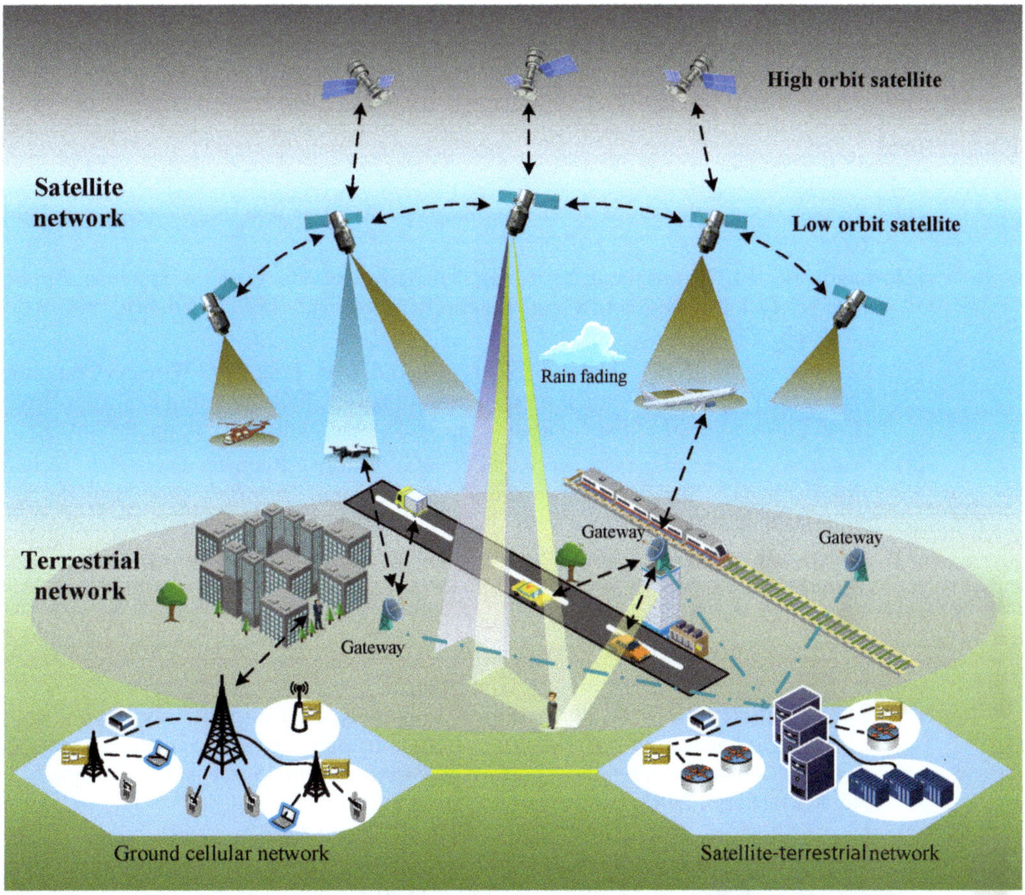

Figure 1.5 Typical satellite communication systems.

networks [14, 15]. To be more specific, the ground network mainly includes ground cellular base stations, satellite gateway stations, and a core network. The near-space network includes drones and air-access platforms, while the space-based network mainly includes high-, medium-, and low-orbit satellite communication payloads and platforms. The satellite-terrestrial integrated communication network employs a unified network architecture and standard transmission technologies, which is capable of providing broadband or narrowband access services for a wide range of communications equipment, thus meeting the needs of space-based, sea-based, and land-based users for anytime and anywhere communications. Satellite-terrestrial integrated communication enables the deep integration of satellite and ground mobile communication in terms of communication technology, components, communication equipment, communication networks, and communication services and applications, greatly reducing costs and improving user experience, and promoting the healthy development of the entire industry.

The 6G-enabled satellite-terrestrial integrated network is expected to truly realize seamless access and seamless coverage, providing global seamless coverage and uninterrupted services. Users will be able to access both ground mobile networks and satellite networks at anytime and anywhere and switch between them without perception.

References

[1] Wander, T., *Guglielmo Marconi: Building the Wireless Age*, New Generation Publishing, 2015.

[2] Saad, W., M. Bennis, and M. Chen, "A Vision of 6G Wireless Systems: Applications, Trends, Technologies, and Open Research Problems," *IEEE Network*, Vol. 34, No. 3, 2020, pp. 134–142.

[3] Chowdhury, M. Z., M. Shahjalal, S. Ahmed, and Y. M. Jang, "6G Wireless Communication Systems: Applications, Requirements, Technologies, Challenges, and Research Directions," *IEEE Open Journal of the Communications Society*, Vol. 1, 2020, pp. 957–975.

[4] Wang, C. X., et al., "On the Road to 6G: Visions, Requirements, Key Technologies, and Testbeds," *IEEE Communications Surveys and Tutorials*, Vol. 25, No. 2, 2023, pp. 905–974.

[5] Andrews, J. G., et al., "What Will 5G Be?," *IEEE Journal on Selected Areas in Communications*, Vol. 32, No. 6, June 2014, pp. 1065–1082.

[6] Agiwal, M., A. Roy, and N. Saxena, "Next Generation 5G Wireless Networks: A Comprehensive Survey," *IEEE Communications Surveys and Tutorials*, Vol. 18, No. 3, 2016, pp. 1617–1655.

[7] Radhakrishnan, R., W. W. Edmonson, F. Afghah, R. M. Rodriguez-Osorio, F. Pinto, and S. C. Burleigh, "Survey of Inter-Satellite Communication for Small Satellite Systems: Physical Layer to Network Layer View," *IEEE Communications Surveys and Tutorials*, Vol. 18, No. 4, 2016, pp. 2442–2473.

[8] Heo, J., S. Sung, H. Lee, I. Hwang, and D. Hong, "MIMO Satellite Communication Systems: A Survey from the PHY Layer Perspective," *IEEE Communications Surveys and Tutorials*, Vol. 25, No. 3, 2023, pp. 1543–1570.

[9] Khammassi, M., A. Kammoun, and M.-S. Alouini, "Precoding for High-Throughput Satellite Communication Systems: A Survey," *IEEE Communications Surveys and Tutorials*, Vol. 26, No. 1, 2024, pp. 80–118.

[10] Hosseinian, M., J. P. Choi, S.-H. Chang, and J. Lee, "Review of 5G NTN Standards Development and Technical Challenges for Satellite Integration with the 5G Network," *IEEE Aerospace and Electronic Systems Magazine*, Vol. 36, No. 8, August 2021, pp. 22–31.

[11] Kodheli, O., et al., "Satellite Communications in the New Space Era: A Survey and Future Challenges," *IEEE Communications Surveys and Tutorials*, Vol. 23, No. 1, 2021, pp. 70–109.

[12] Zedini, E., A. Kammoun, and M.-S. Alouini, "Performance of Multibeam Very High Throughput Satellite Systems Based on FSO Feeder Links with HPA Nonlinearity," *IEEE Transactions on Wireless Communications*, Vol. 19, No. 9, Sept. 2020, pp. 5908–5923.

[13] Abdelsadek, M. Y., G. Karabulut-Kurt, H. Yanikomeroglu, P. Hu, G. Lamontagne, and K. Ahmed, "Broadband Connectivity for Handheld Devices via LEO Satellites: Is Distributed Massive MIMO the Answer?" *IEEE Open Journal of the Communications Society*, Vol. 4, 2023, pp. 713–726.

[14] Liu, J., Y. Shi, Z. M. Fadlullah, and N. Kato, "Space-Air-Ground Integrated Network: A Survey," *IEEE Communications Surveys and Tutorials*, Vol. 20, No. 4, 2018, pp. 27142741.

[15] Chen, S., S. Sun, and S. Kang, "System Integration of Terrestrial Mobile Communication and Satellite Communication: The Trends, Challenges and Key Technologies in B5G and 6G," *China Communications*, Vol. 17, No. 12, Dec. 20, pp. 156–171.

CHAPTER 2

The Evolution for Satellite-Terrestrial Integrated Communication

Integrating satellite and terrestrial communication is a crucial technology for achieving global coverage and seamless communication and has become an inevitable trend in the future development of communication. This chapter introduces the demands of integrated satellite-terrestrial communication, typical application scenarios, the evolution of international standards, as well as the possible challenges.

2.1 Demand for Integrated Communication

Over the past several decades, the relentless pursuit of enhanced spectral efficiency and increased transmission rates has led to significant optimizations in terrestrial wireless communication systems. These systems have been tailored to better serve individual mobile users and Internet of Things devices. Simultaneously, the drive for improved power efficiency and greater throughput has guided the evolution of satellite communication systems, which now provide specialized services for professional mobile users and VSAT users. Due to the distinct needs of these different user groups, as well as varying application scenarios and channel environments, the technological advancements in terrestrial and satellite communication systems have evolved along relatively independent trajectories [1, 2].

In recent years, LEO satellite internet has become a new competitive track and focus in the wireless communication field, triggering a global wave of development. The United States Starlink is currently the largest low-orbit broadband satellite communications system, with more than 1 million VSAT users. AST SpaceMobile is actively conducting related technology trials in the realm of direct mobile-to-satellite connections. SpaceX and T-Mobile are collaborating to build a satellite communication system that supports direct mobile phone connections. Ericsson, Thales, and Qualcomm have jointly announced their collaboration to develop a satellite communication system based on 5G NTNs. Apple is already providing satellite to mobile emergency SOS services in some countries. Relevant Chinese enterprises, research organizations, and universities are also conducting research and validation of 5G/6G satellite-terrestrial communication technologies.

The 3rd Generation Partnership Organization (3GPP) is converging satellite communications and terrestrial mobile communications in the 5G NTN protocol in terms of radio transmission technology, network architecture and security, terminal access, and authentication [3]. This integration creates the conditions for constructing a satellite-terrestrial integrated communication system that provides seamless global coverage to meet the needs of the Internet of Everything and limitless communication. As wireless communication technology advances, particularly with the increasing clarity of 6G technological requirements, the demand for satellite-terrestrial integrated communication has become more vital from the perspectives of user needs, operator expectations, and industry development. This demand is primarily reflected in four aspects, as shown in Figure 2.1.

Popular demand for satellite communication users: The traditional users of satellite communications were governments, the military, and specialized institutions and professionals. With the expansion into space for human activities, satellite communications are gradually moving toward ordinary individuals. Personal mobile communication services and applications centered on cell phones increasingly require direct connection of cell phones to satellites to avoid various problems and inconveniences caused by dedicated satellite communication terminals. Cell phones can flexibly access satellite and earth networks, eliminating the need for dedicated satellite terminals, reducing user costs and enhancing the cell phone user experience, which is undoubtedly an important market driver for integrated satellite-terrestrial communications.

Demand for seamless coverage: While terrestrial mobile communication systems can provide broadband mobile access services for everyone and everything on land, they struggle to cover oceans, mountains, forests, deserts, high altitudes, and near-space regions. While satellite communication systems can cover the entire surface of the Earth, medium- and high-altitude airspace and adjacent space, it is difficult to provide broadband satellite access services for cell phone users. Therefore, it is necessary for them to complement each other to provide seamless coverage as well as anytime, anywhere access services.

Demand for smart connections for everything: The smart Internet of Everything, which is centered on information transfer and application processing, requires not only low-cost local-area IoT connectivity, but also economical wide-area IoT

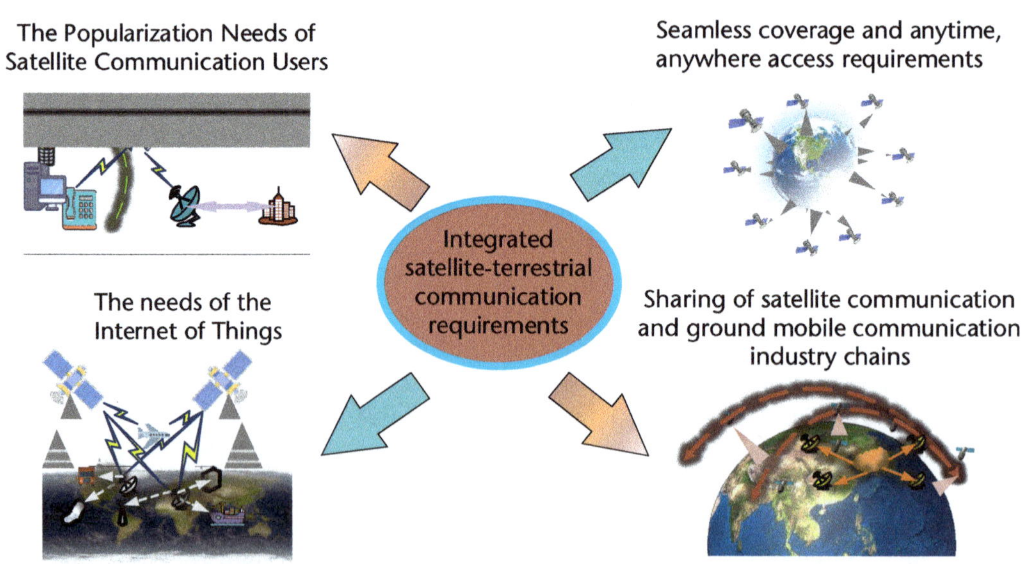

Figure 2.1 Demand for satellite-terrestrial integrated communication.

connectivity to support the rapid interaction and sharing of information and the realization of converged services and applications. Integrated satellite-terrestrial communication provides more convenient and economical technical means for the intelligent connection of everything in remote areas on the ground, medium- and high-airspace, and adjacent space.

Shared industrial chain for integrated communication: Terrestrial mobile communications have billions of users, a strong technical team, and a complete industrial chain. However, satellite communication participants and industrial scale are far lower than that of terrestrial mobile communication, which is also an important reason for the high cost of satellite communication equipment and operation and the slow update of technology. How to fully utilize the industry chain of terrestrial mobile communication and how to develop ordinary mobile communication users into satellite communication users are two important problems faced by the operators, and integrated satellite-terrestrial communication provides a technical path for the solution of these two problems.

2.2 Typical Application Scenarios

The critical value of integrated satellite-terrestrial lies in providing users with continuous, stable, and reliable communication network accessibility anytime and anywhere. The combined provision of ubiquitous communication services by terrestrial networks (TNs) and NTNs ensures that user accessibility to communication services is unaffected by geographical environments [4]. This joint approach offers the only viable connectivity solution for regions with weak communication infrastructure, such as rural and remote areas.

The combined provision of continuous communication services by TNs and NTNs ensures smooth transitions across different geographical environments and effectively reduces costs. Users are encouraged to utilize low-cost, high-capacity terrestrial communication means while demand-based switching between satellites and terrestrial stations to ensure continuous connectivity. Typical scenarios include nearshore coverage areas and regions with overlapping satellite-terrestrial coverage for direct mobile-to-satellite connections.

The combined provision of robust communication service by TNs and NTNs ensures that user accessibility is unaffected or minimally affected during emergencies. These services enhance the resilience of terrestrial communication systems against disasters through space-based and air-based support, thereby providing higher reliability. The typical application scenarios are shown in Figure 2.2 and introduced as follows.

Personal communication: Mobile phones, the most common personal communication tool, can access the network via terrestrial mobile networks when available. In places where there is no terrestrial mobile network signal, or when the terrestrial mobile network fails due to force majeure, the cell phone can be accessed to the network via satellite, realizing truly seamless coverage and anytime, anywhere access.

Telecommunications: Satellite-terrestrial integrated communication systems can supply base station backhaul, broadband access, and satellite relay services

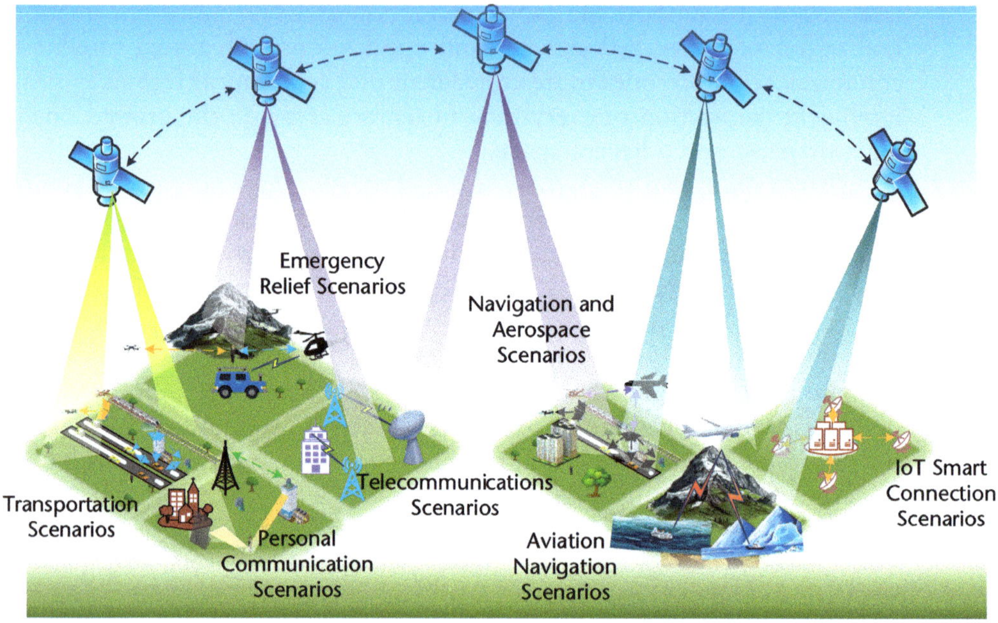

Figure 2.2 Typical application scenarios.

for remote areas, islands, offshore platforms, and mobile platforms, offering satellite connectivity and data relay services without modifying existing communication networks.

Transportation communication: Professional vehicles and trains on land as well as other transportation require professional communication, data transmission, or passenger network access. These can be provided by terrestrial networks where available and by satellite access when in deserts, uninhabited areas, or the centers of lakes and seas.

Low-altitude coverage: A satellite-terrestrial cooperative network can empower net-connected unmanned aerial vehicles (UAVs), utilizing NTN ubiquitous coverage in conjunction with terrestrial cellular networks, in which control commands for control surface and high reliability requirements are carried by non-terrestrial networks, and broadband data services such as images, videos, and so on are transmitted through terrestrial cellular networks, enhancing the control continuity and reliability of the net-connected UAVs while taking into account the performance of the broadband services.

Navigation and aerospace: Satellite-terrestrial integrated communication's satellite network enhances navigation capabilities, providing localization and navigation enhancement services for personal and professional users. It also supports aerospace-related information transmission services, such as high-speed data transmission and Internet Protocol (IP) telemetry and command services.

IoT intelligent connectivity: On fixed or mobile platforms, IoT terminals are provided with direct connection or indirect connection services, enabling applications such as grid monitoring and maintenance, geological monitoring, forest monitoring, UAV data transmission and control, offshore buoy information collection, ocean container information collection, crop monitoring, and rare wildlife monitoring, along with appropriate emergency handling. These services can fulfill the need for large-scale, low-cost terminal access.

Emergency: Satellite-terrestrial integrated communication systems can provide natural disaster warnings for events like earthquakes, floods, typhoons, tsunamis, forest fires, and volcanoes to ordinary users through mobile phones. After disasters, survivors and rescuers can use mobile phones to maintain necessary contact through satellite communication. Additionally, satellite communication can also be used for everyday emergency communication where terrestrial emergency networks or terrestrial mobile networks are unavailable.

2.3 Integration Models

After a long exploration period, satellite and terrestrial communication integration has substantially progressed, which is shown in Figure 2.3. Specifically, the first-generation satellites served as broadcast relay satellites when terrestrial networks predominantly used fixed-line telephones. In the 1990s, China's Emergency Mobility Bureau began using VSAT, signifying the small, intelligent ground stations with tiny aperture antennas. Usually, many such small stations work together with a large station to establish a satellite communication network [5].

The second-generation satellites were digital communication satellites. Satellite and terrestrial mobile communication systems attempt to integrate during the 2G era, creating satellite mobile communication systems. In 1989, the International Maritime Satellite Organization (Inmarsat) expanded its scope from maritime and aeronautical satellite communication to terrestrial satellite communication, renaming itself the International Mobile Satellite Organization. In 1994, it was restructured and transformed into an international commercial company, becoming the earliest satellite mobile communication system service provider. Maritime satellite phone is its service product.

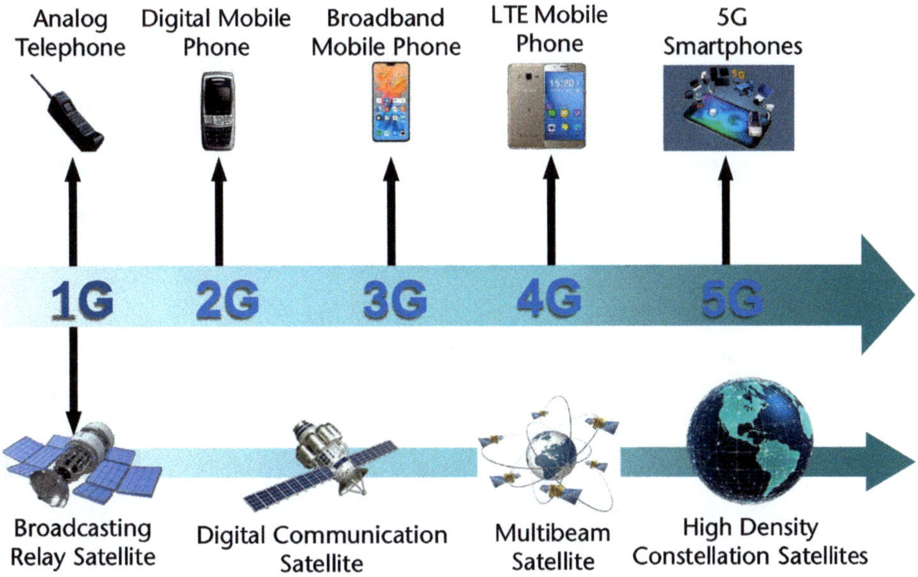

Figure 2.3 Directions for satellite-terrestrial integrated development.

Third-generation satellites are characterized by high throughput and multiple beams. In 2019, China Satellite Communications successfully completed an integration test of 5G data transmission via a high-throughput satellite, which involved building a satellite base station backhaul test system between the portable station of high-throughput satellites and the terrestrial 5G network. Based on this, the information can be transmitted between the 5G base station and the satellite-terrestrial link, marking substantial progress in a satellite-terrestrial network.

Fourth-generation satellites are high-density constellation satellites characterized by large-caliber and multibeam capabilities. In the B5G as well as 6G visions, these satellites will support direct connection for all types of terminals. International Telecommunication Union (ITU) and 3GPP standardization organizations have set up special working groups to study the standardization of satellite-terrestrial integrated communication. Explicitly, the business model, networking model, and terminal development model are the main integrated directions for the future.

2.3.1 Service Models

A NTN's three typical business models include high-speed relay, broadband direct connect, and narrowband direct connect, as shown in Figure 2.4 [6].

To be more specific, the high-speed relay model of NTNs provides services such as base station backhaul and satellite Wi-Fi. For example, by using the high throughput link of high-throughput satellites, it can effectively supplement or replace costly terrestrial optical fiber or wireless backhaul, achieving high-speed relay transmission of video, IoT, and other data to the core network, thereby better serving local cell sites.

Due to the benefits of broad coverage, large capacity, independence from geographical constraints, and the ability to broadcast information, the broadband service of NTNs is capable of addressing the limitations of terrestrial networks (such as coverage restrictions, inability to support high-speed mobile user applications,

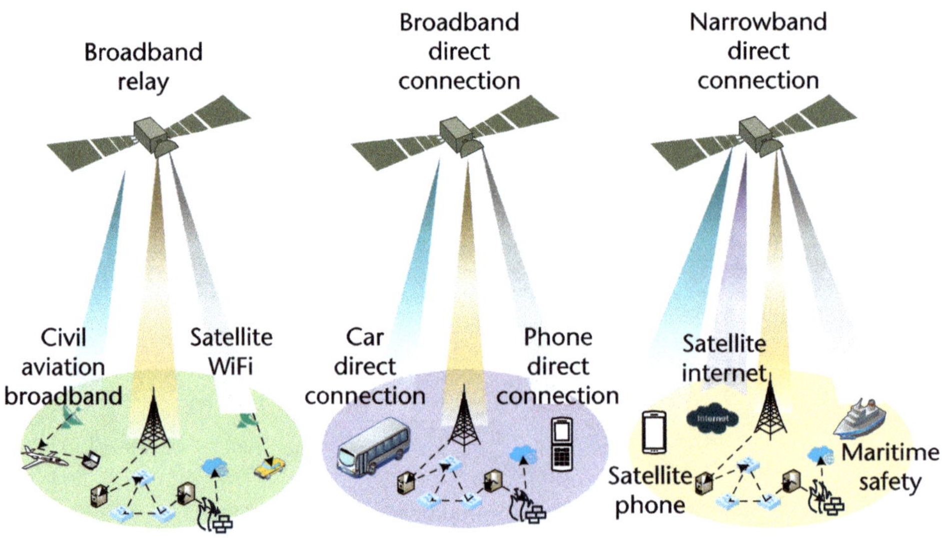

Figure 2.4 Three service models.

high network resource consumption by broadcast services, and vulnerability to natural disasters).

Compared to broadband service, narrowband service is relatively simple, typically offering only phone service and Short Message Service (SMS). However, it is easier to implement at this stage and has a substantial market demand, especially in maritime, emergency rescue, and polar exploration environments.

2.3.2 Networking Models

The overall architecture is divided into NTNs and TNs. NTNs include communication systems that operate independently of terrestrial infrastructure, mainly through high-orbit satellites, medium- and low-orbit satellites, and high-altitude platforms. High-orbit satellites are employed to deploy lightweight core networks (including control-plane functional network elements and user-plane functional network elements), access networks, edge computing service units, and AI-enabled platform functions. Medium- and low-orbit satellites are utilized to deploy core network user-plane functional network elements, access network, edge computing service units, and AI-enabled platforms. TNs refer to communication systems that rely on terrestrial infrastructure, including cellular networks, Wi-Fi and fiber-optic networks used for deploying terrestrial full-featured core networks, terrestrial gateways, and terrestrial mobile base stations. Similar to terrestrial communication system framework, integrated communication systems comprise three primary components: access networks, transmission networks, and core networks [7].

As shown in Figure 2.5, the integrated network model can take the form of loose coupling discrete networking or tight coupling integrated networking.

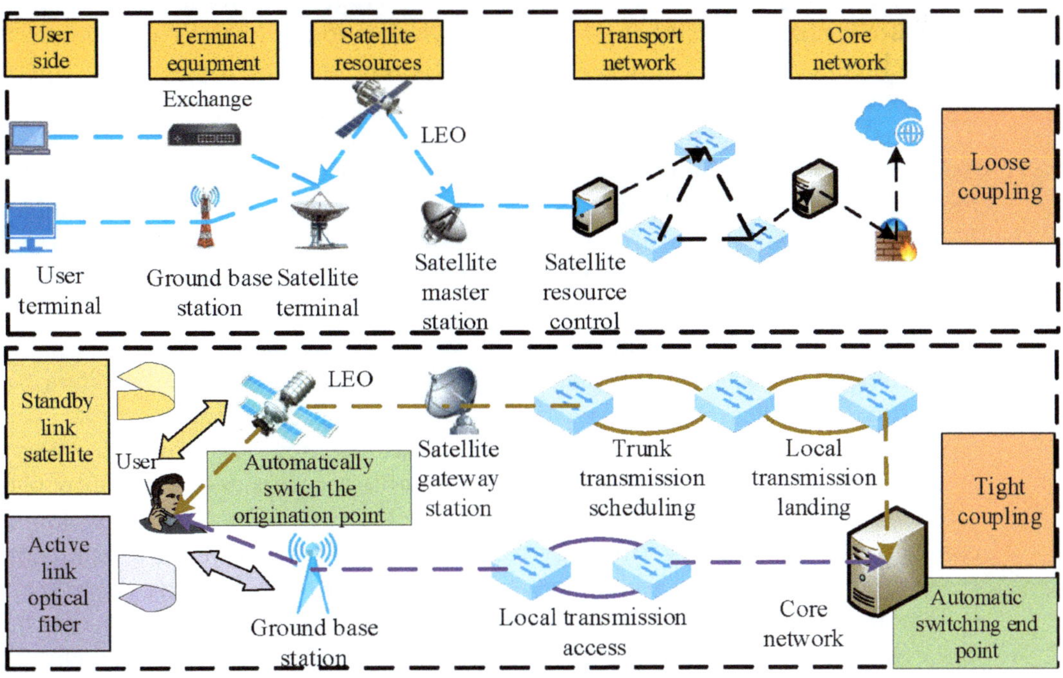

Figure 2.5 Networking models.

In the loosely coupling model, satellites and terrestrial networks operate independently. Explicitly, the satellite sends data to its core network, and data interaction is realized via interfaces between the satellite and terrestrial core networks. Satellites are used as a complement to terrestrial networks, and terrestrial links are used as the primary connection method for end users to access the internet. Satellites are used as an auxiliary connection method, (i.e., as a backup link to provide services when the terrestrial link is unable to do so), thus improving service availability and reliability. At this point, although the quality of service may be degraded, it may still play an important role in extreme scenarios. In natural disaster events, satellite connectivity is the only option for end users to connect to the internet.

Under the tightly coupled mode, the satellite network and ground network employ a compatible air interface design, and the terminal adopts a dual-mode design. The satellite and terrestrial network provide cooperative services and unified management so that users can realize seamless switching between the satellite network and terrestrial network via one terminal.

Satellite-terrestrial integrated networks provide users with more reliable, consistent service and connect air, space, land, and sea, forming an integrated ubiquitous network pattern. Different coverage forms and networking modes are adopted for scenarios with different user densities, such as cities, counties and towns, villages, the air, and the sea. Cellular three-dimensional coverage is adopted in cities to support user density and diverse and massive service characteristics under the trend of three-dimensionalization of cities, and ensure user service experience. Counties and towns use cellular planar coverage to ensure coverage. Rural areas are covered by a cellular line plus point mode, focusing on areas where users are concentrated. At the same time, NTNs provide ubiquitous coverage in all scenarios, acting as a relay satellite, transparent base station, and wireless router in areas where users are relatively concentrated. In areas where users are randomized, dispersed, and sparse, they provide direct connection satellite functions for user terminals.

Satellite-terrestrial integrated networks need synchronized services across low-, medium-, and high-orbit satellites to enrich network capabilities. More specifically, high-orbit satellites offer high-capacity coverage with mature technology and stable industries suitable for broadcast and multicast services. A single satellite can provide extensive service. Integrated with TN in a base station plus satellite broadband relay mode, it extends the coverage of existing base stations. Medium-orbit satellites are able to provide global, rapid, low-cost, and ubiquitous coverage with eight satellites. This system efficiently serves mobile carriers like ships, vehicles, and aircraft with seamless global low-cost and moderate-speed communication.

Low-orbit satellites are characterized by low latency, large overall capacity, and low free-space path loss, making them suitable for small terminals. Low-orbit satellites are more suitable for building satellite internet. 5G has officially entered commercial use, and the technology maturity is high, and so the low-orbit satellite system can reuse the technology and features of 5G standards. In terms of system architecture, satellite internet can be regarded as a kind of 5G access network through the integration of the access network, sharing core network with the ground, and realizing air interface protocol processing and route forwarding by deploying signal processing, link layer, network layer switching routing, and other functional modules on the star. At the same time, the ground equipment of satellite internet can inherit the current 5G base station baseband processing and related terminal chip results, shorten the research and development cycle, and reduce the research and development cost.

2.3.3 Terminal Development Models

The mobile terminals of the integrated network are becoming smaller. As shown in Figure 2.6, the rapid popularization of mobile communications is attributable to the maintenance of miniaturized terminals while continuously improving capacity and fostering the user habit of mobile plus high-speed. The user demand brought about by the rapid development of mobile communications has driven the development of satellite communications toward terminal miniaturization and network broadband [8]. Among them, the miniaturization of mobile terminals refers to the gradual miniaturization of various terminal equipment for mobile satellite communications. Miniaturized mobile terminals, including ground personal mobile terminals, vehicle-mounted terminals, airborne terminals, ship-mounted terminals, and internet of things terminals, provide diversified medium-rate and low-rate communication services.

2.4 Evolution of International Standards

Over the past three decades, satellite and terrestrial cellular mobile communication have gradually transitioned from a competitive relationship to a more integrated one. In the 1990s, LEO satellite and terrestrial cellular mobile communication were

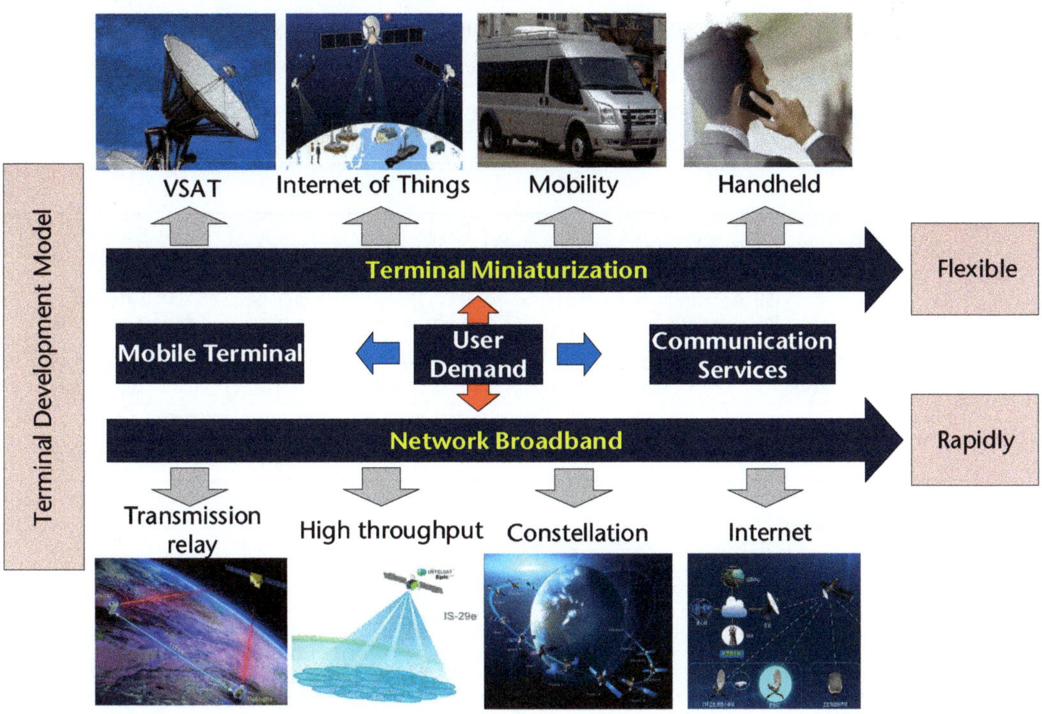

Figure 2.6 Terminal development models.

highly competitive. As mentioned, the earliest Iridium LEO satellite system consisted of 66 satellites, pioneering the field of LEO satellite communication. The Iridium system employed onboard processing and intersatellite link technology aimed at consumers, allowing handheld satellite terminals (satellite phones) to provide seamless communication anywhere on Earth and resolving cross-protocol roaming between satellite and terrestrial cellular networks. However, traditional satellite communication has many limitations. The Iridium system filed for bankruptcy due to massive research and development and system construction costs, later restructuring and shifting to industrial applications. Meanwhile, terrestrial cellular mobile communication evolved significantly from the 2G to the 5G era, with expanding user bases and notable commercial success. As the satellite constellation deployment plans of companies such as Starlink, Telesat, OneWeb, and AST advance, LEO satellite communication is experiencing a resurgence and gradually integrating with terrestrial cellular communication [9].

The 3GPP organization began to investigate integrating satellite networks with terrestrial 5G networks starting from Release 14, aiming to formulate unified standards to regulate terrestrial mobile communication networks and satellite communication networks [10]. As shown in Figure 2.7, an initial definition of satellite 5G network architecture models based on transparent forwarding mode and satellite-based base station mode was provided in Release 15's TS 38.821, along with an evaluation of end-to-end protocol stack segmentation issues under different network architectures. Release 16 further enhanced the flexibility of network element deployment within service-based architecture (SBA) to establish a standard for integrated architectures. As the third phase of the 5G standard, Release 17 aims to further refine the specific technologies defined in Releases 15 and 16 based on current framework and functions, addressing key issues such as software-defined satellite components and new air interface requirements for NTNs.

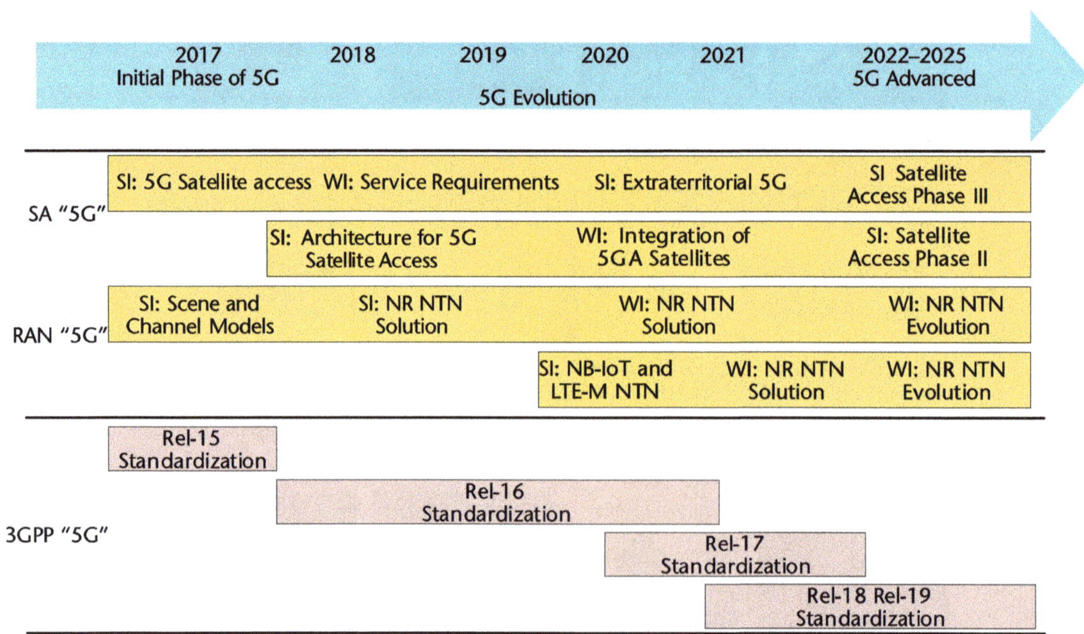

Figure 2.7 Standardization evolution process.

2.4.1 Release-15 for NR NTN

The work of 3GPP on NR NTN commenced in 2017, with Rel-15 emphasizing the definition of deployment scenarios and channel models [11]. In 3GPP TR 38.811, the primary objectives were to identify reference deployment cases for NTNs and reach a consensus on essential parameters, including system architecture, orbital altitudes, and operational frequency bands. The key aspects covered were:

1. Two frequency ranges of S-band and Ka-band;
2. Fixed beams (beams targeted at specific terrestrial areas for extended periods) and moving beams (beams adjust their coverage as the satellite orbits);
3. High-attitude platform systems (HAPS), LEO, and GEO deployments;
4. Two types of NTN terminals: VSAT with paraboloidal reflector typically mounted on buildings or vehicles and handheld terminals;
5. Antenna models for satellites and HAPS.

The second primary goal of Rel-15 was the creation of an NTN-specific channel model, building upon the existing 3GPP terrestrial channel models. The developed channel model supported a variety of environments, from urban to rural settings. Multipath propagation is a common occurrence. For NTNs, due to the vast distances involved, signal paths tend to be nearly parallel, resulting in minimal angular spread. Consequently, large-scale fading characteristics, such as line-of-sight probability, angular dispersion, and delay spread, exhibit significant differences compared to terrestrial conditions, varying with the elevation angle of the satellite. For path loss modeling, while free-space path loss forms the basis, additional considerations for clutter loss and shadowing effects are incorporated to accurately represent signal attenuation caused by obstacles and terrain. The channel model also considers atmospheric gas absorption parameters and ionospheric and tropospheric scintillation losses. These losses are significant under specific conditions, such as low elevation angles, intense solar activity periods, and low-latitude regions. Shadow regions indicate additional losses due to scintillation, assuming moderate scintillation intensity.

Release 15 presented two fast-fading models. One of these is a broad, frequency-selective model that originates from terrestrial channel models but is adjusted to fit the geometry of satellite communications. It takes into account different delay and arrival angles, along with their interrelationships, in a manner similar to the adjustments for shadow fading as well as clutter loss. Moreover, in the version of Release 15, channel models with tapped delay line and the clustered delay line are formulated for NTN link-level simulations.

2.4.2 Release-16 for New Radio NTN

After completing Rel-15's studies on New Radio (NR) support for NTN scenarios and channel models, 3GPP continued with further studies in Rel-16 to adapt NR for NTN solutions. The main goal was identifying the minimal necessary functions that enable NR to support NTN. These functions encompassed considerations for system architecture, higher-layer protocols, and physical layer aspects. The results were detailed in 3GPP TR 38.821.

The Next Generation Radio Access Network (NG-RAN) supports the split of 5G base (gNB) stations into a Distributed Unit (DU) and a Central Unit (CU). Figure 2.8 presents some NTN-based NG-RAN architecture options. The NR higher-layer protocol stack is divided into the user plane (UP), which handles data transmission, and the control plane, which manages signaling. For the user plane, the primary challenge arises from the long propagation delays characteristic of NTNs. Consequently, Release 16 investigated the impact of these long delays on several protocols, including Medium Access Control (MAC), Radio Link Control (RLC), Packet Data Convergence Protocol (PDCP), and Service Data Adaptation Protocol (SDAP) [12]. Enhancements to the MAC layer were identified as necessary to address issues related to random access, discontinuous reception (DRX), scheduling requests, and Hybrid Automatic Repeat Request (HARQ).

Mobility management was a primary focus for the CP due to the NTN platform, particularly for the movement of LEO satellites. In idle mode, NTN-specific system information had to be introduced. Earth-fixed tracking areas can help avoid frequent tracking area updates. Moreover, it is beneficial to add auxiliary information for cell selection. Efficient handover schemes are necessary for the rapid movement of satellites.

From a physical layer perspective, extensive link-level and system-level evaluations were carried out within S-band and Ka-band. Results concluded that LEO and GEO satellites could serve handheld user equipment (UE) in the S-band with appropriate satellite beam layouts. Conversely, other UEs with higher transmit and receive antenna gains, such as VSATs and UEs equipped with suitable phased array antennas, could be served by LEO and GEO satellites in both S-band and Ka-band.

Despite challenges such as long propagation delays, significant Doppler shifts, and moving cells in NTN, Rel-15 and Rel-16 NR features have laid a robust foundation for supporting NTNs.

2.4.3 Release-17 for NR NTN

Building on the work from Release 16, 3GPP has furthered its investigation of NTNs in NR Release 17. The goal is to define the required enhancements for NTNs, including those based on LEO and GEO satellites, as well as support for HAPS and air-to-ground networks. The scope of this research encompasses various aspects, such as the physical layer, protocols, system architecture, radio resource management, RF requirements, and the frequency bands used [13]. The primary focus is on transparent payload architectures with geostationary tracking areas and frequency division duplex (FDD) systems, under the assumption that all UE is equipped with Global Navigation Satellite System (GNSS) capabilities.

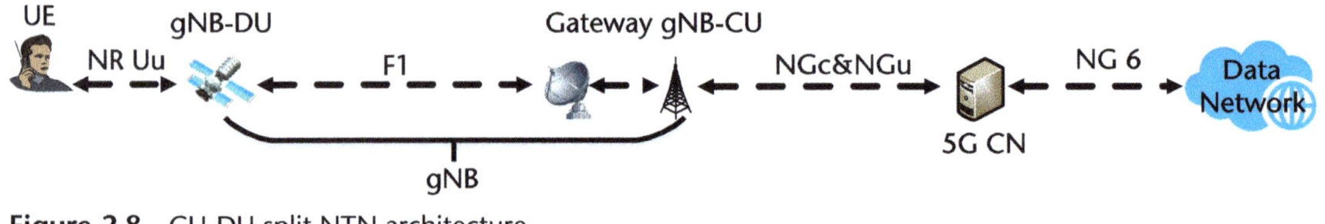

Figure 2.8 CU-DU split NTN architecture.

In terrestrial NR networks, uplink timing is synchronized based on the downlink reception timing, and the propagation time is typically much shorter than the transmission time slot. However, in NTN environments, the propagation time is significantly longer than the transmission time slot. UEs equipped with GNSS capabilities can determine the relative speed between the UE and the satellite, as well as the round-trip time (RTT) between them, by using their own position and the NTN ephemeris data. Using the calculated relative speed, the UE can apply Doppler precompensation to ensure that its uplink signals are received at the correct frequency by the satellite or gNB. The gNB provides a common timing advance (TA) to the UE, which represents the RTT between the satellite and the gNB. The UE then adds the RTT between itself and the satellite to this common TA to derive the complete TA. This complete TA serves as the offset between the downlink timing received by the UE and the uplink transmission timing, ensuring proper synchronization.

In the transmission mechanism of Release 16 NR, up to 16 stop-and-wait HARQ processes are employed for continuous data transmission. These HARQ processes cannot be reused for new transmissions until feedback for the previous transmission is received. In scenarios with long RTTs and the use of the stop-and-wait protocol, transmissions can stall when all HARQ processes are waiting for feedback, leading to reduced communication throughput. To address this issue, the number of HARQ processes was increased to 32, which helps mitigate stalling in some air-to-ground scenarios. However, even 32 HARQ processes are insufficient to handle the RTT in NTNs based on LEO and GEO satellites. Since further expanding the number of HARQ processes is not desirable, mechanisms must be implemented to reuse the same HARQ process before the full RTT has elapsed to prevent stalling. For downlink transmissions, if a HARQ process is reused before the RTT has completed, HARQ feedback is not required and is disabled. In the uplink, no HARQ feedback is used, and the gNB can dynamically decide whether to reuse the HARQ process before the RTT has passed. This decision is made by sending authorization for new data or retransmissions, or by deciding to send retransmission authorization after decoding the uplink transmission.

For HARQ processes with feedback disabled, the UE does not need to monitor for retransmission assignments after a certain period, which helps conserve energy. When HARQ is not used for retransmissions, link adaptation can aim for a lower block error rate. However, this approach requires higher RLC retransmission rates and more frequent RLC status reports to ensure robustness.

To accommodate the long RTT in NTNs, specific MAC and RLC timers have been extended. As satellites move, UEs must select new satellites based on both existing standards and new criteria, such as when a satellite no longer covers the UE's area. Conditional handover has been enhanced with new criteria that take into account the UE's location and the timing of satellite coverage, thereby improving the measurement and handover process.

2.4.4 Release-18 for NR NTN

In June 2022, the international standards organization 3GPP completed the first standard version for NTN with low-frequency bands and a transparent forwarding mode in 5G Rel-17. For the current R18 phase, network-side efforts are advancing the space-based deployment of core network user plane functions (UPF), and the radio access network side is enhancing link transmissions required for mobile phone direct-to-satellite connections [14]. Looking forward to R19, it is anticipated that 3GPP will further consider the standardization of onboard processing modes and beam-hopping technology enhancements. This progression aims to support diverse networking modes for satellite-terrestrial integrated communications and to achieve 6G satellite-terrestrial integrated network capabilities by 2030. Concurrently, the technical validation and networking of satellite-terrestrial integrated communications will proceed in three phases:

1. 2023–2025: Technical validation and commercial networking of direct mobile-to-satellite connections based on 5G Rel-17 NTN;
2. 2025–2030: Technical validation and commercial networking of direct mobile-to-satellite connections based on later versions of 5G NTN following the expansion of satellite capabilities;
3. After 2030: Realization of technical validation and commercial networking applications based on 6G integrated networks.

2.4.5 Other Initiatives

Several organizations and projects have also discussed integrating satellite communication networks with terrestrial 5G networks. For instance, the International Telecommunication Union (ITU), the European Union (EU) H2020-funded Shared Access Terrestrial Satellite Backhaul Network enabled by Smart Antennas (SANSA) project, and the European Space Agency-initiated SATis5 project are actively engaged in these topics.

The ITU launched a research project in 2016 aimed at integrating satellite communication systems into Next Generation Access Technologies (NGAT). The project focuses on the key elements of satellite network integration and proposes four application scenarios for network convergence: relay transmission, hyperlink communication, cell backhaul and broadcast distribution, and hybrid multimedia services.

The EU H2020 5GPPP funded the Sat5G project in its second phase to evaluate the architecture of satellite access networks integrated with 5G networks and validate key technologies. To achieve these goals, Sat5G defined four application scenarios for satellite 5G: multimedia content distribution and offloading based on satellite broadcast services, satellite-based 5G fixed backhaul services, satellite-based 5G to the home services, and satellite-based 5G mobile platform backhaul services. In June 2019, the Sat5G project team successfully demonstrated the integrated architecture of satellite and terrestrial networks at the European Conference on Networks and Communications (EUCNC).

2.5 Possible Challenges

At present, there are three mismatches in satellite communications: a mismatch between satellite capacity and communication capacity, a mismatch between radio-frequency capacity and baseband capacity, and a mismatch between cost and commercial value [15]. The cost, capability, and scale of satellites constrain each other. The cost of satellite manufacturing is relatively high, and the manufacturing efficiency still has a distance in meeting the demand for bulk commercialization. The cost of satellite launching is relatively high, the current number of shifts is relatively small, and the demand for batch deployment is still to be met. The supply chain system mainly adopts the major project system model, and the maturity still needs to be improved. The closed loop of a business model has not yet been formed, and there are certain difficulties in sustainable investment. With the technological and industrial integration of space and ground communications, it is expected to further enhance capacity, expand scale, and reduce costs. The specific challenges are introduced as follows.

Air interface design: Unlike terrestrial mobile communication, which typically operates under multipath fading channels, satellite communication channels are mainly characterized by strong line-of-sight (LoS) channels. Significant differences exist in terms of Doppler shift, channel attenuation, propagation delay, and time drift. Additionally, the linearity of RF channels (including power amplifiers) and variations in signal strength are fundamentally different. These discrepancies present considerable challenges for unified air interface design, directly impacting the design of critical technical solutions such as waveform, synchronization, random access, and beamforming coordination. Furthermore, technological innovations are required to optimize the performance of phased array antennas and power amplifier efficiency.

Seamless handover: The integrated network is characterized by a large spatial-temporal scale, high dynamics, and multilayer heterogeneity. It is centered around services/users and needs to adapt flexibly to various dynamic services and space network environments with limited onboard payload resources. These factors not only affect the logical architecture of the network but also the implementation and deployment of network functions, as well as resource management and mobility management associated with the architecture. As a result, seamless handover becomes a key challenge.

For scenarios with multiple coverage layers involving satellite and terrestrial networks, the handover process includes at least the transition between satellite and terrestrial networks and between high- and low-orbit multilayer satellite networks. The handover requirements primarily consider the continuity of data transmission and network load balancing. For terminals, both soft and hard handover methods exist. Suppose a terminal has dual- or multiconnection capabilities. In that case, zero-latency handover is possible, requiring the terminal to connect with the new network before disconnecting from the existing one. Handover prediction and data forwarding techniques are needed without dual-connection capabilities to minimize the impact on user data transmission.

Dynamic frequency allocation: In integrated networks, a multilayer-aided three-dimensional network is formed among high-orbit, low-orbit satellites and terrestrial base stations. As user demands grow, spectrum resources become increasingly scarce, and traditional rigid frequency segmentation significantly reduces transmission efficiency. To improve the utilization efficiency of frequency resources, it is necessary to study the transmission characteristics of multilayer space networks, leverage the differences in beams and coverage, and explore soft frequency reuse methods for satellite-terrestrial communication. By predicting interference and coordinating resources, further research into dynamic frequency sharing and reuse technologies and methods is essential, which is capable of enhancing the transmission efficiency at the cell edge and reducing interference at the cell edge.

Integration security: In satellite networks, security and privacy are significantly more important than in traditional terrestrial networks. Satellite communications are carried out in a broadcast mode and can cover large geographical areas of the Earth. Therefore, the confidentiality of communications can be exposed if appropriate measures are not taken. In addition, satellites are characterized by special architectural features such as extended propagation time, limited computing power, and unstable dynamics of the network topology compared to traditional cellular networks. Extremely unstable wireless links and long propagation delays will significantly increase the delay in authentication. The limited computing power and storage capacity of the entities involved in LEO satellites are not suitable for realizing high-complexity algorithms [16]. The design of secure and efficient integrated security protocols is necessary.

References

[1] Ding, R., et al., "5G Integrated Satellite Communication Systems: Architectures, Air Interface, and Standardization," in *2020 International Conference on Wireless Communications and Signal Processing (WCSP)*, Nanjing, China, 2020, pp. 702–707.

[2] White Paper on Integrated Terrestrial-Satellite Communication, CICT Mobile, 2024, https://www2.sofarsolar.com/upload/file/20240207/1707243094170026993.pdf.

[3] Vanelli-Coralli, A., N. Chuberre, G. Masini, A. Guidotti, and M. El Jaafari, "NR NTN Architecture and Network Protocols," in *5G Non-Terrestrial Networks: Technologies, Standards, and System Design*, IEEE, 2024, pp. 91–109, doi:10.1002/9781119891185.ch4.

[4] Zhang, Z., H. Guo, and W. Xie, "Research of NTN Technical Scheme Based on 5G Network," in *2023 IEEE International Symposium on Broadband Multimedia Systems and Broadcasting (BMSB)*, Beijing, China, 2023, pp. 1–6.

[5] Zhong, P., T. Tan, and Y. Yu, "Enlightenment for Chinas LEO Internet Satellite Industry from Typical Development Model of European Commercial Satellite," in *2022 International Symposium on Networks, Computers and Communications (ISNCC)*, Shenzhen, China, 2022, pp. 1–6.

[6] Harounabadi, M., and T. Heyn, "Toward Integration of 6G-NTN to Terrestrial Mobile Networks: Research and Standardization Aspects," *IEEE Wireless Communications*, Vol. 30, No. 6, December 2023, pp. 20–26.

[7] Ge, M., R. Zhu, K. Li, J. Wei, H. Sang, and X. Hou, "Convergence-Efficient Satellite-Ground Federated Learning for LEO Mega Constellations Optical Networks," *2023 21st International Conference on Optical Communications and Networks (ICOCN)*, Qufu, China, 2023, pp. 1–3.

[8] Wang, S., R. Li, Y. Han, and M. Yao, "Opportunities and Challenges of Antenna Design for Future 5G Mobile Terminals," in *2021 Cross Strait Radio Science and Wireless Technology Conference (CSRSWTC)*, Shenzhen, China, 2021, pp. 115–116.

[9] Naous, T., M. Itani, M. Awad, and S. Sharafeddine, "Reinforcement Learning in the Sky: A Survey on Enabling Intelligence in NTN-Based Communications," *IEEE Access*, Vol. 11, 2023, pp. 19941–19968.

[10] Le, T.-K., U. Salim, and F. Kaltenberger, "An Overview of Physical Layer Design for Ultra-Reliable Low-Latency Communications in 3GPP Releases 15, 16, and 17," *IEEE Access*, Vol. 9, 2021, pp. 433–444.

[11] Ghosh, A., A. Maeder, M. Baker, and D. Chandramouli, "5G Evolution: A View on 5G Cellular Technology Beyond 3GPP Release 15," *IEEE Access*, Vol. 7, 2019, pp. 127639127651.

[12] Inoue, T., "5G NR Release 16 and Millimeter Wave Integrated Access and Backhaul," *2020 IEEE Radio and Wireless Symposium (RWS)*, San Antonio, TX, USA, 2020, pp. 56–59.

[13] Liu, W., X. Hou, J. Wang, L. Chen, and S. Yoshioka, "Uplink Time Synchronization Method and Procedure in Release-17 NR NTN," *2022 IEEE 95th Vehicular Technology Conference: (VTC2022-Spring)*, Helsinki, Finland, 2022, pp. 1–5.

[14] Vanelli-Coralli, A., N. Chuberre, G. Masini, A. Guidotti, and M. el Jaafari, "Release 18 and Beyond, 5G Non-Terrestrial Networks: Technologies, Standards, and System Design," *IEEE*, 2024, pp. 261–279.

[15] Judice, A., J. Livin, and K. Venusamy, "Research Trends, Challenges, Future Prospects of Satellite Communications," *2022 2nd International Conference on Advance Computing and Innovative Technologies in Engineering (ICACITE)*, Greater Noida, India, 2022, pp. 1140–1143.

[16] Ahmad, I., J. Suomalainen, P. Porambage, A. Gurtov, J. Huusko, and M. Hyhty, "Security of Satellite-Terrestrial Communications: Challenges and Potential Solutions," *IEEE Access*, Vol. 10, 2022, pp. 96038–96052.

CHAPTER 3

Constellation Design for Satellite-Terrestrial Integrated Communication

With the development of communication technology and influenced by factors such as the surge in mobile users, the dramatic increase in communication data and the frequent occurrence of natural disasters, traditional terrestrial communication systems have struggled to meet the modern communication needs of people. Compared to terrestrial communication systems, LEO satellite communication systems have broader communication coverage, more reliable transmission, larger transmission capacity, stronger resistance to destruction, and less restriction by ground conditions. These advantages effectively compensate for the shortcomings of terrestrial communication systems, thus attracting the attention of communication researchers worldwide [1–3].

Since each LEO satellite's coverage area is unstable, a single satellite cannot meet the communication requirements for global coverage. The idea of using multiple satellites to operate together as a system, known as a satellite constellation, has been proposed to address this shortcoming. Satellite constellations have significant technical advantages and a wide range of application scenarios, so the construction of satellite constellation systems has become the main program for satellite communications, making it a hot research topic for communication-related researchers in various countries [4–6]. As shown in Figure 3.1, the service scenario of a LEO satellite communication system is formed by multiple satellites with intersatellite links. Under shared control, these satellites can provide various communication services. This chapter systematically overviews the definition, types, and design methods of satellite internet constellations and introduces their development history and the design methods of several classic constellations.

3.1 Overview of Satellite Constellations

3.1.1 Definition of Satellite Constellation

A satellite constellation, also known as a satellite system, is a communication system formed by a group of artificial satellites operating together. Since a single satellite can only cover a small part of an area and is unstable while a satellite constellation can provide stable global or near-global coverage, building satellite constellations has become the primary trend in satellite communication. In a satellite constellation, each satellite is usually positioned in a complementary

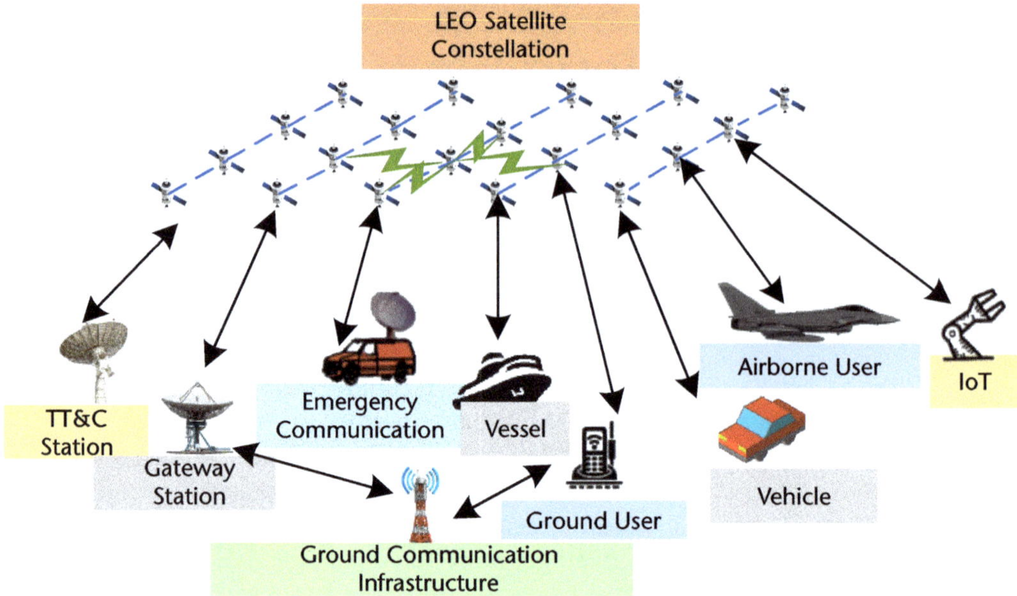

Figure 3.1 Schematic diagram of a LEO satellite constellation service scenario.

set of orbital planes and establishes connections with ground stations on Earth. Furthermore, communication between satellites is also possible.

A satellite constellation is a collection of satellites organized according to certain rules, working together to achieve specific functions. It is the primary form of multiple satellites working together. At the same time, a satellite constellation can also be defined as satellites of similar types and functions distributed in similar or complementary orbits and working together under shared control to accomplish specific tasks. In practical applications, satellite constellations have significant advantages over single satellites and are increasingly receiving widespread attention and application.

3.1.2 Development of Satellite Constellations

The satellite internet communication system has overcome many limitations of traditional ground-based communications, playing a crucial role in emergency rescue, live global broadcasting, remote communication, and marine fisheries. In 1957 [7], the successful launch and orbit of the first artificial satellite marked a new era in satellite communication. Subsequently, the concept of using multiple satellites to create a network for global coverage began to take shape. This idea can be traced back to 1945 when Arthur C. Clarke published an article in *Wireless World* [8], where he first proposed the concept of a satellite constellation. In the article, Clarke pointed out that placing three artificial satellites evenly in the GEO could achieve long-distance wireless communication excluding the poles. His idea of forming a system by networking multiple satellites laid a solid foundation for the subsequent research and development of satellite constellation systems.

The exploration of satellite internet began in the 1980s with Motorola's development of the Iridium system. Later, scholars from various countries proposed numerous satellite constellation deployment plans. Among the globally renowned satellite constellations are Starlink, OneWeb, and Hongyan, with their deployment plan details shown in Table 3.1.

Iridium constellation: In 1987, the Iridium system [9] was proposed as the world's first satellite constellation system with intersatellite links, and it was launched into operation in November 1998. The Iridium system has undergone updates and iterations, and it became commercially available in 2018.

Orbcomm constellation: In 1998, the Orbcomm system [10] was launched into operation, providing short data communication and positioning services to users worldwide through heterogeneous orbits. The Orbcomm constellation was the world's first bidirectional short data LEO microsatellite constellation system, providing a valuable reference for developing subsequent satellite constellation systems.

GlobalStar constellation: The GlobalStar system [11] mainly offered mobile communication services such as voice, fax, data, short messaging, and positioning. The satellite communication services provided by GlobalStar have served as an essential supplement to global communication.

OneWeb constellation: The OneWeb constellation [12] system was proposed in 2013. The OneWeb satellite network is utilized for internet access and various applications, including enterprise, government, maritime, aviation, and land mobile services.

Starlink constellation: The Starlink constellation [13] is currently one of the most popular constellations. Starlink's service has now been activated in multiple countries and regions worldwide, providing a new high-speed internet access option for users globally.

Hongyan constellation: Hongyan constellation system, independently developed by the China Aerospace Science and Technology Corporation (CASC), is a global satellite communication system primarily aimed at advancing the integration of ground networks and space-based systems, providing users with real-time data communication and various information services.

Table 3.1 Global Satellite Constellation Deployment Plans

Names of Constellations	*Country*	*Company*	*Initial Launch Date*	*Total Number*	*Orbital Altitude (km)*	*Orbital Inclination (°)*	*Orbit Type*
Iridium	USA	Iridium Communications Inc.	1997	66	780	86.4	Polar orbit
Orbcomm	USA + CAN	Orbcomm Inc.	1999	78	740+825	45	Inclined circular orbit
GlobalStar	USA	Globalstar Inc.	1998	56	1,414	52	Inclined circular orbit
OneWeb	UK	OneWeb	2019	720	1,200	87.9	Polar orbit
Starlink	USA	SpaceX	2019	11943	340–1,325	42–81	Inclined circular orbit
Hongyan	CHN	CASC	2018	324	1,100	50	Circular orbit

3.2 Classification of Satellite Constellations

Satellite constellations can be classified into various types based on their coverage performance (ground coverage area, constellation coverage density, coverage time distribution rate), orbital distribution (satellite spatial distribution), and satellite application functions [14]. For example, based on coverage performance, satellite constellations can be divided into global coverage constellations (Starlink, Iridium, OneWeb) and regional coverage constellations (O3b, Galileo, Eutelsat). According to the orbital distribution, satellite constellations can be divided into LEO constellations (Starlink, OneWeb), MEO constellations (O3b, Galileo), GEO constellations (Eutelsat), and so on. According to application functions, satellite constellations can be divided into communication constellations (Starlink Iridium, OneWeb), navigation constellations (GPS, Galileo), remote sensing constellations (Landsat), and so on. Each type of constellation has specific functions that can accomplish different communication tasks. Furthermore, classification standards are not mutually exclusive so a satellite constellation can belong to multiple types simultaneously. Many mature solutions exist for global coverage satellite constellations, such as the Walker, star, and flower. Next, these satellite constellations will be introduced in detail.

3.2.1 Walker Constellation

The Walker constellation is an inclined circular orbit satellite constellation proposed by John Walker [15] of the British Aerospace Institute. The Walker constellation has two unique configurations: the delta and the Rosette. Initially, the Walker constellation referred to the delta constellation, whose structure is shown in Figure 3.2. It consists of three orbital planes, each with satellite ground track intersections that form a pattern similar to the Greek letter delta when viewed from the pole. Later, American scholar Ballard conducted optimization research on the Walker constellation and designed the rose constellation [16]. The Rose constellation consists of more orbital planes.

Below, we will provide a brief overview of the characteristic and configuration parameter model for the delta constellation within the Walker constellation. The

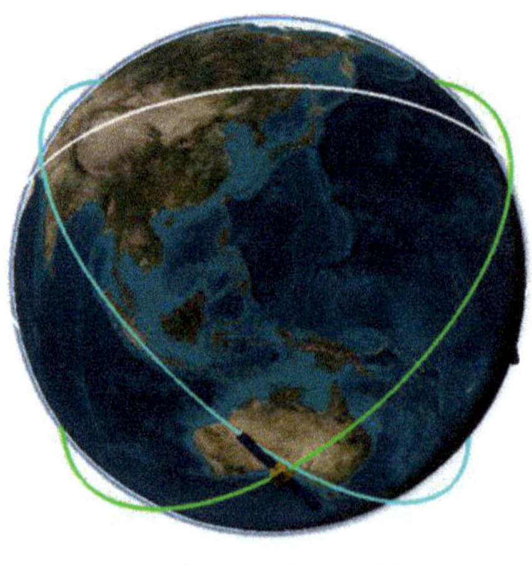

Figure 3.2 Delta constellation structure diagram (viewed from the pole).

delta constellation consists of multiple orbital planes with the same altitude and inclination. Satellites are evenly spaced and distributed on each orbital plane, providing stable communication services for coverage areas. Given the specific satellite orbital altitude and inclination, the Delta constellation only requires three parameters, N, P, and F, to determine the distribution of satellites within the constellation. Here, N represents the total number of satellites in the constellation, P represents the number of orbital planes, and F represents the phase factor (with values ranging from 0 to $P - 1$). The delta constellation configuration parameter model can be described as follows:

$$\begin{cases} i_k = i \\ a_k = a \\ e_k = 0 \\ \omega_k = 0^\circ \\ \Omega_k = \Omega_0 + \frac{360}{P}(k-1) \\ \theta_{k,m} = \theta_{0,0} + F(k-1)\frac{360}{N} + P(m-1)\frac{360}{N} \end{cases} \tag{3.1}$$

where $k(k = 1, 2, \ldots, P)$ represents the index number of the orbital plane in the constellation, m $(m = 1, 2, \ldots, N_{sat} - 1)$ denotes the index number of the satellite within the orbital plane, N_{sat} is the number of satellites in each orbital plane, and $[k, m]$ is the mth satellite in the kth orbital plane.

3.2.2 Star Constellation

The characteristics of the star constellation are as follows: all orbits in the constellation share a standard pair of nodes, and adjacent orbits in the same direction have approximately equal or equal inclinations. The ground tracks are most densely packed at the nodes and most sparse in between the nodes. The relative positions of satellites in codirectional orbits remain relatively stable, while those in reverse orbits experience significant changes. Polar orbit constellations are typical star constellations with codirectional orbits, counterdirectional orbits, and counterdirectional gaps at the poles.

A polar orbit constellation is a circular orbit constellation, with the design concept proposed by R.D. Luders [17]. As shown in Figure 3.3, the polar orbit constellation has an orbital inclination of 90°, enabling seamless coverage of the Earth. Similar to the Walker constellation, in a polar orbit constellation, each orbit has the same altitude, and each orbital plane contains an equal number of satellites distributed at equal intervals. The configuration parameter model of the polar orbit constellation can be represented as follows:

$$\begin{cases} i_k = 90^\circ \\ a_k = a \\ e_k = 0 \\ \omega_k = 0^\circ \\ \Omega_k = \frac{360}{P}(j-1)(j = 1, 2, \ldots, P) \\ \theta_k = (N_k - 1)\frac{360}{N_{sat}}(N_k = 1, 2, \ldots, N_{sat} - 1) \end{cases} \tag{3.2}$$

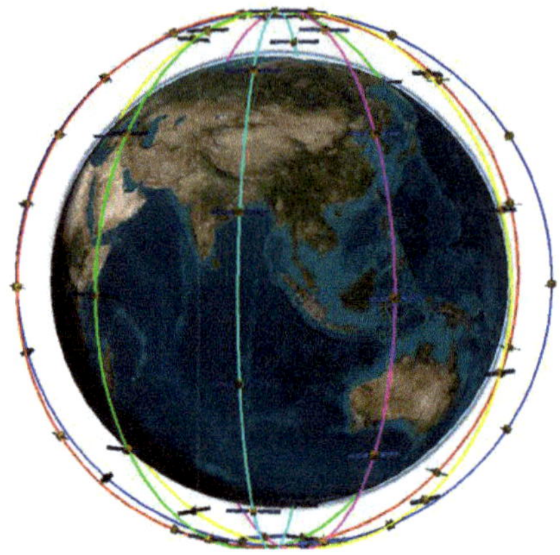

Figure 3.3 Polar orbit constellation structure diagram (viewed from the equator).

3.2.3 Flower Constellation

A flower constellation is a type of repeating ground track satellite constellation where multiple satellites share the same ground track. This concept was proposed by Professor Mortari [18] from the United States. In a flower constellation, each satellite has the same orbital altitude, argument of perigee, and orbital inclination. The satellites' ground tracks form a pattern resembling a flower when observed from the pole, thus earning the "Flower" designation.

The parameters of a flower constellation include two types: orbital parameters of individual satellites and constellation configuration parameters [18]. There are six most critical orbital parameters: semimajor axis (SMA) α, inclination (INC) i, perigee altitude h_ρ, right ascension of ascending node (RAAN) Ω, argument of perigee (AoP) ω, and true anomaly θ.

The flower constellation uses quasi-repeating orbits to achieve global coverage. Additionally, the flower constellation can use the number of orbits per repeating cycle N_p and the number of days per repeating cycle N_d to replace the semimajor axis a. For a simple two-body problem, these three parameters have the following relationship:

$$\begin{cases} n_e \cdot N_d = n \cdot N_p \\ n^2 \cdot a^3 = \mu \end{cases} \tag{3.3}$$

where n_e represents the angular velocity of Earth's rotation, n denotes the angular velocity of the satellite's orbit, and μ is Earth's gravitational constant.

The flower constellation has only one important configuration parameter, which is the number of orbital planes F_d. However, the following points need to be considered when determining the number of orbital planes:

1. In a flower constellation, the maximum number of satellites in each orbital plane is equal to the number of days per repeating cycle N_d. Therefore, the maximum number of satellites in the flower constellation is $N = F_d * N_d$.
2. The distribution characteristics of the orbital slots need to be considered. For example, for orbital planes with satellites evenly spaced, it is only necessary to design the RAAN of the first orbital plane Ω_0 and the initial TA of the first

satellite θ_0 [19]. Similarly, for the two-body problem, the orbital positions of the first and the kth satellites in adjacent orbital planes can be calculated using the following equation:

$$\begin{cases} \Delta\Omega = \dfrac{2\pi}{F_d} \\ \Delta\theta(0) = -\Delta\Omega \cdot \dfrac{n}{n_e} \\ \Delta\theta(k) = \Delta\theta(0) + k \cdot \dfrac{2\pi}{N_d} \end{cases} \tag{3.4}$$

A restrictive constellation configuration is a special type of constellation configuration that restricts the RAAN of all orbital planes within a certain range of longitudes. Adopting a restrictive flower constellation configuration can meet the requirements of regional navigation or coverage, ensuring that each orbital plane is evenly distributed within a specific range of longitudes (instead of over the entire range), while reducing the total number of satellites N to lower costs [19]. Therefore, when using a restrictive flower constellation configuration, the configuration parameters also include the range of RAAN. In this case, only the first equation in (3.4) needs to be modified:

$$\begin{cases} \Delta\Omega = \dfrac{\Omega_e xtend}{F_d} \\ \Delta\theta(0) = -\Delta\Omega \cdot \dfrac{n}{n_e} \\ \Delta\theta(k) = \Delta\theta(0) + k \cdot \dfrac{2\pi}{N_d} \end{cases} \tag{3.5}$$

The determination of the range of RAAN Ω_{extend} for a restrictive flower constellation configuration needs to be done manually, and the steps to determine this range are as follows [20–22]:

1. Determine the four orbital parameters shared by all satellites in the constellation, namely $\{a, e, i, \omega\}$.

Table 3.2 Summary and Comparison of Three Types of Constellation Configurations

Constellation Configuration	*Walker Constellation*	*Flower Constellation*	*Star Constellation*
Basic principles	The orbit type and orbit altitude are the same, with all orbits utilizing circular orbits. Satellites are evenly spaced within each orbital plane.	The orbit type and orbit altitude are identical, with all orbits utilizing retrograde orbits.	The orbit type and orbit altitude are the same, with all orbits utilizing polar orbits. Satellites are evenly spaced within each orbit.
Coverage capability	There are coverage blind spots in high-latitude regions, while stable communication services can be provided in mid-to-low-latitude areas.	The satellites have the characteristic of repeating ground tracks, resulting in a more compact coverage of specific areas.	Coverage is better in high-latitude regions, while coverage in mid-to-low-latitude regions is relatively poorer.
Application	Regional coverage.	Telemetry and control.	Global coverage.

2. Examine the distribution of RAAN and argument of ascending node separation $\{\Omega, u = \omega + \Omega\}$. Here, $\Omega \in [0, 2\pi)$, $u \in [0, 2\pi)$, with a search step typically set to 5°.
3. Calculate the visibility distribution $\{\Omega, u\}$ of satellites from each region, and then make a judgment. The specific determination rule is as follows: If the elevation angle between the region and the satellite is greater than the set elevation cutoff angle, it is determined that the satellite is visible from the region and is marked accordingly. Based on these marked points, a coverage map of satellite visibility can be plotted (usually with Ω on the x-axis and u on the y-axis), and the range of RAAN is determined accordingly.

3.2.4 Classical Satellite Constellation Design Solution

3.2.4.1 Design Solutions for Polar and Near-Polar Orbit Constellations

American scientist R. D. Luders [17] first proposed the concept of polar orbit constellations to achieve global coverage. Subsequently, D. C. Beste [23] further optimized Luders' polar orbit constellation concept, minimizing the number of satellites required for the constellation system. An optimization method for polar orbit constellation design proposed by L. Rider [24] has been widely adopted in the academic community. This polar orbit constellation consists of multiple orbital planes with the same altitude and inclination, and each orbital plane is evenly populated with the same number of satellites. The phase difference between each satellite and the satellites in the same orbital plane is twice that between the satellites in the adjacent orbital planes.

Polar orbit constellations belong to the typical star constellation type, the characteristics of which were introduced in Section 3.2.3. An illustration of the polar orbit satellite constellation as observed from the pole is shown in Figure 3.4. The constellation includes same-direction orbits, opposite-direction orbits, and opposite-direction gaps. The shaded area in the figure represents the opposite-direction gap. The arrows on both sides of the boundary of the shaded area indicate

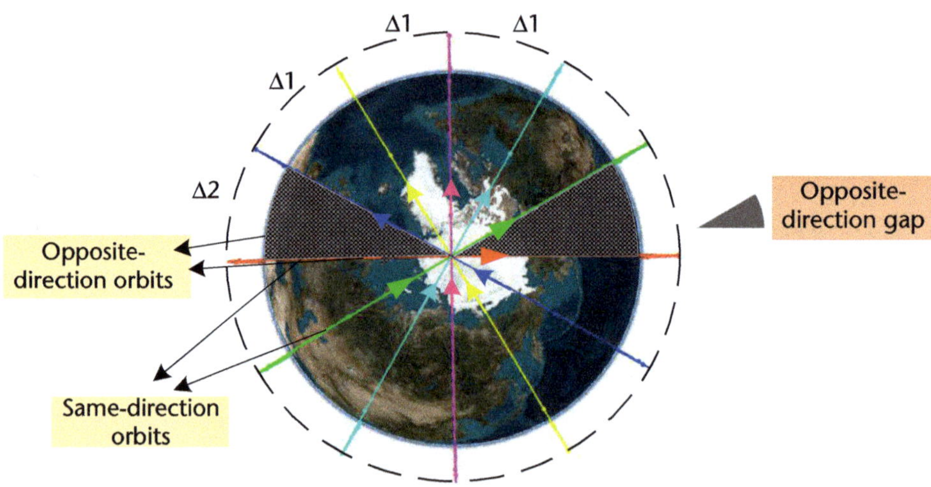

Figure 3.4 Polar orbit satellite constellation as observed from the pole.

the direction of satellite orbits around the Earth. The opposite-direction gap occurs because adjacent satellites in two orbital planes fly in opposite directions at high speeds, making it impossible to establish intersatellite links.

The design method for near-polar orbit satellite constellations is similar to that of polar orbit satellite constellations. By appropriately adjusting the orbital inclination angle and optimizing the coverage performance, near-polar orbit satellite constellations ensure that any two satellites in the constellation do not intersect at the poles, thereby avoiding satellite collisions at the North and South Poles. This optimization is based on the design theory of polar orbit satellite constellations. Near-polar orbit satellite constellations consist of multiple circular orbital planes with inclinations close to but less than 90°. Due to the different RAAN intervals between adjacent codirectional orbital planes, any two satellites in the constellation do not intersect at the poles, thereby preventing satellite collisions.

The design of polar orbit satellite constellations can use the coverage band method. The coverage band of a satellite constellation is shown in Figure 3.5. The primary method is to determine the parameters of the constellation based on the continuous coverage band formed by satellites in the same orbit. In the figure, θ is the angle formed between the satellite coverage area and the subsatellite point, α is the half-width of the coverage band, and N_p is the number of satellites on each orbit.

To achieve global coverage of the constellation, the parameters need to satisfy

$$\alpha = \arccos\left(\frac{\cos\theta}{\cos(\pi/N_p)}\right) \tag{3.6}$$

Assuming $\Delta 1$ is the RAAN difference between each codirectional orbital plane and $\Delta 2$ is the RAAN difference between each counterdirectional orbital plane. In conjunction with Figure 3.5, the adjacent coverage bands should satisfy:

$$\begin{cases} \Delta 1 = \theta + \alpha \\ \Delta 2 = 2\alpha \end{cases} \tag{3.7}$$

At the same time, the RAAN differences need to satisfy

$$(P - 1)\Delta 1 + \Delta 2 = \pi \tag{3.8}$$

where N is the total number of orbital planes in the constellation.

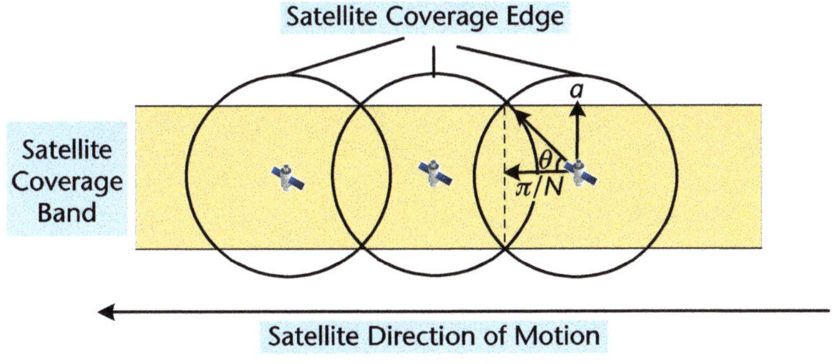

Figure 3.5 The coverage band of a satellite constellation.

Combining (3.6) to (3.8), the relationship between parameters P, N_p, and θ can be obtained:

$$(P-1)\theta + (P+1)\arccos\left(\frac{\cos\theta}{\cos(\pi/N_p)}\right) = \pi \tag{3.9}$$

3.2.4.2 Design Scheme for Inclined Circular Orbit Constellation

Among the extensive research on the optimal design of inclined circular orbit constellations, the schemes proposed by the British scientist Walker [25] and the American scientist Ballard [16] have gained recognition from researchers and are widely used. All the RAANs of the orbital planes in the Walker constellation are evenly distributed over the equatorial plane from 0° to 360°. The Walker constellation is further divided into two types: the delta (Δ) constellation and the rosette constellation. The structure of the delta constellation is shown in Figure 3.6(a), and the structure of the rosette constellation is shown in Figure 3.6(b). Although the structures of the two constellations differ, both belong to the Walker constellation. The Walker constellation consists of multiple circular orbital planes with the same altitude and inclination, each uniformly populated with the same number of artificial satellites.

Assuming the Walker constellation consists of P orbital planes, each uniformly populated with N_p satellites of the same orbital altitude h and inclination i, the total number of satellites is $N = P \times N_p$. As shown in Figure 3.7, the distribution of satellites in the Walker constellation can be determined by three parameters: N, P, and F. The phase factor F is related to the total number of satellites, and $0 \leq F \leq P-1$.

Due to the uniform symmetry of the Walker constellation, the phase angle difference between adjacent satellites in the same orbit can be expressed as

$$\Delta\omega = 2\pi/N_p \tag{3.10}$$

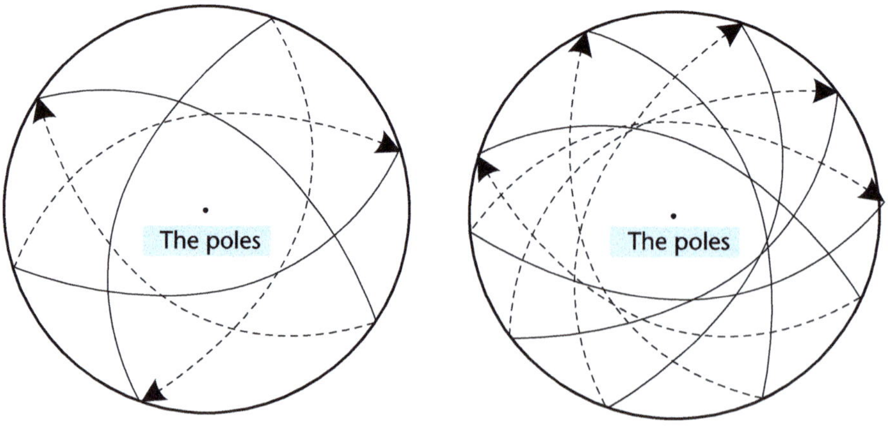

(a) (b)

Figure 3.6 Walker constellation diagram observed from the pole: (a) δ constellation, and (b) rosette constellation.

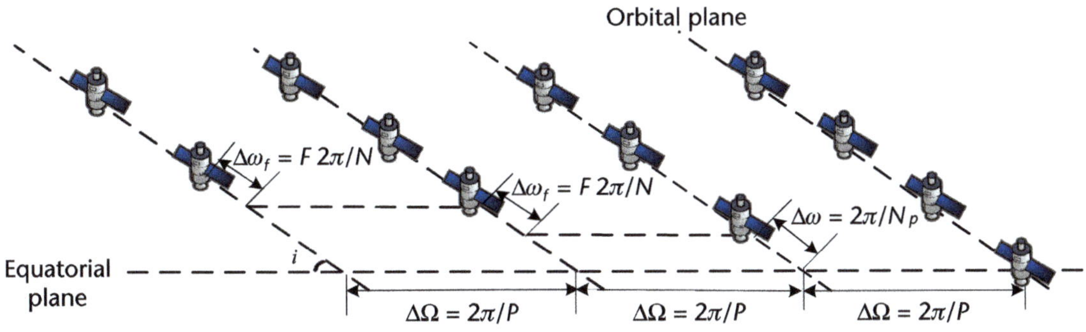

Figure 3.7 Walker constellation satellite distribution plane diagram.

The difference in RAAN between adjacent orbital planes is

$$\Delta\Omega = 2\pi/P \tag{3.11}$$

The phase difference between two adjacent satellites in adjacent orbital planes is

$$\Delta\omega_f = F \cdot 2\pi/N = 2\pi \cdot F/(P \cdot N_p) \tag{3.12}$$

As observed, the spatial configuration of the Walker constellation is influenced by the parameters of the number of orbital planes P and the phase factor F. These parameters also affect the coverage performance of the satellite constellation. As shown in the text and Figure 3.7, this formula represents the calculation of the phase difference between adjacent satellites in two neighboring orbital planes.

3.3 Satellite Constellation Design

The primary purpose of satellite constellation design is to ensure global coverage and reliability of satellite communication. Constellation design is a complex process where any parameter change directly or indirectly affects its communication performance. Factors such as the large number of satellites, the lengthy construction time, and the high management costs make satellite constellation design and construction increasingly complex and challenging. Therefore, designing appropriate satellite constellations to optimize system performance and reduce costs is a pressing issue in development. This section will introduce the basic theoretical knowledge of constellation design, evaluation metrics for assessing constellation configuration superiority, and two classical design methods for constellations.

3.3.1 Configuration Design

3.3.1.1 Coordinate and Time Systems

- Coordinate Systems

a. Earth-Centered Earth-Fixed (ECEF) Coordinate System

The ECEF coordinate system is defined with its origin at the Earth's center. The x-axis extends toward the point where the equator intersects the prime meridian. The y-axis, which completes the right-hand rule with the XOZ plane, is directed toward the North Pole. The z-axis aligns with the Earth's rotation axis, pointing toward the North Pole as well. This three-dimensional system is commonly used to describe positions in space, particularly for determining satellite ground tracks.

b. Longitude Latitude Altitude (LLA) Coordinate System

The LLA coordinate system is the most widely used Earth-based system, representing positions through longitude, latitude, and altitude values. It offers a more intuitive and tangible understanding compared to the Earth-centered coordinate system and is frequently employed after converting from that system. The LLA coordinate system is commonly utilized to describe the distribution of visible satellites across the Earth's surface.

c. Earth-Centered Inertial (ECI) Coordinate System

The ECI coordinate system is a celestial reference frame with its origin located at the Earth's center. The x-axis is directed toward the vernal equinox, while the z-axis points toward the North Pole. The y-axis completes a right-handed system with the XOZ plane. This coordinate system is primarily used to describe the motion of celestial bodies and is commonly applied in satellite orbit propagation calculations.

d. Radial, Tangential, Normal (RTN) Coordinate System

The RTN coordinate system is centered at the satellite's center of mass. The R-axis extends from the Earth's center to the satellite's center of mass, while the T-axis points along the direction of the satellite's motion within the orbital plane and is perpendicular to the R-axis. The N-axis completes the right-handed system with the RT plane. This coordinate system is used to describe the satellite's position and motion at any given moment in its orbit.

- Time Systems

a. Sidereal Time (ST)

Sidereal time is a timekeeping system that bases its measurements on the Earth's rotation relative to the fixed stars. Sidereal time is location-specific and is also referred to as local sidereal time.

b. International Atomic Time (TAI)

TAI is a highly precise time standard derived from the fundamental properties of time on Earth. It uses the extremely stable frequency of cesium atomic transitions to define the atomic second. TAI started at 0 hours on January 1, 1958, with a difference from Universal Time (UT) of 0.0039 seconds, represented as

$$(UT - TAI)_{1958.0} = 0.0039s \tag{3.13}$$

c. Coordinated Universal Time (UTC)

However, TAI's precision and stability does not account for the Earth's rotational variations. Hence, Coordinated Universal Time was introduced. UTC is based on Atomic Time but is adjusted to stay within 0.9 seconds of UT through leap seconds. This ensures synchronization with the Earth's rotation.

d. Terrestrial Time (TT)

Terrestrial Time is a modern time standard used on the Earth's surface, reflecting the mean solar time at the geoid level. TT is a dynamical time standard for geocentric reference frames, with a conversion relationship to TAI as

$$TT = TAI + 32.184s \tag{3.14}$$

Additionally, the cumulative relationship between TAI and UTC due to leap seconds is expressed as

$$UTC = TAI + \text{leap second} \tag{3.15}$$

In satellite constellation design, UTC is typically used instead of TT for orbit propagation due to the minimal difference and the broader acceptance of UTC.

3.3.1.2 Constellation Basic Parameters

Only the ideal two-body interaction between the satellite and the Earth is typically considered in satellite orbit design. The Keplerian orbital elements, also known as the six orbital parameters, describe the state of motion for a single satellite. These parameters include the semimajor axis α, eccentricity e, inclination i, right ascension of the ascending node (RAAN) Ω, the argument of perigee ω, and true anomaly υ. Each of these six parameters serves different functions in satellite design and can be categorized as follows:

- Defining Orbital Size and Shape
  Semi-major axis α: Describes the size of the satellite's orbit. Kepler's first law states that a satellite follows an elliptical orbit, where the semimajor axis represents half of the orbit's longest axis. For a circular orbit, α is equivalent to the orbit's radius.
  Eccentricity e: Describes the shape of the satellite's orbit calculated as the ratio of the distance between the ellipse's two foci to the length of the central axis. The value of e ranges from 0 to 1. When $e = 0$, the satellite's path is a circle centered on the Earth's center of mass; when $0 < e < 1$, the orbit is an ellipse with the gravitational body at one of the foci.
- Defining Orbital Orientation
  Inclination i: The angle between the Earth's equatorial plane and the satellite's orbital plane determines the orbit's tilt. When $i = 0°$, the orbital plane coincides with the Earth's equatorial plane, forming a geostationary orbit; when $0 < i < 90°$, the orbit is inclined; when $i = 90°$, the orbital plane is perpendicular to the equatorial plane, forming a polar orbit.
  RAAN Ω: The angle is measured along the equatorial plane from the vernal equinox to the ascending node, where the satellite moves from south to north as it crosses the equatorial plane. The value of Ω ranges from 0 to 360 degrees.
  Argument of perigee ω: The angle from the ascending node to the perigee is measured in the direction of the satellite's orbit. The value of ω ranges from 0 to 360 degrees. For circular orbits, ω is undefined. The spatial relationships of these three parameters are shown in Figure 3.8:

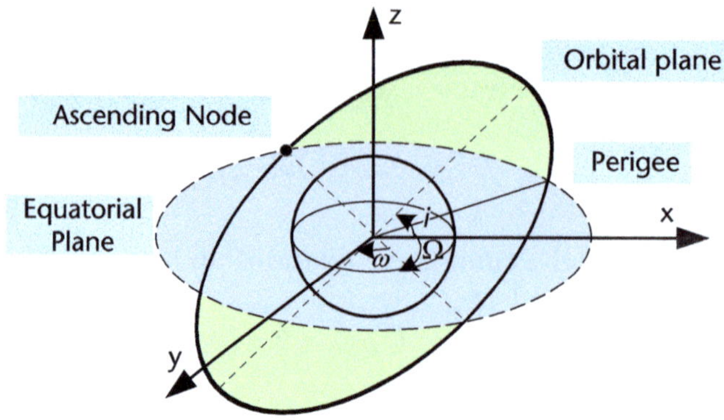

Figure 3.8 Spatial relationships of orbital parameters.

- Defining Satellite Position in the Orbit
  True Anomaly θ: The angle between the direction of the perigee and the current position of the satellite measured at the Earth's center. θ varies over time as the satellite moves along its orbit and is typically represented by the mean anomaly for simplicity in calculations.

These parameters collectively describe the complete state of a satellite's orbit, enabling precise orbit determination and prediction essential for constellation design and satellite operation.

3.3.2 Coverage Design

3.3.2.1 Single Satellite Coverage Characteristics

Each satellite has its coverage area. For a specific satellite, its coverage area refers to the range where signals transmitted from the satellite can propagate in a straight line and be received on the ground. In other words, it represents the area on the ground where satellite signals can be directly received. The coverage area for a single satellite is illustrated in Figure 3.9.

In the figure α is the ground angle; β is the half-angle, which is the angle formed between the line connecting the satellite to the center of the Earth and the line connecting the satellite to the observation point; E is the elevation angle representing

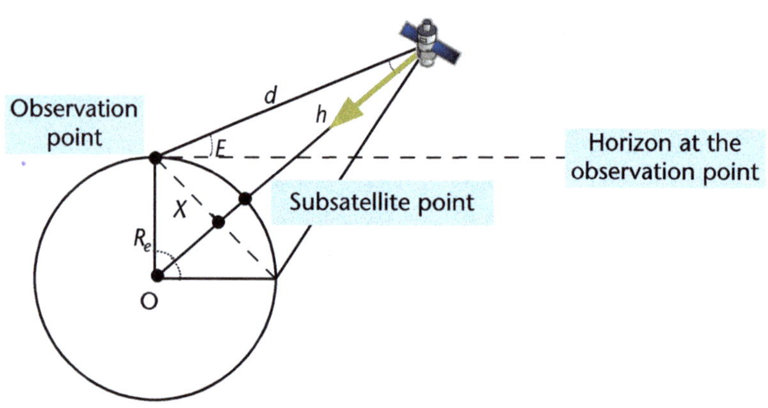

Figure 3.9 Illustration of single satellite coverage characteristics.

the angle between the observer's horizon and the line connecting the observer to the satellite; X is the radius of the satellite coverage area; d is the distance from the satellite to the ground station or user; R_e is the radius of the Earth; and h is the height of the satellite orbit.

Based on Figure 3.9, mathematical geometric knowledge can be utilized to derive the conversion relationships between various parameters (E, β, R_e, and h) that are easier to measure. The relationships are as follows:

The ground angle between the satellite and the observation point:

$$\alpha = \arccos\left[\frac{R_e}{h+R_e} \cdot \cos E\right] - E = \arcsin\left[\frac{h+R_e}{R_e} \cdot \sin\beta\right] - \beta \tag{3.16}$$

Half-angle of the satellite:

$$\beta = \arcsin\left[\frac{R_e}{h+R_e} \cdot \cos E\right] = \arctan\left[\frac{R_e \cdot \sin\alpha}{(h+R_e) - R_e \cdot \cos\alpha}\right] \tag{3.17}$$

Elevation angle at the observation point:

$$E = \arctan\left[\frac{(h+R_e)\cdot\cos\alpha - R_e}{(h+R_e)\cdot\sin\alpha}\right] = \arccos\left[\frac{h+R_e}{R_e} \cdot \sin\beta\right] \tag{3.18}$$

Distance between the observation point and the satellite:

$$\begin{aligned} d &= \sqrt{R_e^2 + (h+R_e)^2 - 2R_e(h+R_e)\cos\alpha} \\ &= \sqrt{R_e^2 \sin^2 E + 2hR_e + h^2} - R_e \sin E \end{aligned} \tag{3.19}$$

Coverage radius:

$$X = R_e \cdot \sin\alpha \tag{3.20}$$

Coverage area:

$$A = \pi R_e^2(1 - \cos^2\alpha) \tag{3.21}$$

It is important to note that, theoretically, the elevation angle ranges from 0 to $\pi/2$. However, when the elevation angle approaches 0, signal propagation becomes more susceptible to terrain and ground noise interference, leading to significant signal disruption. To avoid such occurrences, a parameter known as the minimum elevation angle is typically set in satellite communication system design. The satellite system cannot communicate effectively if the elevation angle is below this threshold.

3.3.3 Design Factor Analysis

Satellite constellation assessment is essential for satellite constellation design and optimization, primarily focusing on a comprehensive evaluation of the constellation's performance. Setting appropriate evaluation metrics is the prerequisite and basis for quantitatively assessing the satellite constellation. This section will introduce five performance evaluation metrics for satellite constellations: coverage performance, communication capacity, communication quality, reliability, and system cost.

Coverage performance evaluation metrics: Coverage performance evaluation metrics are typically divided into spatial domain and temporal domain indicators. In the spatial domain, indicators characterize the area covered by the satellite constellation, including coverage redundancy, coverage rate, and coverage percentage. In the temporal domain, indicators represent the coverage of target points and areas over time, system responsiveness, and the probability of coverage at a given time, including the average coverage gap, maximum coverage gap, and average response time.

Communication capacity evaluation metrics [26]: Communication capacity evaluation metrics are typically divided into space satellite- and ground user-side indicators. Satellite-side indicators mainly focus on theoretical capacity concepts, including channel capacity and throughput, reflecting the communication capability of the satellite constellation. User-side indicators mainly focus on engineering capacity concepts, including the number of users and channels, reflecting the number of users the satellite constellation can accommodate.

Communication quality evaluation metrics [27]: According to ETSI TS 102 250-1 specifications, communication quality evaluation metrics can be divided into three categories: service access quality indicators, service holding quality indicators, and service integrity quality indicators. Service access quality indicators include blocking rate and delay, service integrity quality indicators refer to the quality of service during user application service usage, mainly including packet loss rate and bit error rate, and service holding quality indicators are measured primarily by interruption probability.

System cost evaluation metrics [28]: Based on the cost composition of satellite constellations, system cost evaluation metrics are typically divided into four evaluation indicators: satellite construction cost, launch cost, insurance cost, and maintenance cost.

Reliability evaluation metrics [29]: The reliability of satellite constellations is closely related to satellite nodes and communication links and is typically divided into satellite survivability and link resilience indicators. Satellite survivability indicators refer to the survival capability of satellite nodes under system random failures, including interference resistance, intrusion resistance, and durability. Link resilience indicators characterize the stability of network topology under emergency conditions, including link connectivity and link robustness.

References

[1] Kodheli, O., E. Lagunas, N. Maturo, et al., "Satellite Communications in the New Space Era: A Survey and Future Challenges," *IEEE Communications Surveys and Tutorials*, Vol. 23, No. 1, 2020, pp. 70–109.

[2] Su, Y., Y. Liu, Y. Zhou, et al., "Broadband LEO Satellite Communications: Architectures and Key Technologies," *IEEE Wireless Communications*, Vol. 26, No. 2, 2019, pp. 55–61.

[3] Abo-Zeed, M., J. B. Din, I. Shayea, et al., "Survey on Land Mobile Satellite System: Challenges and Future Research Trends," *IEEE Access*, Vol. 7, 2019, pp. 137291–137304.

[4] Leyva-Mayorga, I., B. Soret, M. Röper, et al., "LEO Small-Satellite Constellations for 5G and Beyond-5G Communications," *IEEE Access*, Vol. 8, 2020, pp. 84955–184964.

[5] Marcuccio, S., S. Ullo, M. Carminati, et al., "Smaller Satellites, Larger Constellations: Trends and Design Issues for Earth Observation Systems," *IEEE Aerospace and Electronic Systems Magazine*, Vol. 34, No. 10, 2019, pp. 50–59.

[6] Qu Z., G. Zhang, H. Cao, et al., "LEO Satellite Constellation for Internet of Things," *IEEE Access*, Vol. 5, 2017, pp. 18391–18401.

[7] https://www.aps.org/apsnews/2007/10/soviets-launch-first-satellite-orbit.

[8] https://web.mit.edu/m-i-t/science_fiction/jenkins/jenkins_4.html.

[9] Leopold, R. J., and A. Miller, "The IRIDIUM Communications System," *IEEE Potentials*, Vol. 12, No. 2, 1993, pp. 6–9.

[10] Ilcev, S. D., "Orbcomm Space Segment for Mobile Satellite System (MSS)," in *2011 10th International Conference on Telecommunication in Modern Satellite Cable and Broadcasting Services (TELSIKS)*, 2011, pp. 689–692.

[11] Dietrich, F. J., P. Metzen, and P. Monte, "The Globalstar Cellular Satellite System," *IEEE Transactions on Antennas and Propagation*, Vol. 46, No. 6, 1998, pp. 935–942.

[12] De Selding, P. B., "Virgin, QUALCOMM Invest in OneWeb Satellite Internet Venture," *Space News*, January 15, 2015.

[13] De Selding, P. B., "SpaceX to Build 4,000 Broadband Satellites in Seattle," *Space News*, January 19, 2015.

[14] Magan, V., "Samsung Exec Envisions LEO constellation for Satellite Internet Connectivity," *Via Satellite*, August 18, 2015.

[15] Walker, J. G., "Satellite Constellations," *Journal of the British Interplanetary Society*, Vol. 37, 1984, p. 559.

[16] Ballard, A. H., "Rosette Constellations of Earth Satellites," *IEEE Transactions on Aerospace and Electronic Systems*, Vol. 5, 1980, pp. 656–673.

[17] Luders, R. D., "Satellite Networks for Continuous Zonal Coverage," *ARS Journal*, Vol. 31, No. 2, 1961, pp. 179–184.

[18] Mortari, D., M. P. Wilkins, and C. Bruccoleri, "The Flower Constellations," *Journal of Astronautical Sciences*, Vol. 52, No. 1, 2004, pp. 107–127.

[19] Wilkins, M. P., "The Flower Constellations: Theory, Design Process, and Applications," PhD Dissertation, Texas A&M University, 2004.

[20] Zhang, T. J., H. X. Shen, Z. Li, et al., "Restricted Constellation Design for Regional Navigation Augmentation," *Acta Astronautica*, Vol. 150, 2018, pp. 231–239.

[21] Ulybyshev, Y., "Satellite Constellation Design for Complex Coverage," *Journal of Spacecraft and Rockets*, Vol. 45, No. 4, 2008, pp. 843–849.

[22] Wilkinson, C. K., "Coverage Regions: How They Are Computed and Used," *Journal of the Astronautical Sciences*, Vol. 42, No. 1, 1994, pp. 47–70.

[23] Beste, D. C., "Design of Satellite Constellations for Optimal Continuous Coverage," *IEEE Transactions on Aerospace and Electronic Systems*, Vol. 3, 1978, pp. 466–473.

[24] Rider, L., "Optimized Polar Orbit Constellations for Redundant Earth Coverage," *Journal of the Astronautical Sciences*, 1985, Vol. 33, pp. 147–161.

[25] Walker, J. G., "Some Circular Orbit Patterns Providing Continuous Whole Coverage," *J. British Interplanetary Society*, 1971, Vol. 24, pp. 369–384.

[26] Abass, E. S., J. V. M. Halim, and H. M. El-Hennawy, "Coded-Beam Strategy for High Throughput Satellites—A Path Toward Terahertz Communications-System Model, Realization and Performance Assessment," in *2020 12th International Conference on Electrical Engineering (ICEENG)*, IEEE, 2020, pp. 181–185.

[27] ETSI, Speech and Multimedia Transmission Quality (STQ); QoS Aspects for Popular Services in Mobile Networks; Part 2: Definition of Quality of Service Parameters and Their Computation, ETSI TS 102 250-2 V2.7.1, 2019.

[28] Landis, G. A., S. G. Bailey, and R. Tischler, "Causes of Power-Related Satellite Failures," in *2006 IEEE 4th World Conference on Photovoltaic Energy Conference*, IEEE, Vol. 2, 2006, pp. 1943–1945.

[29] Fu, X., H. Yao, and Y. Yang, "Modeling Cascading Failures for Wireless Sensor Networks with Node and Link Capacity," *IEEE Transactions on Vehicular Technology*, Vol. 68, No. 8, 2019, pp. 7828–7840.

CHAPTER 4
Intersatellite Free-Space Optical Communication

Relative to conventional satellite microwave communications, free-space optical communications offer superior channel throughput. It also provides enhanced transmission bandwidth and robust anti-interference features. Additionally, it ensures increased security and privacy, while being lightweight and energy efficient. These attributes render it a promising alternative for next-generation satellite communications. This chapter presents the fundamental concepts and crucial technologies of laser communications and explores new challenges and prospective strategies to address the evolving dynamics of intersatellite connections.

4.1 Fundamentals

The free-space optical communication system should have the ability to autonomously calibrate while in orbit, form equivalent point-to-point connections among various satellite terminals, and maintain link stability. According to these specific functions, a laser communication system can be categorized into four primary subsystems: communication, control, optical, and acquisition pointing and tracking (APT), as depicted in Figure 4.1. Each subsystem is composed of various functional modules responsible for performing different tasks. These modules will coordinate and interact with other modules to ensure smooth, organized, and reliable system operation.

The communication subsystem serves as the core of the entire laser terminal, primarily responsible for the transmission and processing, modulation and demodulation, as well as sending and receiving of communication signals. The transmitting module is chiefly involved in the generation, modulation, amplification, and transmission of laser signals, including components in terms of laser transmitter, electro-optical modulator, amplifier, and driver. The receiving module is tasked with detecting and capturing laser signals. Its primary functions encompass laser signal detection, photoelectric conversion, amplification, and initial filtering. The message processing unit is able to amplify the electrical signal from the detector and accomplish tasks in terms of filtering, electro-optical conversion, coding, and decoding.

The optical subsystem is an integral part of the laser terminal and is vital for beam transmission, reception, and adjustment. It consists of two components: optical antennas for both transmitting and receiving, and a beam processing unit. The antenna system, effectively a telescope, comprises transmitting and receiving optical antennas. The transmitting antenna can increase the size of the beam spot, reduce

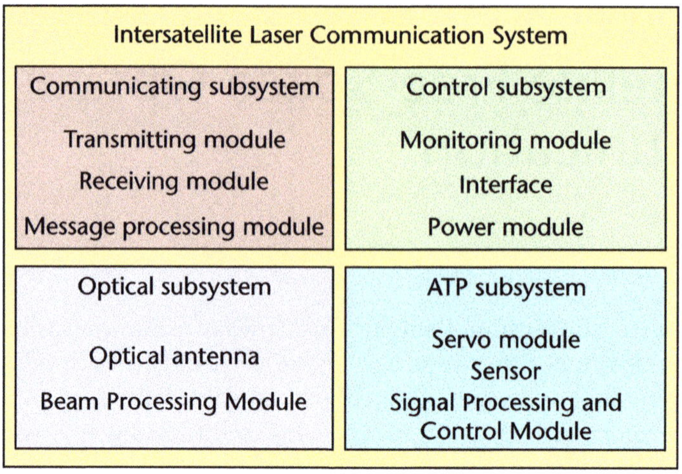

Figure 4.1 Diagram of system components.

the beam divergence angle, and thus enhance the transmission gain. As for the receiving antenna, it collects the laser light from the opposite terminal, utilizes the energy from the increased spot size, and boosts both the optical power and gain. The beam processing unit focuses on managing transmitted and received beams with processes such as beam shaping, collimation, filtering, splitting, and combining. Its key components include beam shapers, beam expanders, beam splitters, narrowband filters, optical path folding mirrors, and diaphragms to manage the field of view and eliminate stray light.

The main function of the control subsystem is to supervise, control, and administer the working modules of the laser communication system according to its established operational principles and procedures. This guarantees organized operation throughout the system, adherence to specific roles by each module, and realization of the desired functionalities.

The APT subsystem primarily handles the capture, tracking, and alignment of the target terminal while also establishing a stable laser connection. This subsystem includes modules in terms of beam detection, signal processing, control, and a servo mechanism. The beam detection module, essential for initiating the link, captures the laser beam at the launching end and relays the spot's position within the view of the detector. The signal processing and control module aims to process the received signals, ascertain the errors in acquisition and tracking, and generate control commands for the actuators. The function of the servo mechanism is to implement these control commands, correct any errors in capturing and tracking, and achieve precise alignment, which typically consists of a two-dimensional frame and a rapid response reflector.

From the start to the end of a communication task, the complete workflow of the system can be segmented into four stages according to the system structure and underlying principles: the initial pointing, the scanning and acquisition, the tracking, and the communication, as depicted in Figure 4.2.

1. *Initial pointing*: At this stage, the terminal's laser, optical modulator, sighting mirror, fast reflector, servo turntable, and detector perform self-checks

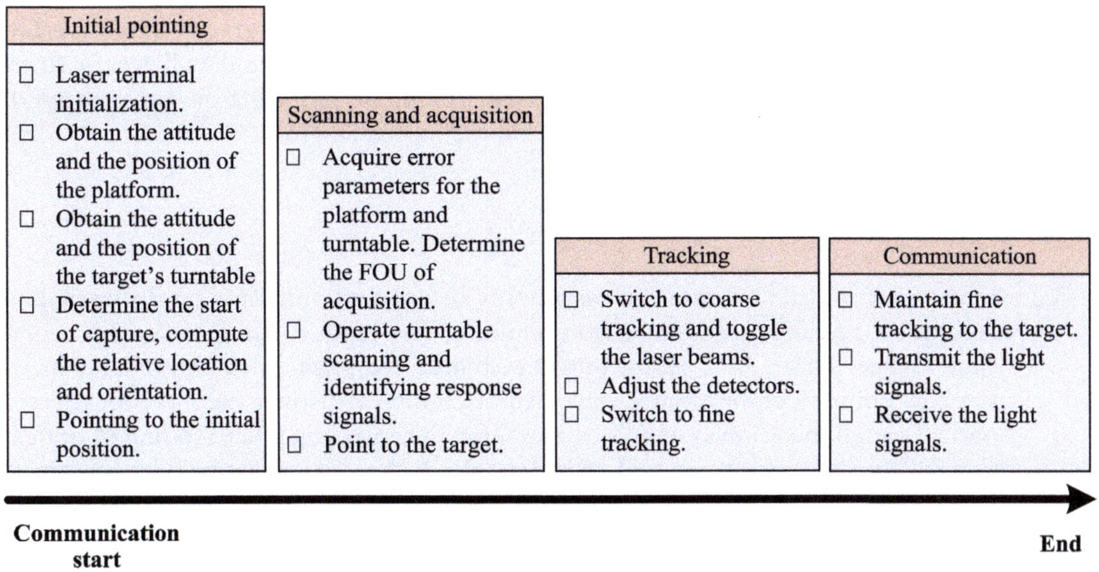

Figure 4.2 Flowchart of free-space optical communication systems.

and enter calibration mode, while the laser terminal initializes. The control subsystem gathers data on attitude, ephemeris, and local time to calculate the initial pointing direction. It then adjusts the servo turntable to align the optical axis with the cooperative target's anticipated position.

2. *Scanning and acquisition*: After adjusting the servo turntable, the visual axis aligns with the area of uncertainty (FOU) where the target is located, triggering the system to begin scanning. Using injected data on satellite attitude and orbit predictions, the system calculates the FOU size and determines the scanning path. The servo mechanism then systematically scans the FOU. As the transmitter emits a beacon light, the detector captures the response signal. Upon detection, the system recalibrates the visual axis within the detector's field of view, preparing for the tracking phase.
3. *Tracking*: After the system establishes an initial connection between the terminals, it engages in coarse tracking, continuously adjusting the alignment of the optical beam. Once the tracking precision reaches the required level for fine tracking, the system transitions to this stage, enabling real-time control and ensuring a stable connection between the satellites. If fine tracking fails, the system must revert to coarse tracking or initiate scanning to restore the connection and resume the tracking process.
4. *Communication*: After achieving successful fine tracking, the system stabilizes the laser link and initiates communication. The transmitting terminal then encodes and modulates the data before transmitting the signal light. At the receiving terminal, the signal light is detected, demodulated, and decoded to complete the communication process.

During the entire procedure, interruptions can occur in any of the four stages due to various reasons including cancellation of the task, the cooperative target becoming obscured, alterations in the intersatellite link network, or extensive calibration requirements. Once a task is completed, the terminal will return to its original standby mode and wait for a new command, requiring the reexecution of the aforementioned workflow to perform the next task [15].

4.2 Key Techniques

Free-space optical communications employ beams with microradian divergence to facilitate long-range communication spanning thousands of kilometers. This necessitates a specialized APT beam control system to establish quick and reliable intersatellite point-to-point connections. Additionally, a distinct incoherent/coherent optical communication system is essential for the encoding and decoding of optical signals. In this section, we will introduce the technologies involved in constructing intersatellite optical links along with the typical modulation techniques used in intersatellite incoherent/coherent optical communications.

4.2.1 Link Construction

Creating an intersatellite optical link faces numerous challenges due to the harsh conditions in space. These challenges include limited load and space on satellite platforms, unpredictable platform vibrations, and inaccuracies in satellite orbit and attitude predictions. Additionally, the narrow width of the laser beam, long communication distances, and the detector's limited sensing capabilities further complicate the process [15]. The establishment of a stable free-space optical link comprises three stages: initial pointing, beam scanning and acquisition, and beam tracking, as illustrated in Figure 4.3. Terminal A acts as the scanning endpoint that emits a beacon light, and terminal B serves as the gazing endpoint.

4.2.1.1 Initial Pointing

During the initial pointing stage, the position, velocity, and orientation of the satellite, as well as the position of the satellite to be communicated with, are known in advance. In addition, synchronization has been achieved between these two satellites. Initially, the onboard computer determines the pointing directions for both terminals using the navigation satellite's ephemeris, creates the navigation data, and then uploads these data to the onboard laser terminal. The two terminals then used the satellite platform's attitude information and feedback from the APT rotary encoder to determine the azimuth and elevation angles required for aligning their visual axes through coordinate transformation. This data instructs the rotary encoder to modify the visual axes from their initial position to aim at the opposite terminal. Generally, the onboard terminal gathers information such as position, attitude, velocity, and time from the ephemeris and its internal attitude sensors [4, 15].

However, considering the large distance and relative speed between the two terminals, the angular advance of their relative motion during the beam's travel time cannot be neglected. Therefore, initial pointing includes applying an overtargeting fast steering mirror (FSM) to adjust the direction of the beam within the field of view to offset the alignment errors caused by relative movement. The precession

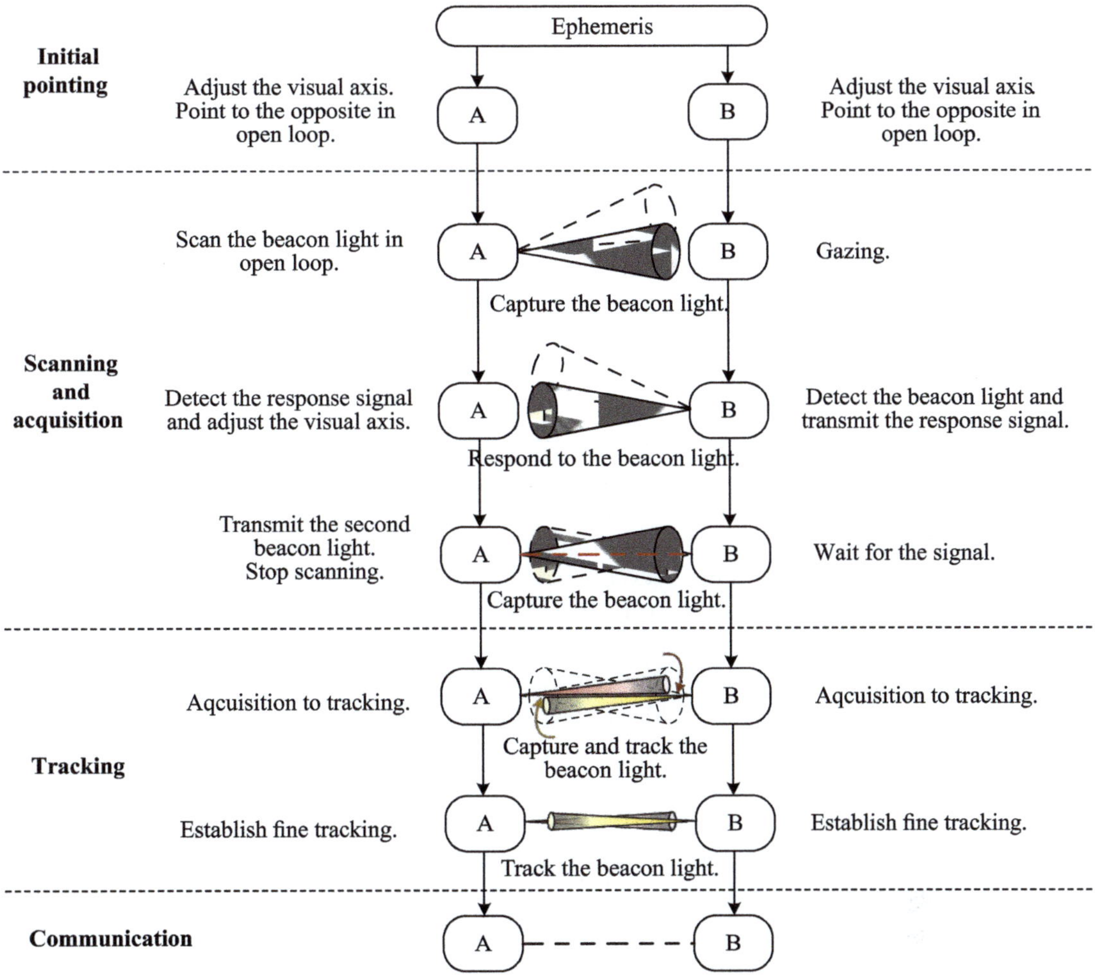

Figure 4.3 Flowchart of APT systems.

angle at which the FSM should be deflected is calculated from the ephemeris and the velocity of the satellite.

4.2.1.2 Scanning and Acquisition

Throughout the scanning and acquisition phase, the ATP subsystems on both terminals utilize the obtained pointing azimuth and altitude angles of the satellite platforms. These angles are used to direct their own two-dimensional (2D) tracking turntables toward each other's satellites in an open-loop manner. Meanwhile, the subsystems calculate and adjust the advance deflection angles for their advanced vibrating mirrors. Terminal A illuminates the FOU of terminal B with a beacon light. Terminal B determines the pointing angle error of its satellite platform from the received spot position and adjusts its 2D tracking turntable in real time to follow terminal A's beacon light. Terminal B then sends a response signal to terminal A to achieve initial single-end capturing. Once terminal A detects the returned beacon light from terminal B, it corrects its visual axis pointing based on the spot position and transmits the beam back to terminal B. When terminal B's detector identifies the incoming beam again, the initial capturing is completed at both ends.

4.2.1.3 Tracking

Once the coarse detector identifies the opponent's beam, it transmits the spot's position signal on the coarse detector to the coarse tracking controller. This controller initiates coarse tracking and adjusts the coarse pointing mechanism to align the spot as closely as possible to the center of the coarse detector. Subsequently, using the wide beam beacon light, both A and B begin emitting the narrow beam beacon light based on the control command. The centers of the coarse and fine detectors are superimposed and the fine detector's field of view exceeds the coarse tracking system's tracking error. Thus, after the coarse tracking stabilizes in a closed loop, the fine detector can identify the light spot's position. This position signal is then sent to the fine-tracking controller, which starts fine tracking and adjusts the fast reflector's deflection. This further minimizes the deviation between the system's visual axis and the received light's axis, ensuring that the system's visual axis accurately tracks the other terminal. As such, the system's optical axis is precisely aligned with the terminal on the other side.

Typically, the detector employs an open-window technique for fast switching between frame rate and field of view. Once the spot enters the view of the fine tracking detector, fine tracking with a narrow beam can be established. However, the spot may escape from the fine-tracking field of view, in which case the terminals have to reconstruct fine tracking by turning back to coarse tracking or even open-loop scanning. When fine tracking is stabilized, A and B can turn off the emission of the captured beacon light, and the intersatellite link is successfully established.

4.2.2 Signal Modulation Technique

Optical modulation formats, determined by the physical parameters utilized within the optical domain, are generally categorized into four main types: intensity, frequency, phase, and polarization modulation [18]. Furthermore, optical communication systems are generally classified into two types according to the modulation techniques and detection strategies applied at the receiver: direct detection systems and coherent systems. The direct detection system relies on intensity modulation combined with direct detection techniques. Conversely, the coherent system adjusts the amplitude, phase, or frequency of the optical carrier using an external modulation technique, which is applied after the optical signal generation. This system employs optical coherence detection, leveraging the intrinsic oscillatory characteristics of light to accomplish signal demodulation.

For laser-based intersatellite communication, either the intensity modulation/direct detection (IM/DD) approach or coherent optical communication techniques can be utilized. In IM/DD systems, as shown in Figure 4.4, the information is directly modulated onto optical pulses, with the receiver employing a direct detection method. This method is known for its simplicity and cost-efficiency; however, it relies on photodetectors operating under the square-law detection principle, which restricts the system to capturing only amplitude information.

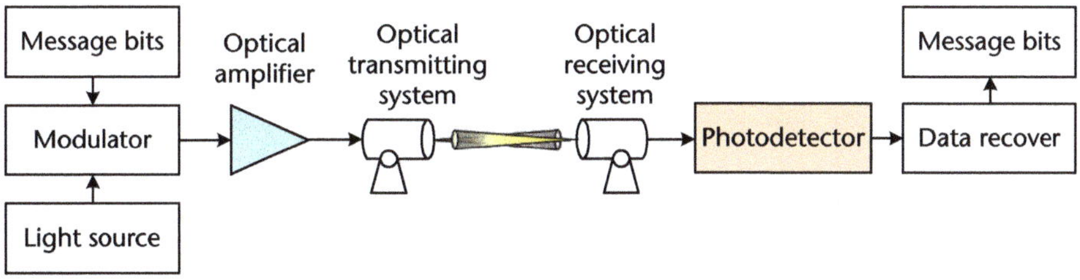

Figure 4.4 Block diagram of an IM/DD system.

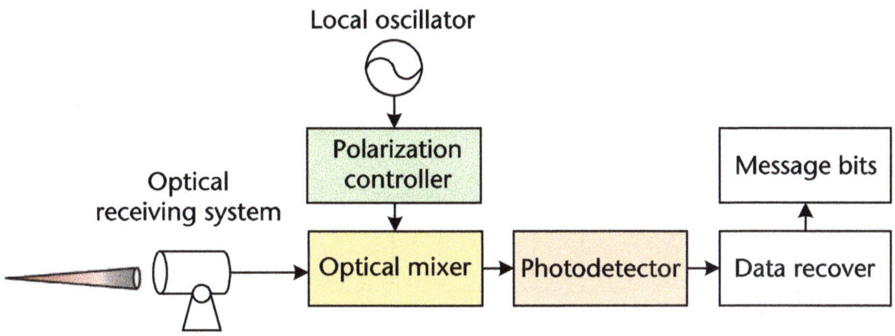

Figure 4.5 Block diagram of the receiver in a coherent system.

In coherent optical communication systems, the transmitter's structure is similar to that of the IM/DD system; however, a significant difference lies in the use of an external modulation method after the optical signal is generated to alter the optical carrier frequency, phase, or amplitude. Additionally, the detection mechanisms between the coherent and IM/DD systems differ considerably. The receiver configuration in the coherent system, as depicted in Figure 4.5, is distinct due to the incorporation of a local oscillator. This setup involves combining the polarization-controlled primary light with the incoming signal in an optical mixer. The optical mixer output power encompasses the full range of information related to the signal intensity, frequency, or phase, thereby ensuring that any modulation applied by the transmitter is accurately retained and conveyed in the output [18]. Consequently, coherent detection exhibits a versatile and robust application across various optical communication systems [5, 19].

4.2.2.1 Noncoherent Modulation

On-off keying (OOK) stands out as a simple form of noncoherent modulation, utilizing a digital pulse signal where each bit's optical pulse that is either present or absent, a state referred to as the "switch." An OOK signal can be represented as

$$s(t) = \sum_{n} a_n g\left(t - nT_s\right), \tag{4.1}$$

where T_s denotes the duration of the codeword, $g(t)$ is the baseband pulse waveform, and a_n represents the amplitude level for the nth symbol. The peak communicating speed of an OOK system is contingent on the light source's switching speed. Although semiconductor lasers' rapid switching rates suffice for certain high-speed OOK applications, the optical pulse intensity diminishes as the

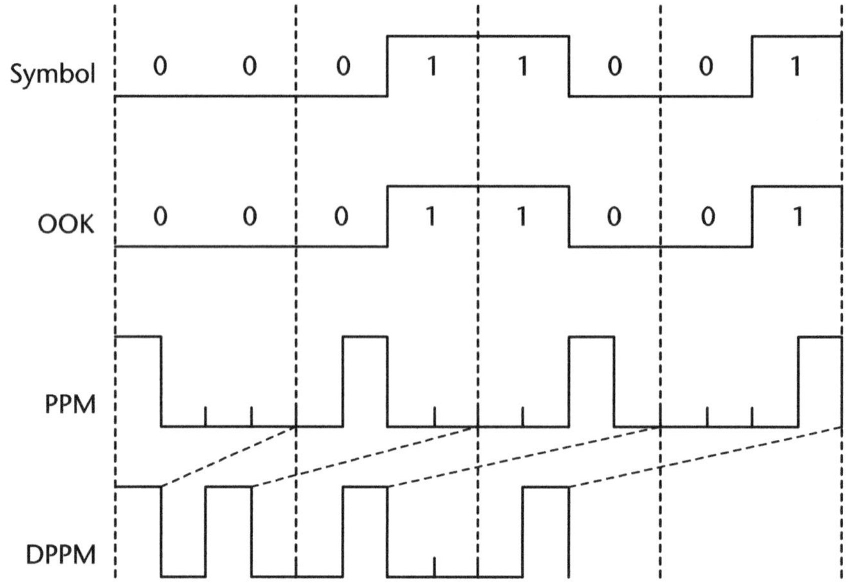

Figure 4.6 Comparison of different modulation timings [1].

modulation rate increases. This makes OOK unsuitable for particularly high-speed and long-distance laser communications [6].

Pulse position modification (PPM) employs the location of an optical pulse to transmit data. In detail, PPM allocates each frame into L time slots, each with a duration of T_s, and the optical pulse also has a width of T_s. The specific position of the optical pulse within these slots indicates the data being transmitted, with the bit length determined by $log_2 L$. The use of low bandwidth in PPM modulation does not favor high-speed communications, but the modulation's low duty cycle for signal light pulse slots facilitates achieving higher peak powers when average laser detecting power is constrained. Additionally, the ability of this modulation to use superconducting nanowire single-photon detectors allows for reception sensitivity at the single-photon level. As such, PPM is more suitable for lunar-terrestrial and deep-space laser communications that demand lower code rates and involve extensive link distances [13].

Differential pulse position modulation (DPPM) offers an enhancement over conventional PPM by having unspecified bit counts in its code groups. The primary distinction between DPPM and PPM lies in the elimination of the "0" following the "1" in the signal's time slot, thereby maintaining the same transmission rate for both L-DPPM and L-PPM code groups at $\log_2 L$ bits. With identical transmission rates, DPPM benefits from reduced bandwidth usage and improved power efficiency compared to PPM. However, DPPM requires strict symbol synchronization; otherwise, a single erroneous code can compromise the interpretation of subsequent signals, thus limiting DPPM for widespread application [14].

4.2.2.2 Coherent Modulation

Coherent modulation can be achieved using an optoelectronic modulator, predominantly the Mach-Zehnder modulator (MZM), which utilizes the electro-optic properties of a lithium niobate (LiNbO3) substrate. Exhibiting strong optoelectronic effects, this modulator alters the refractive index of the LiNbO3 crystals when an external voltage is applied, thereby modifying the amplitude, frequency, or phase of the light signal that travels through it.

Binary phase shift keying (BPSK) modulation employs binary digital baseband signals to encode information about carrier phase changes using a 180° phase shift, which can be realized by an MZM. Specifically, BPSK translates a unipolar signal into a bipolar electrical signal with nonzero values, subsequently modulating this electrical signal onto the laser carrier. BPSK modulation operates through either homodyning detection or heterodyne detection, and in either case, it requires the use of a local oscillator laser to mix the incoming light signals. Due to the Doppler shift encountered in free-space optical communications, the modulation technique must adequately compensate for these shifts. However, current technologies for Doppler shift compensation still face challenges in correcting frequency shifts in the 30-GHz range.

Differential phase shift keying (DPSK) modulation utilizes the phase difference between successive signal elements to encode information, implementing differential precoding based on BPSK. This modulation scheme demonstrates detection sensitivity that is intermediate between direct modulation detection and conventional phase modulation systems while providing strong resilience against background noise. DPSK is widely adopted in modern high-speed free-space optical communication systems, particularly those linking satellites and ground stations. In such systems, atmospheric turbulence causes both intensity scintillation and phase fluctuations in the laser beam. To counteract these effects, distributed antenna array receiving technology is employed to perform automatic coding compensation, accounting for intensity variations and phase disturbances induced by turbulence on both large and small scales [12].

Quadrature phase shift keying (QPSK) modulation employs a quadrature-based phase modulation technique that uses four primary carrier phases: 45°, 135°, 225°, and 315°. This can be realized using an optical vector modulator that includes two MZMs and an optical phase modulator. In comparison to BPSK, QPSK modulation offers higher communication rates primarily utilized for high-rate intersatellite or satellite-earth communications, but it also demands greater frequency stability.

4.2.3 Laser Antenna Technology

The intersatellite laser communication system comprises four subsystems that collectively facilitate the completion of the laser communication task.

1. *Optical mechanical subsystem*: The optical-mechanical subsystem constitutes the fundamental component of the laser communication terminal. It encompasses the optical telescopic unit, the optical relay unit, and the precision mechanical base unit. The optical telescopic unit and the optical antenna represent the central elements of the optical system. They are typically designed with integrated transceivers, with the performance of the optical telescopic unit directly influencing the establishment of links and the quality of communication.

2. *Communication subsystem*: The communication subsystem constitutes the principal functional entity for intersatellite laser communication. This is divided into two principal units: communication transmission and communication reception. Currently, high-efficiency and low-power semiconductor lasers are employed as emission sources. At the receiving end, the communication subsystem can be divided into two categories: direct detection and coherent detection. Among these, coherent detection is receiving increasing attention due to its high sensitivity.
3. *APT subsystem*: The APT (acquisition, alignment, tracking) subsystem is the foundation for establishing intersatellite laser communication links. The capture unit is tasked with rapidly and with high probability capturing targets in uncertain areas, thereby building an optical closed-loop link. The coarse tracking unit is responsible for rapidly realizing dynamic coarse tracking to ensure that the target enters and stabilizes within the field of view of fine tracking. The fine tracking unit subsequently corrects the coarse tracking residuals to achieve the system's high-precision tracking. The leading alignment unit compensates for the leading direction of the field of view caused by the relative motion between satellites, thereby achieving accurate alignment of the communication light.
4. *Total control subsystem*: The terminal's communication process is entirely under the system's control, which also manages thermal control and power management.

The optical antenna represents the fundamental component of an optical machine system within a laser communication terminal. Its principal functions can be delineated in two key areas: First, at the transmitting end, the antenna facilitates the expansion of the transmitted signal, the enlargement of the beam waist radius, and the reduction of the beam divergence angle, thereby reducing the beam divergence loss and the transmission power requirements of the light source. Second, at the receiving end, the antenna increases the receiving area, compresses the receiving field of view, and reduces background light interference, thus enhancing the signal-to-noise ratio of the received signal and extending the communication distance.

The majority of onboard laser communication terminals employ transceiver-integrated optical antenna design. However, the specific optical antenna structure forms employed vary. The most commonly used structure is the Cassegrain form. Additionally, there are transmission structures, off-axis reflection structures, folding hybrid structures, and other forms. This section will present a comparative analysis of various optical antennas.

The transmission optical antenna is comprised of two sets of lenses, which constitute the objective lens and the eyepiece, respectively. The transmission optical antenna is primarily classified into two categories: The Galileo and Kepler types.

The reflective structure is defined as follows: Laser communication terminals are more commonly employed as reflective optical antennas. If masking is present, the reflector antenna can be categorized into two distinct groups: those with and those without masking.

The folding hybrid structure is a novel telescope design that combines the advantages of transmission and reflection systems. It offers high definition and contrasts similar to that of a transmission telescope while maintaining the low color difference of a reflection telescope. Furthermore, it boasts a larger field of view than a simple two-mirror reflection system. These features and its potential for high-quality imaging make the folding hybrid structure an attractive option for future telescope designs. This implies that the telescopic objective incorporates both a reflector and a lens. The lens is positioned at the front of the system as a correction mirror, which corrects aberrations generated by the spherical primary mirror. Introducing the correction lens ensures that the transverse system exhibits optimal image quality. An example of laser antenna technology is SpaceX's Starlink constellation, which uses optical intersatellite links (OISLs) for high-speed data transmission between satellites. This technology allows satellites to communicate directly via laser beams, reducing reliance on ground stations and enabling global coverage with low latency.

4.2.4 Microwave Antenna Technology

Radio waves are typically defined as electromagnetic waves with frequencies within the range of 300 GHz. Radio waves can be further classified into microwave and millimeter-wave links based on the frequency band. Compared to the laser link, the microwave/millimeter-wave link exhibits a broader beam, conferring more excellent reliability and representing a more prevalent technical approach [16]. As illustrated in Table 4.1, the frequency band selection for microwave links is predominantly the L/S band, whereas the millimeter-wave link encompasses a higher Q/V band in addition to the currently prevalent Ka-band. The millimeter wave frequency band offers greater communication capacity and more abundant spectrum resources, which has led to it becoming a research focus and a promising area for broad application.

The design of a high data rate intersatellite link antenna is contingent upon several factors, including full duplex, an up to 2 Gbps data rate, reverse circular polarization, and the position of each satellite, which varies from 25° to 160° from the geosynchronous orbital plane (18,000–83,000 km range, 213–226 dB path loss, 77 GHz). The azimuth angle is between 5 and 10 degrees. Duplex communication links with other tracking and data acquisition system (TDAS) satellites are not in the sun and can operate at total capacity. This is achieved using a 10W RF power amplifier and a 360K low-noise receiver. Based on these considerations, it can be concluded that GEO-LEO links require antenna gains of approximately 63 dB, which equates to 3.2-diameter antennas at both ends of the link. In order

Table 4.1 Part of the Intersatellite Link Frequency Band Planning

Carrier Type	*Frequency Band Range*	*Bandwidth*
Microwave	22.55–23.55 GHz	1000 MHz
	24.45–24.75 GHz	300 MHz
	25.25–27.50 GHz	2250 MHz
Millimeter-wave	32–33 GHz	1000 MHz
	54.25–58.20 GHz	3950 MHz
	59–64 GHz	5000 MHz
	65–71 GHz	6000 MHz

to operate at total capacity when in the sun irradiation area, the antenna must be increased in size, the transmission power must be increased, and both methods may be employed. Given that the solar noise fading of the GEO-GEO link occurs only a few times a year, it is not necessary to adjust the link for this phenomenon.

4.3 Current Status and Possible Challenges

In recent years, various countries have successfully conducted on-orbit demonstrations of laser communication technologies across different orbital platforms, gradually moving toward large-scale applications. These demonstrations, while often utilizing customized laser terminals tailored to specific mission requirements, have also driven the development of next-generation space systems. Despite the successful on-orbit validation of satellite laser communication systems and the breakthroughs in key technologies, several significant challenges remain due to the rapid movement of satellites and the unique space environment. These challenges include issues related to the transmission medium, node mobility, link stability, energy supply, storage, and data processing, among others [9, 10, 15, 16]. This section will analyze the development of satellite laser communication across different countries, explore the associated technical challenges, and suggest potential solutions.

4.3.1 Current Status

This section categorizes and examines key technical validation projects in Europe, the United States, and Asia, focusing on the technological details and development processes in each region. Through this analysis, we summarize the current state and emerging trends in satellite communication, providing a clear understanding of the global advancements in this field.

4.3.1.1 Europe

The European Data Relay System (EDRS) is a satellite relay platform based on a GEO satellite platform equipped with both laser and Ka-band communication payloads. This system facilitates communication between LEO satellites, GEO satellites, and ground stations, providing relay services for LEO satellite users, aviation users, UAVs, and ground terminal devices. The communication range of EDRS is 45,000 km, with a laser transmission power of 5W, a communication rate of 1.8 Gbps, using BPSK modulation at a laser wavelength of 1,064 nm, and it supports bidirectional communication.

In June 2016, EDRS-A introduced intersatellite laser communication with a data rate of 600 Mbps, providing relay services for up to 40 LEO and GEO satellites daily. By August 2019, EDRS-C was successfully launched into geostationary orbit, with its laser intersatellite link mounted on the SmallGEO platform. The third satellite, EDRS-D, scheduled for launch in 2025, will be equipped with three next-generation laser communication terminals. These terminals will enable simultaneous communication with multiple satellites, with a transmission range of up to 80,000 km, and will be compatible with both 1,064-nm and 1550-nm wavelengths.

This expansion aims to relay data from the Asia-Pacific region to Europe, thus enabling global data relay services.

The German Aerospace Center (DLR) has developed the OSIRIS program, which focuses on experimental optical terminals optimized for small satellites. The development of OSIRIS began with two scientific missions launched in 2016 and 2017, OSIRISv1 and BiROS (OSIRISv2), followed by the OSIRIS4 Cubesat launch in the fourth quarter of 2018 and the installation of OSIRISv3 on the Airbus DS Bartolomeo platform aboard the International Space Station in 2019. Currently, the fourth generation, OSIRIS-4, is under development. It is a miniaturized version with dimensions smaller than 10 cm × 10 cm × 3 cm, and with a low power consumption of just 8W during operation, making it compatible with virtually any CubeSat platform [7].

Germany's TESAT company has introduced a range of laser communication terminals designed to meet the demands of multiple tasks. For low Earth orbit missions, TESAT offers the SmartLCT terminal, which can be deployed on smaller, lighter satellites, significantly conserving mass and space. The SmartLCT can transmit data over distances up to 45,000 km, providing high-speed data transfer of 1.8 Gbps, while weighing only around 30 kg. It ensures secure, rapid, and highly reliable performance. In the small satellite sector, TESAT's laser product line includes the TOSIRIS and CubeLCT terminals, which offer lightweight solutions. TOSIRIS transmits Earth data at speeds of up to 10 Gbps and weighs just 8 kg. It employs IM/DD modulation with an adjustable downlink rate. The CubeLCT, with an edge length of only 10 cm and weighing just 0.397 kg, transmits at speeds of up to 100 Mbps. By utilizing these laser terminals to build a global data backbone, TESAT enables near-real-time data transmission worldwide.

4.3.1.2 The United States

The Laser Communications Relay Demonstration (LCRD) is one of NASA's key projects for advancing space laser communication technologies. This initiative is a foundational step toward developing the next generation of space tracking and laser communication relay satellites [3]. As part of the LCRD program, NASA integrated and tested the ILLUMA-T terminal, a satellite laser communication terminal, in October 2023. The ILLUMA-T terminal can establish a two-way communication link between satellites in different orbits, facilitating a multilayered space network. This terminal employs photonic integration technology, replacing traditional electronic components with photonic ones, which significantly reduces the terminal's weight, volume, and power consumption, thereby enhancing reliability.

Previously, the Optical Communication and Sensor Demonstration (OCSD) satellites showcased the capability of small satellites to facilitate high-speed satellite-to-ground communication using laser intersatellite links, thereby addressing earlier constraints related to the size and mass of laser communication systems. The OCSD-A satellite was launched in October 2015, with the subsequent deployment of OCSD-B/C in November 2017. These missions successfully demonstrated the feasibility of high data rate transmission between satellites and ground stations using laser links [8].

The CubeSat Laser Infrared Crosslink (CLICK) system, collaboratively developed by The Massachusetts Institute of Technology (MIT), the University of

Florida, and NASA Ames Research Center, serves as a platform for testing laser communication between satellites as well as between satellites and ground stations. This system features laser terminals with minimal size, weight, and power (SWaP), enabling full-duplex, high-speed data downlinks and intersatellite links, while also facilitating precise ranging and time synchronization. The CLICK-A payload, which comprises a laser transmitter and an advanced pointing, acquisition, and tracking (PAT) system, has successfully completed its assembly and testing phase and has been delivered for integration with the spacecraft. CLICK-B/C, slated for launch no earlier than August 2025, is designed to build upon the CLICK-A mission by incorporating additional elements, such as beacon lights and detector systems essential for communication [8]. These twin CubeSats will be launched together to demonstrate full-duplex intersatellite links with data rates exceeding 20 Mbps, along with 0.5-m ranging capability and 200-ps time synchronization accuracy [3]. In May 2022, the Terabyte Infrared Delivery (TBIRD) project demonstrated an innovative 200-Gbps downlink using a system with a volume of just 1.8U ×1U ×1U and a mass of less than 2.25 kg.

4.3.1.3 Asia

The Japanese Data Relay Satellite System (JDRS) was developed collaboratively by the Japan Aerospace Exploration Agency (JAXA) and the government, with its ownership and management under the Cabinet Satellite Intelligence Center. JAXA oversees the optical data relay operations of the satellite. Positioned in a geostationary orbit at approximately 35,400 km above Earth, JDRS functions as a high-altitude platform, facilitating rapid data transfer between Japanese satellites and ground stations. This system proves particularly beneficial when a direct line of sight between satellites and ground stations is unavailable [36]. JDRS-1, an integral element of the system, supports both military and civilian applications, taking over from the "Kodama" Data Relay Test Satellite (DRTS), which was in service from its launch in 2002 until its decommissioning in August 2017. Featuring two laser terminals, JDRS-1 employs infrared beams to achieve data transmission rates of up to 1.8 Gbps between spacecraft. Its LUCAS payload enhances data transmission capabilities, offering speeds that are seven times greater than those achieved by the earlier S-band and Ka-band systems used in DRTS [8, 17].

The National Institute of Information and Communications Technology (NICT) in Japan has developed several terminals for various missions. Among these is the Small Optical Transponder (SOTA), a compact satellite optical communication terminal weighing only 6 kg. SOTA is currently deployed on the 50-kg-class satellite SOCRATES, which focuses on advanced space optical communication research. Optical communication experiments using SOTA are primarily conducted at the Koganei Earth Station in Tokyo. These experiments include image transmission tests with Low-Density Generator Matrix (LDGM) codes and quantum satellite communication demonstrations, confirming that SOCRATES, aided by SOTA, can achieve photon-level information exchange with ground stations [37].

In 2016, China conducted a successful laser communication experiment between the Tiangong-2 space laboratory and the Nanshan ground station in Xinjiang. The laser terminal achieved a downlink data rate of 1.6 Gbps using IM/DD technology. This experiment marked the first successful daytime laser communication, with the terminal's tracking performance during daylight nearly matching that of nighttime conditions. Following this, in 2017, China launched the Shijian-13 satellite, whose laser terminal, developed by Harbin Institute of Technology, enabled the first high-speed, bidirectional laser communication between a geostationary satellite and the ground. The communication system, based on IM/DD technology, achieved data rates of up to 5 Gbps over distances up to 45,000 km, setting a new record for high-data-rate laser communication between satellites and ground stations at that time.

In 2019, China successfully launched the Shijian-20 satellite from the Wenchang Space Launch Center in Hainan. This satellite is equipped with a laser terminal, designed by the China Academy of Space Technology, which employs coherent modulation techniques. In 2020, the Shijian-20 established a laser communication link with the Lijiang ground station, achieving a downlink transmission rate of up to 10 Gbps utilizing QPSK modulation [8]. Furthermore, the Xingyun satellite series, launched in 2020, included laser communication payloads developed by LaserFleet. This event marked China's inaugural experiment in low Earth orbit satellite-to-satellite laser communication, successfully achieving a link distance exceeding 3,000 km with a data rate of 100 Mbps.

4.3.2 Possible Challenges

Building laser intersatellite links involves selecting various parameters, such as modulation schemes and wavelengths, which are closely tied to the satellite's orbital altitude, distance, payload size, and specific mission requirements. Typically, satellites in medium to high Earth orbits handle high data rate, wide-bandwidth communication tasks. These satellites usually carry larger communication payloads and employ coherent modulation schemes with shorter wavelengths.

In contrast, LEO satellites often support more frequent Earth communication tasks and require the deployment of a greater number of satellites. These tasks typically involve smaller, lighter communication payloads, noncoherent modulation schemes, and longer wavelengths. However, the actual selection of parameters for laser intersatellite links must be flexibly adjusted according to specific mission needs and design objectives. Moreover, free-space optical communications still face numerous technical challenges, particularly due to the rapid movement of satellites and the unique space environment.

4.3.2.1 Transmission Environment

Free-space optical communications operate in a space environment markedly distinct from that of terrestrial fiber optic communications. The space environment is highly intricate, involving solar radiation and reflections from the Moon and other celestial bodies, which may surpass the intensity of the signal light and significantly impact the quality of signal transmission. Consequently, free-space optical communication systems require exceptionally strong resistance to background noise to ensure reliable signal transmission. First, reception of data at very low signal-to-noise ratios can be supported by high-sensitivity detectors that incorporate advanced noise reduction methods such as optical filtering and signal amplification. Moreover, signal immunity can be further improved by complex signal processing algorithms such as digital filtering, adaptive precoding, and error correction coding. These algorithms are capable of identifying valid information in the received data and minimizing noise interference.

4.3.2.2 High Nodes Mobility

Satellites are in high-speed motion and their relative positions are constantly changing, making dynamic alignment and connection of the laser link a major challenge. Satellite vibration can cause jitter in the transmit beam, reducing the stability of the laser link, which in turn leads to jitter in the received power or even error codes. Rapid beam acquisition and precise beam tracking technology are essential to address this issue. This necessitates an optical antenna with high precision, accurate orbital prediction, and robust load control compensation to maintain the resilience of the optical followup system. Furthermore, the robustness and stability of the free-space optical communication system can be further enhanced by employing multisensor fusion technology and advanced control algorithms.

4.3.2.3 Signal Processing with Low Energy Supply

Satellite energy sources are primarily dependent on solar panels and internal batteries and constrained by the size of the solar panels, the battery capacity, and eclipse events. These constraints impact the satellite's ability to gather, store, process, and transmit data. To address these issues, it is essential to implement cutting-edge optoelectronic integration techniques to boost the processing and transmission efficiency of satellite payloads and to decrease their size and energy consumption. In addition, the complex signal transmission, data handling, and network routing algorithms used in ground-based optical communications should be streamlined to suit the limited energy and processing resources of satellites. Furthermore, the use of high-efficiency solar cells, energy recovery systems, and low-power electronics can also markedly improve the energy efficiency of satellites. The use of lightweight yet robust materials such as aluminum-silicon carbide composites, titanium alloys, and beryllium-aluminum alloys not only reduces the weight of the satellite, but also strengthens its structure. This, in turn, increases its capacity to accommodate more energy systems.

4.3.2.4 Complex Routing Protocols

Considering the severe instability of intersatellite links, intersatellite communication systems require robust defenses against link disruptions. Protocols for data links and routing are essential in satellite communications to guarantee ongoing communication and dependability in the event of link failure. In recent years, the approach of using machine learning to discover the best policy has yielded significant outcomes on numerous issues, and has been implemented in a variety of resource scheduling algorithms. Meanwhile, to accommodate the unique requirements of satellite communications, established algorithms from terrestrial optical communications should be modified and simplified. These modifications should align with the satellites' energy and processing limitations. Employing strategies like hierarchical routing and dynamic link management can enhance data transmission's effectiveness and dependability.

4.3.2.5 Lightweight Design

The reduction in size while ensuring longevity and reliability presents a significant technical hurdle for laser communication systems. This miniaturization requires optimization across four dimensions: link, terminal, stand-alone units, and material choices. Initially, at the link level, it is crucial to strike a balance between the reliability of the system and its weight. Traditional payloads, typically expected to last over five years, should include provisions for redundancy, and the resilience of a grouped satellite system's single laser communication terminal must be evaluated from an overall system standpoint. At the terminal level, the design of the terminal influences both its weight and size, with the telescope aperture affecting the channel's weight. Minimizing the terminal's weight hinges on refining the optical magnification, the design of the optical path, and the size of the detector. When selecting materials, a compromise between performance and cost is necessary. Currently, the primary materials used are aluminum alloy, aluminum-silicon carbide, and titanium alloy. For payloads where terminal weight is critically restricted, such as in deep space missions, metals such as beryllium or beryllium-aluminum, known for their high specific stiffness, are preferred.

In the context of designing free-space optical communications for extended durability and reliability, the choice of components, the design process, and the implementation of terminals must adhere to aerospace design standards. Further studies are necessary to improve reliability and adapt to the space environment. Strategic planning of high-level switching routes and minimizing the operating duty cycle of laser terminals can significantly extend their lifespan.

4.3.2.6 Networking and Security

In order to achieve interconnected networking and on-demand access among satellite systems, it is essential to standardize intersatellite optical link techniques and manage the status of links and equipment. This involves unified planning of signal systems, data rates, link-layer data formats, and handshake protocols. Regarding link security, especially for satellite systems with stringent security needs, coherent laser links and end-to-end encryption schemes are necessary to ensure secure and reliable data transmission.

References

[1] Chang, J., C. M. Schieler, K. M. Riesing, J. W. Burnside, K. Aquino, and B. S. Robinson,"Body Pointing, Acquisition and Tracking for Small Satellite Laser Communication," in *Free-Space Laser Communications XXXI*, Vol. 10910 (H. Hemmati and D. M. Boroson, eds.), International Society for Optics and Photonics, SPIE, 2019, p. 109100P.

[2] Chishiki, Y., S. Yamakawa, Y. Takano, Y. Miyamoto, T. Araki, and H. Kohata, "Overview of Optical Data Relay System in JAXA," in *Free-Space Laser Communication and Atmospheric Propagation XXVIII*, Vol. 9739 (H. Hemmati and D. M. Boroson, eds.), International Society for Optics and Photonics, SPIE, 2016, p. 97390D.

[3] Cornwell, D. M., "NASA's Optical Communications Program for 2017 and Beyond," in *2017 IEEE International Conference on Space Optical Systems and Applications (ICSOS)*, 2017, pp. 10–14.

[4] Jiaxin, C., and H. Junfeng, "Research on Initial Pointing of Inter-Satellite Laser Communication," in *2020 12th International Conference on Intelligent Human-Machine Systems and Cybernetics (IHMSC)*, Vol. 1, 2020, pp. 218–221.

[5] Kikuchi, K., "Fundamentals of Coherent Optical Fiber Communications," *Journal of Lightwave Technology*, Vol. 34, No. 1, 2016, pp. 157–179.

[6] Kumar, N., V. K. Jain, and S. Kar, "Evaluation of the Performance of FSO System Using OOK and M-PPM Modulation Schemes in Inter-Satellite Links with Turbo Codes," in *2011 International Conference on Electronics, Communication and Computing Technologies*, 2011, pp. 59–63.

[7] Li, L., X. Zhang, J. Zhang, C. Xu, and Y. Jin, "Advanced space Laser Communication Technology on CubeSats," *ZTE Communications*, Vol. 18, 2021, pp. 45–54.

[8] Li, R., B. Lin, Y. Liu, M. Dong, and S. Zhao, "A Survey on Laser Space Network: Terminals, Links, and Architectures," *IEEE Access*, Vol. 10, 2022, pp. 34815–34834.

[9] McLemore, B., and M. L. Psiaki, "Navigation Using Doppler Shift from LEO Constellations and INS Data," *IEEE Transactions on Aerospace and Electronic Systems*, Vol. 58, No. 5, 2022, pp. 4295–4314.

[10] Mücheberg, S., C. Gal, J. Horwath, H. Kinter, L. Martin Navajas, and M. Soutullo, "Development Status and Breadboard Results of a Laser Communication Terminal for Large LEO Constellations," in *International Conference on Space Optics ICSO 2018* (Z. Sodnik, N. Karafolas, and B. Cugny, eds.), Vol. 11180, International Society for Optics and Photonics, SPIE, 2019, p. 1118034.

[11] Munemasa, Y., Y. Saito, A. Carrasco-Casado, et al., "Feasibility Study of a Scalable Laser Communication Terminal in NICT for Next-Generation Space Networks," in *International Conference on Space Optics ICSO 2018* (Z. Sodnik, N. Karafolas, and B. Cugny, eds.), Vol. 11180, International Society for Optics and Photonics, SPIE, 2019, p. 111805W.

[12] Popoola, W. O., E. Poves, and H. Haas, "Spatial Pulse Position Modulation for Optical Communications," *Journal of Lightwave Technology*, Vol. 30, No. 18, 2012, pp. 2948–2954.

[13] Fu,Q., H.-L. Jiang, X.-M. Wang, Z. Liu, S.-F. Tong, and L.-Z. Zhang, "Research Status and Development Trend of Space Laser Communication," *Chinese Optics*, Vol. 5, No. 2, 2012, pp. 116–125.

[14] Shiu, D.-S., and J. M. Kahn, "Differential Pulse-Position Modulation for Power-Efficient Optical Communication," *IEEE Transactions on Communications*, Vol. 47, No. 8, 1999, pp. 1201–1210.

[15] Wang, G., F. Yang, J. Song, and Z. Han, "Free Space Optical Communication for Inter-Satellite Link: Architecture, Potentials and Trends," *IEEE Communications Magazine*, Vol. 62, No. 3, 2024, pp. 110–116.

[16] Xie, J., G. Huang, R. Liu, et al., "Design and Data Processing of China's First Spaceborne Laser Altimeter System for Earth Observation: Gaofen-7," *IEEE Journal of Selected Topics in Applied Earth Observations and Remote Sensing*, Vol. 13, 2020, pp. 1034–1044.

[17] Yamakawa, S., Y. Chishiki, Y. Sasaki, Y. Miyamoto, and H. Kohata, Jaxa's Optical Data Relay Satellite Programme," in *2015 IEEE International Conference on Space Optical Systems and Applications (ICSOS)*, 2015, pp. 1–3.

[18] Zhang, J., and J. Li, "InterSatellite Link Laser Modulation Mode," in *Laser Inter-Satellite Links Technology*, IEEE, 2023, pp. 167–185, doi:10.1002/9781119910749.ch9.

[19] Zhang, Y., R. Ding, Z. Qian, M.Liu, Y. Liang, and H. Jiang, "Research on High-Speed Clock Synchronization Technology for Inter-Satellite Coherent Laser Communication Link," in *2023 3rd International Conference on Communication Technology and Information Technology (ICCTIT)*, 2023, pp. 80–83.

CHAPTER 5

Channel Models for Satellite-Terrestrial Integrated Communication

5.1 Wireless Channel Fundamentals

Due to the presence of occlusions, signals propagate through different paths. Consequently, signals arriving at the receiver exhibit varying time delays consisting of the overall received signal. The maximum delay spread $\tau_{\max}$ of the channel is defined as the difference between the paths with the largest and smallest transmission delays. Moreover, the coherence bandwidth of the channel B_c is defined as follows;

$$B_c \approx \frac{1}{\tau_{\max}}.$$

When the bandwidth of the signal surpasses the channel's coherence bandwidth, the attenuation across different parts of the signal becomes dependent on frequency, leading to waveform distortion. This channel is termed as a frequency-selective channel. This type of channel is referred to as a frequency-selective channel. On the other hand, if the signal bandwidth is less than the coherence bandwidth, the entire signal undergoes consistent attenuation, maintaining the waveform, and such a channel is called a nonfrequency-selective (or flat) channel.

For the case of mobile propagation environment, relative motion between the transmitter and receiver induces the Doppler shift phenomenon. When multipath propagation exists concurrently, the Doppler shift transforms into Doppler spread. Assuming the transmitted signal has a frequency f_c, Doppler spread results in the received signal's power spectrum expanding from $f_c - f_m$ to $f_c + f_m$, and the received signal's power spectrum spread B_D is referred to as f_m. If the signal bandwidth B_s is greater than this power spectrum spread B_D, the channel is classified as a slow-fading channel. Conversely, if the signal bandwidth B_s is smaller than the power spectrum spread B_D, the channel is regarded as a fast-fading channel. Additionally, coherence time T_c is defined as

$$T_c \approx \frac{1}{B_D}.$$

Based on the relationships among signal bandwidth B_s, coherence bandwidth B_c, and power spectrum spread B_D, fading channels can be categorized as illustrated in Figure 5.1.

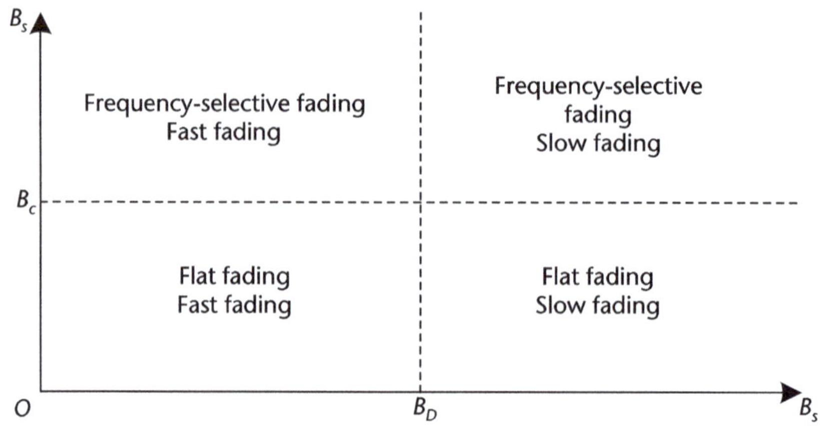

Figure 5.1 Schematic diagram of fading channels classification.

5.2 Satellite-Terrestrial Channel Characteristics

As shown in Figure 5.2, satellite communication signals tend to suffer from various attenuation fading such as the nonlinear effects of RF devices, scintillation effects caused by rain and clouds, the shadow fading effect caused by obstacles, as well as the multipath effect caused by reflection, scattering and bypassing phenomena. Additionally, the complex relative motion between LEO satellites and the

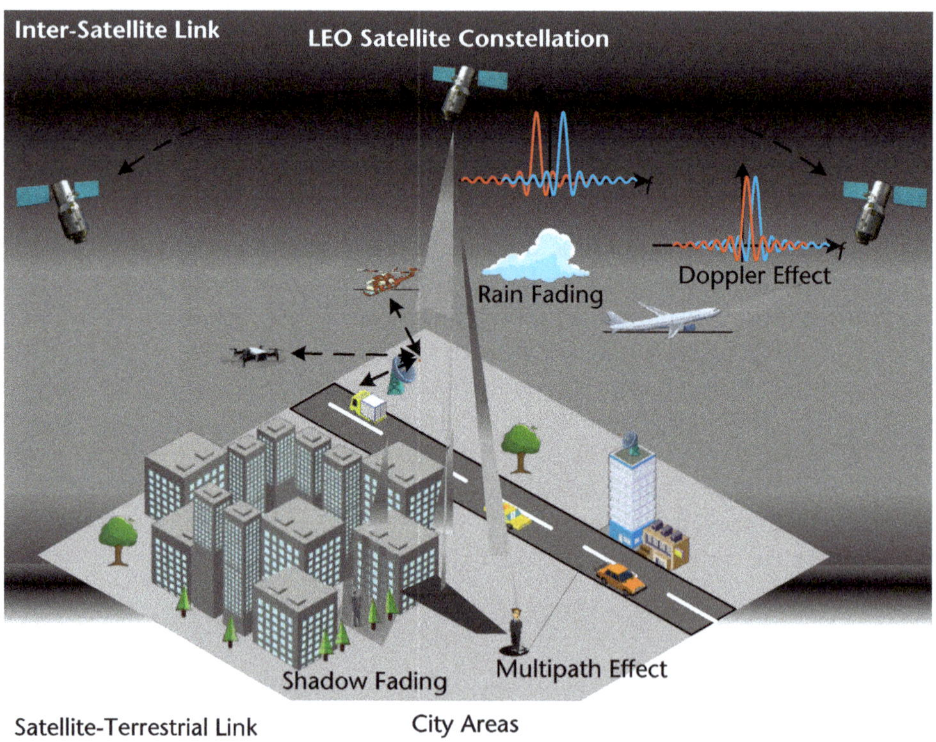

Figure 5.2 Satellite-terrestrial channel characteristics.

receiving end contributes to a high dynamic time-varying characteristic in the above attenuation, accompanied by a significant Doppler effect [2].

5.2.1 Free-Space Loss

Signal attenuation occurs as radio waves propagate through free space, with its magnitude depending on the communication frequency and propagation distance [3]. The free-space path loss can be described by the following equation:

$$L_{FS} = \left(\frac{4\pi d}{\lambda}\right)^2 = \left(\frac{4\pi d f_c}{c}\right)^2,$$

where d represents the distance in kilometers between the satellite and the terminal, λ is the wavelength of the carrier wave used for communication in m, f_c is the communication carrier frequency in MHz, and is the speed of light, with a value of $3 \times 10^8 m/s$. The above equation can be expressed in logarithmic form as follows:

$$L_{FS} = 32.45 + 20 \lg f_c + 20 \lg d,$$

At this time, the free space loss L_{FS} is in dB. This equation reveals that as the carrier frequency and the distance between the satellite and the terrestrial station increase, the free-space path loss correspondingly increases.

5.2.2 Ionospheric Scintillation

Ionospheric scintillation occurs due to the nonuniformity of the concentration of the medium within the troposphere and ionosphere, causing signal scattering as electromagnetic waves pass through. This redistribution of electromagnetic energy in space and time leads to sharp fluctuations in signal phase and amplitude over short periods, which is called ionospheric scintillation [4]. This effect can lead to random changes in the amplitude and phase of the received signal of the satellite mobile communication system, affecting the system performance and hindering the normal communication process. While ionospheric scintillation is a significant factor for signals with a carrier frequency below 6 GHz, it particularly affects those below 3 GHz traveling through the ionosphere. Strong scintillation is rarely observed in mid-latitude regions but can occur frequently in low-latitude regions shortly after sunset. High-latitude regions generally witness medium to high-intensity scintillation phenomena.

For LEO satellites operating below 6 GHz, PL_c must be considered. However, in mid-latitude degree regions of [$\pm 20^\circ$, $\pm 60^\circ$], strong scintillation is typically undetectable, and PL_c can be assumed negligible. In other latitudes, PL_c can be calculated as [5]

$$PL_c = \frac{1}{\sqrt{2}}\left(27.5 \cdot \psi^{1.26} \cdot (f_c/4)^{-1.5}\right).$$

The above formula is an PL_c empirical formula based on the gigahertz scintillation model. ψ represents the amplitude scintillation index.

5.2.3 Shadow Fading and Clutter Loss

Shadow fading, a form of signal loss caused by the obstruction of terrestrial obstacles, occurs when satellite signals are blocked by shadowing objects such as rugged terrain, buildings, and trees [5]. Behind these obstacles, a shadow area with a weak radio signal field is formed, resulting in the attenuation of the received signal energy. The fading depth caused by the shadow effect is mainly related to the size of the shadow area and the signal frequency. The larger the shadow area, the higher the communication frequency and the greater the fading depth. For GEO satellites, the south mountain effect occurs where visibility between the GEO satellites and terrestrial stations might be impaired by southern mountains or buildings, obstructing the communication link for users in the northern hemisphere. Conversely, LEO satellites can lead to the urban canyon effect, where some signals at low elevation angles are more likely to be blocked by shadow objects such as urban structures in the propagation path, rendering communication infeasible.

Clutter loss is one of critical factors in the loss of the satellite-terrestrial links, which denotes the degradation of electromagnetic wave signal caused by the occlusion and scattering of nearby mountains, buildings, and other structures. Clutter refers to various objects such as buildings or vegetation on Earth's surface rather than the actual terrain. The propagation effect is significantly influenced by the clutter around a radio transmitter/receiver terminal, with nearer objects having more substantial effects. The actual impact will depend on the characteristics of the clutter and the radio parameters [6].

5.2.4 Rain Fading

Satellite-terrestrial links often traverse regions with heavy rainfall. Rain fading refers to the reduction of electromagnetic wave signal energy due to the absorption and scattering by raindrops. The degree of rain fading is determined by the wavelength and radius of the raindrops. The larger the wavelength in relation to the diameter of the water particles, the greater the signal distortion encountered by the electromagnetic wave. The diameter of raindrops is generally between 0.25–8 mm, the wavelength of the C-band is between 37.5–75 mm, and the wavelength of the Ka-band is between 7.5–11.1 mm. Therefore, compared with satellite communication using the C-band, the satellite communication using the Ka-band is more significantly affected by rain fading. Among them, LEO satellites, due to their shorter transmission distances and lower path loss, more frequently utilize the Ka-band, making them more susceptible to rain fading [7].

5.2.5 Multipath Fading

During the propagation of electromagnetic wave signals between satellites and terrestrial, reflections, and scattering caused by various structures, mountains, and clouds transform direct signals into multipath signals, each with different delays and attenuations [8]. The received signal consists of multipath signals with varying amplitudes, delays, and phases, leading to intersymbol interference. Orthogonality between subcarriers is disrupted, thereby decreasing the stability of the signal transmission link. In the case of GEO satellite communication and MEO satellite communication, the propagation distance is long, and the multipath signal suffers from severe path loss; the power of the multipath interference signal thus becomes small. Hence, the multipath effect of high-Earth orbit satellite communication can be neglected. However, the propagation distance of LEO satellite communication is relatively short, and the power of the multipath interference signal remains large, which gives rise to serious intersymbol interference.

5.2.6 Doppler Effect

5.2.6.1 Doppler Shift

In a high-mobility communication environment, the carrier frequency of the received signal will shift in relation to the transmitted signal. This phenomenon is referred to as the Doppler effect. The corresponding frequency shift is referred to as the Doppler shift. The Doppler distribution of a satellite specifically includes three parts: the Doppler fixed frequency offset caused by the satellite's movement; the Doppler spread caused by local scattering around the terrestrial equipment and the pure Doppler peak caused by the movement of the terrestrial equipment. To be more specific, the Doppler frequency offset can be expressed as

$$f_d = \frac{v}{c} \cdot f_c,$$

where f_c represents the carrier frequency of the transmitting end, v represents the relative movement speed between the transmitting and the receiving end, and c represents the speed of light.

The phenomenon of carrier frequency offset caused by the Doppler effect is shown in Figure 5.3. Assuming that the carrier frequency of the transmitted signal is f_c, an oscillator is usually used at the receiving end to generate the same carrier frequency as that of the transmitted signal f_c to achieve carrier matching and recover the signal. However, due to the Doppler effect, the carrier frequency of the

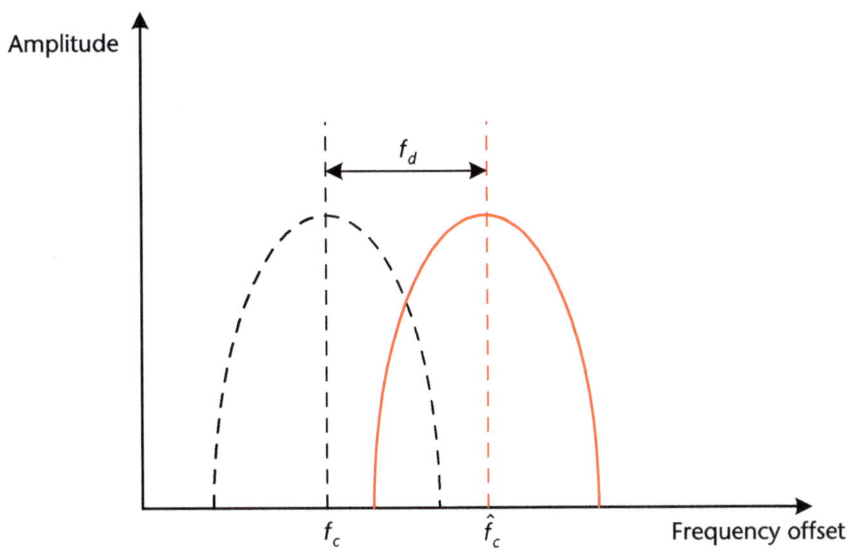

Figure 5.3 Schematic diagram of Doppler frequency offset.

signal is shifted to the right by f_d, and the carrier frequency becomes $\hat{f_c} = f_c + f_d$, which prevents the signal carrier from being matched correctly, resulting in demodulation errors. The normalized frequency offset ε is used to measure the degree of impact of the frequency offset on the modulation system, defined as the ratio of the frequency offset to the subcarrier interval as follows:

$$\varepsilon = \frac{\Delta f}{\Delta f_0},$$

where Δf is the frequency offset and Δf_0 is the subcarrier interval. The integer part of the normalized frequency offset is called integer frequency offset (IFO) ε_i and the fractional part is called fractional frequency offset (FFO) ε_f.

5.2.7 Atmospheric Absorption

Electromagnetic wave signals will be affected by atmospheric absorption, rainfall, fog, and other factors in the process of passing through the atmosphere. As the frequency band used by the satellite-terrestrial link gradually increases, especially in the context of terahertz communication, which is one of the key technologies of 6G, the impact caused by atmospheric absorption cannot be ignored. The characteristic atmospheric attenuation of radio waves at the 1,000 GHz frequency is mainly due to dry air and water vapor. The value of this attenuation is related to the satellite elevation angle, communication frequency, pressure, temperature, and other factors, which can be formulated as

$$PL_a = \frac{L_{zenith}(f_c)}{\sin\theta},$$

where θ is the elevation angle, f_c is the carrier frequency (in GHz), h is the satellite altitude, and $L_{zenith}(f_c)$ is the zenith attenuation (i.e., atmospheric absorption loss at an elevation angle of 90° for different altitudes and environments on earth). The effect of $L_{zenith}(f_c)$ is mainly caused by the resonant absorption lines of oxygen and water vapor, which is usually less than 10 dB based on the reference values for different weather conditions given by ITU-R.

3GPP provides a simplified methodology and sets all the required parameters to their annual average global values based on the original methodology of the ITU that considers only UEs placed at sea level. Atmospheric absorption loss should be considered only for frequencies above 10 GHz, or for any frequency with an elevation angle of less than 10 degrees. We apply the piecewise cubic hermite interpolating polynomial (PCHIP) algorithm to model $L_{zenith}(f_c)$ illustrating the relationship between atmospheric absorption and frequency. The PCHIP algorithm performs piecewise cubic polynomial interpolation, ensuring that the data's monotonicity is preserved and preventing the introduction of spurious oscillations during the interpolation process. The values of $L_{zenith}(f_c)$ are shown in Figure 5.4.

5.2.8 Building Penetration Loss

Today, the majority of mobile network users are indoor users, which makes the most communication data services generated from indoors, whereas indoor users

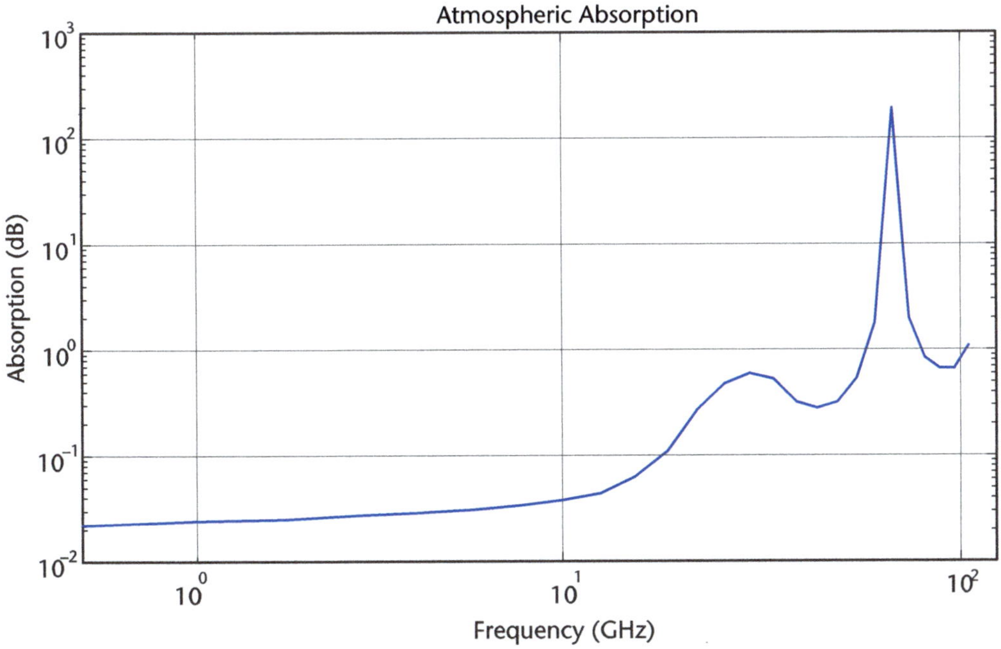

Figure 5.4 Atmospheric absorption loss of 1–100 GHz (90° elevation angle).

usually utilize outdoor base stations for access. Therefore, building penetration loss is also an important factor to consider. When modern, thermally efficient building materials are used (metalized glass, foil backed panels), building penetration losses are usually significantly higher than in buildings without such materials. L_h is the median loss for horizontal paths, given by [11]

$$L_h = r + s\log(f) + t(\log(f))^2.$$

The path elevation angle on the building fade is expressed as

$$L_e = 0.212|\theta|,$$

where f is the frequency in GHz and θ is the elevation angle on the building facade in degrees. In a nutshell, the propagation effects and impacts present in the channel transmission environment are summarized as follows. We summarize the above channel transmission environment influences in Table 5.1.

5.3 Classical Satellite-Terrestrial Channel Models

Conventional methods for satellite mobile channels can be divided into three categories: empirical models, statistical models, and geometrically stochastic models. In recent years, extensive research has also been conducted on satellite-to-terrestrial channel modeling based on machine learning.

One of the classic models within empirical models is the empirical roadside shadowing (ERS) model, which can calculate the fade margin needed to account for signal attenuation caused by the environment surrounding the receiver. Due to

Table 5.1 Propagation Effects in the Channel Transmission Environment

Transmission Characteristics	*Physical Causes*	*Effect*
Nonlinear effect	Nonlinear operating characteristics of onboard critical RF devices	Received signal's amplitude and phase are distorted, causing group delay
Free-space loss	Electromagnetic wave loss	Signal-to-terrestrial power, beam coverage
Ionospheric scintillation	Changes in ionospheric electron density	Random jumps in signal amplitude and phase
Rain fading	Rainfall attenuation of electromagnetic signals	Signal power attenuation, depolarization effect
Shadow fading and clutter loss	Obstruction to obstacles	Dynamic easing of signal envelope level values
Multipath fading	Signal reflection, scattering	Rapid fluctuations in signal amplitude and phase
The doppler effect	Relative motion	Received signal frequency offset
Atmospheric absorption	Absorption of electromagnetic signals by dry air and water vapor	Signal power attenuation
O2I penetration loss	Building attenuation of electromagnetic signals	Signal power attenuation

the narrow usable frequency range and elevation angle range of the ERS model, many modifications have been made to it. However, because empirical models can only establish simple mathematical relationships based on measured data, they can describe the characteristics of a particular parameter but cannot reveal the underlying signal propagation characteristics. Therefore, recent research on empirical models for satellite channels has been limited [12].

In the study of satellite channel models, probability distribution functions are commonly used to describe channel characteristics. This approach is not only straightforward and intuitive but also provides an accurate representation of the statistical properties of channel models. The currently widely researched satellite channel models primarily include the C.Loo model, Corazza model, LutZ model, and tapped delay line (TDL) satellite mobile channel model.

5.3.1 The C.Loo Model

The C.Loo model [14] was proposed by Chun Loo in 1985, and is suitable for channels with shadowing in rural or suburban environments. The received signal is composed of both direct and multipath components. While the direct signals are influenced by shadowing, the multipath signals remain unaffected. As a result, the C.Loo model is also referred to as the partial shadowing model, and its mathematical representation is shown in Figure 5.5.

Specifically, the received signal can be expressed as

$$r(t) = z(t)s(t) + d(t),$$

where $z(t)$ denotes the direct signal obeying a lognormal distribution, $s(t)$ is the shadowing effect on the direct signal, and $d(t)$ denotes the multipath signal component obeying a Rayleigh distribution. The probability density function of the

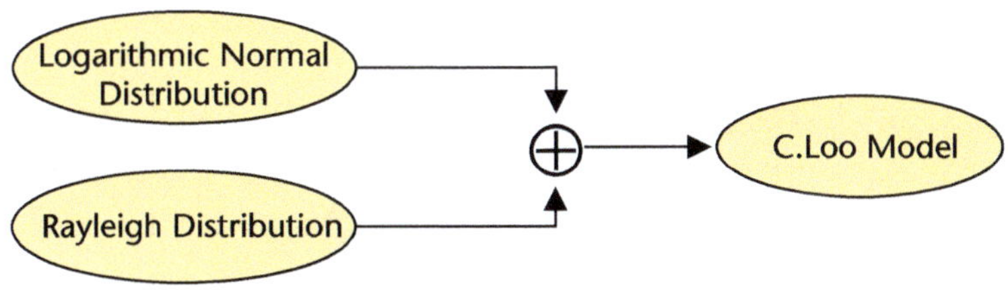

Figure 5.5 C.Loo mathematical model.

direct signal component $z(t)$ can be expressed as

$$f_z(z) = \frac{1}{z\sigma_1\sqrt{2\pi}} \exp\left[-\frac{(\ln z - \mu)^2}{2\sigma_1^2}\right],$$

whereμ, σ_1^2 are the mean and variance of $\ln z$, respectively.

When the direct signal component is fixed, the envelope of the received signal r follows the Rice distribution, and its probability density function is expressed as

$$f_r(r|z) = \frac{r}{\sigma_2^2} \exp\left[-\frac{r^2+z^2}{2\sigma_2^2}\right] I_0\left(\frac{rz}{\sigma_2^2}\right),$$

where σ_2^2 is the variance of the Rice distribution , I_0 is modified Bessel function of the first kind, order zero.

The probability density function of the received signal envelope r is

$$\begin{aligned} f_r(r) &= \int_0^\infty f_r(r|z) f_z(z) dz \\ &= \frac{r}{\sigma_1\sigma_2^2\sqrt{2\pi}} \int_0^\infty \frac{1}{z} \exp\left[-\frac{r^2+z^2}{2\sigma_2^2} - \frac{(\ln z - \mu)^2}{2\sigma_1^2}\right] dz \end{aligned}$$

This equation is the theoretical formulation of the C.Loo model.

5.3.2 Corazza Model

The Corazza model [15] is applicable to rural suburban as well as urban areas, with the received signal composed of direct and multipath signal components. Unlike the C.Loo model, both the direct and multipath components in the Corazza model are affected by shadowing, so it is also referred to as the full shadowing model. Its mathematical model is shown in Figure 5.6.

The received signal of Corazza model can be expressed as

$$r(t) = [z(t) + d(t)] * s(t),$$

where $d(t)$ is the multipath signal following a Rayleigh distribution, $s(t)$ is the shadow fading following a lognormal distribution, and $z(t)$ is the diameter component at the receiving end. When the shadow fading is considered as a constant, $r(t)$

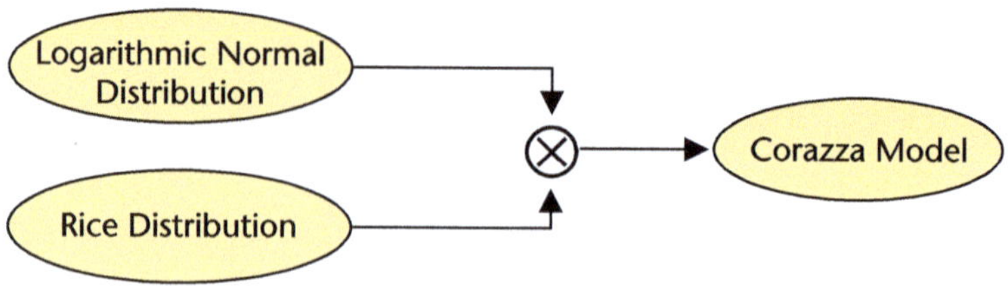

Figure 5.6 Corazza mathematical model.

follows the Rice distribution, and the probability density function of $r(t)$ is

$$f(r \mid s) = \frac{r}{s\sigma_0^2} \exp\left[-\frac{r^2/s^2 + z^2}{2\sigma_0^2}\right] I_0\left(\frac{zr}{s\sigma_0^2}\right).$$

According to the total probability theorem, we can obtain $f(r)$

$$f(r) = \int_0^{\infty} \frac{1}{s} f(r \mid s) f(s)\, ds,$$

where $f(s)$ is

$$f(s) = \frac{1}{s\sigma_s\sqrt{2\pi}} \exp\left[-\frac{(\ln s - \mu_s)^2}{2\sigma_s}\right].$$

5.3.3 Lutz Model

The Lutz model [16] divides the channel into two states: good state and bad state based on whether there are obstacles obstructing the signal transmission process. Thus, the model is also known as the two-state channel model, whose mathematical model is shown in Figure 5.7.

The good state corresponds to the case where the signal is not affected by shadowing. In this case, the received signal's envelope obeys the Rice distribution, and its probability density function $r(t)$ can be expressed as

$$f_r(r) = K \cdot exp[-K(r+1)] \cdot I_0(2K\sqrt{r}),$$

where K is the Rice factor, I_0 is the first class of zero-order Bessel functions.

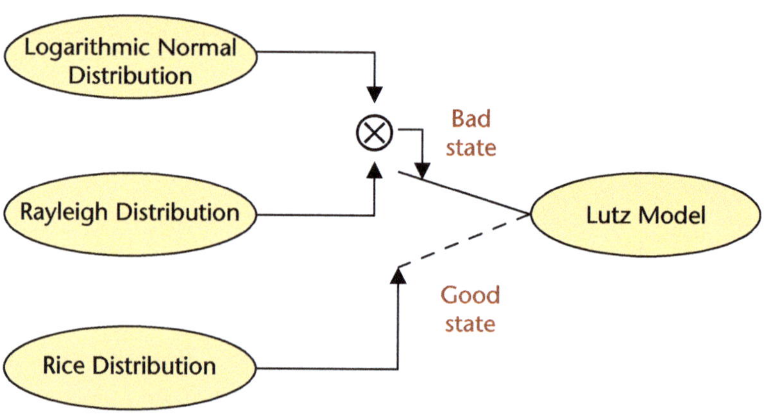

Figure 5.7 Lutz mathematical model.

The bad state corresponds to the situation where the signal is impacted by shadowing. At this time, there is no direct signal and the multipath is affected by shadow fading, so that the received signal obeys the Rayleigh-lognormal distribution and the probability density function $r(t)$ is expressed as

$$f_r(r|s) = \frac{r}{\sigma_1^2} \exp\left[-\frac{r^2}{2\sigma_1^2}\right].$$

Shadow fading $s(t)$ follows the lognormal distribution and the probability density function is expressed as

$$f_s(s) = \frac{1}{h\sigma_2 s\sqrt{2\pi}} \exp\left[-\frac{(\ln s - \mu)^2}{2h^2\sigma_2^2}\right],$$

where $h = (\ln 10)/20$, μ and $(h\sigma_2)^2$ are the mean and variance of, respectively.

5.3.4 TDL Satellite Mobile Channel Models

The TDL is applied to single antenna channels and is considered to consist of taps, each having distinct fading and delay coefficients. The number of taps can be set according to the demand, making it suitable for satellite mobile communication channel simulation. In the channel model simulation, each tap represents each path of the signal, and different configurations of the tap coefficients can realize modeling under different scenarios. The TDL model divides the scenarios of the LEO satellite communicating with the terrestrial terminals into non-line-of-sight (NLOS) and LOS scenarios. TDL-A and TDL-B are the simulation models for NLOS scenarios, while TDL-C and TDL-D are for LOS scenarios. For the case of NLOS scenarios, there is no direct path, and the multipath follows Rayleigh fading. For the case of LOS scenarios, the direct path at the receiving end follows Rice fading, while the rest of the multipath obeys Rayleigh fading.

For the most cases of wireless channels, the resolvable paths can be viewed as fixed values that do not change with time. Similarly, in the TDL model, each resolvable path has a fixed delay phase as well as a Doppler shift. If the transmitted signal at the satellite is $s(t)$, the received signal at the terrestrial $y(t)$ can be expressed as

$$y(t) = \sum_{n=1}^{N} s\left(t - \sum^{n} \tau_n\right) g_n(t),$$

where N is the number of resolvable paths, τ_n is the time delay of the nth path, and $g_n(t)$ is the channel impulse response of the nth path denoted by

$$g_n(t) = \alpha_n \exp\left(j\theta_n + j2\pi f_{Dn} t\right),$$

where α_n is the fading factor of the nth tap, θ_n is the initial phase of the nth tap, and f_{Dn} is the Doppler frequency offset.

The geometric randomness channel model is a reasonable assumption about the distribution of obstacles and scatterers in the vicinity of the receiving end. Specifically, the effective scatterers in the environment are abstracted into one or more specific geometric models and their distributions are modeled. Studying channel variations by varying the scatterer distribution function has the advantages of strong model generalization and low computational complexity. Currently there are fewer studies on geometric randomness modeling for satellite-terrestrial links, which can be categorized into 2D models, 3D models, and so on.

Machine learning based channel modeling usually uses long short-term memory (LSTM), convolutional neural networks (CNNs), recurrent neural network (RNN), and other methods to predict fading. LSTM is one of widely used methods with good prediction accuracy. Machine learning-based modeling can take advantage of the superior self-learning and prediction capabilities to extract channel features from complex channel data and improve the generalization ability of the model. However, the prediction accuracy of current machine learning-based models is highly dependent on the richness of the training set data and consumes more computational resources. Therefore, its trained models are only applicable to some specific scenes or frequency bands.

5.4 Evolution of Satellite-Terrestrial Channel Standards

The integration of satellite-terrestrial networks began to be studied by 3GPP from Release 15 [13]. The deployment scenarios of NTN are defined in Technical Report (TR) version 38.811, including eight enhanced mobile broadband scenarios and two large-scale machinelike communication scenarios. In addition, some key system parameters, such as channel model, orbit height, carrier frequency point, and bandwidth, are also defined. During Release 16, 3GPP detailed the protocol layer improvements required for the convergence of NTN networks with 5G in the TR 38.821, and defined the parameters for satellite-terrestrial link-level simulation in the S and Ka bands. During Release 17, 3GPP started to develop NTN-related standards, which were further improved in Release 18.

According to TR 38.811 [5], NTN LEO satellite channels are divided into two types: LOS and NLOS propagation. The channel of LOS scenarios can be modeled by the Rice distribution, while that of NLOS scenarios can be modeled by the Rayleigh distribution due to the zero Rice K-factor. The TR 38.811 standard provides four different typical scenarios based on empirical data: TDL-A/B/C/D. In particular, TDL-A and TDL-B correspond to different NLOS environments, while TDL-C and TDL-D refer to different LOS environments. The simulation parameters for these scenarios are shown in Tables 5.2 to 5.5.

Table 5.2 Scene 1 NLOS-TDL-A

Channel Type	*Tap Serial Number*	*Normalization Delay*	*Power [dB]*	*Fading Distribution Characteristics*
NTN-TDL-A	1	0	0	Rayleigh
	2	1.0811	−4.675	Rayleigh
	3	2.8416	−6.482	Rayleigh

Table 5.3 Scene 2 NLOS-TDL-B

Channel Type	*Tap Serial Number*	*Normalization Delay*	*Power [dB]*	*Fading Distribution Characteristics*
NTN-TDL-B	1	0	0	Rayleigh
	2	0.7249	−1.973	Rayleigh
	3	0.7410	−4.332	Rayleigh
	4	5.7392	−11.914	Rayleigh

Table 5.4 Scene 3 NLOS-TDL-C

Channel Type	*Tap Serial Number*	*Normalization Delay*	*Power [dB]*	*Fading Distribution Characteristics*
NTN-TDL-C	1	0	0.394	LOS path
	2	0	10.618	Rayleigh
	3	14.8124	−23.373	Rayleigh

Table 5.5 Scene 4 NLOS-TDL-D

Channel Type	*Tap Serial Number*	*Normalization Delay*	*Power [dB]*	*Fading Distribution Characteristics*
NTN-TDL-D	1	0	−0.284	LOS path
	2	0	−11.991	Rayleigh
	3	0.5596	−9.887	Rayleigh
	4	7.3340	−16.771	Rayleigh

References

[1] Fang, K., "Research and Key Technology Verification on Satellite-Earth Communication Transmission of 800 Mbps Throughput," Master's dissertation, University of Electronic Science and Technology, 2018.

[2] Liu, H., "Research on Key Technology of Channel Modelling and Simulation Towards, Low Orbit Satellite Communication System," PhD dissertation, Beijing University of Posts and Telecommunications, 2022.

[3] ITU-R P.618, Propagation Data and Prediction Methods Required for the Design of Earth-Space Telecommunication Systems [S/OL], 2017–12.

[4] ITU-R P.531, Ionospheric Propagation Date and Prediction Methods Required for the Design of Satellite Networks and Systems [S/OL], 2019-08.

[5] 3GPP, TR 38.901: Study on Channel Model for Frequencies from 0.5 to 100 GHz (Release 16) [R/OL], 2020-01-11.

[6] ITU-R P P.2108, Prediction of Clutter Loss [S/OL], 2021-09.

[7] Huang, M., "OCDM and Its Transmission Performance in NTN Low-Orbit Satellite Scenario," Master's dissertation, Harbin Institute of Technology, 2022.

[8] ITU-R P. 678, Representation of Natural Variability of Communication Phenomenon [S/OL], 2015-07.

[9] Qiao, Y., "Modelling and Simulation Implementation For LEO Mobile Channel," Master's dissertation, Beijing University of Posts and Telecommunications, 2023.

[10] ITU-R P. 676, Attenuation by Atmospheric Gases and Related Effects [S/OL], 2007-02.

[11] ITU-R P, 2109, Prediction of Building Entry Loss [S/OL], 2023-08.

[12] Zhaoyang, S., L. Liu, A. Bo, et al., "A Review of Satellite-Terrestrial Channel Models for Low-Orbit Satellites," *Journal of Electronics and Information*, 2023.

[13] 3GPP, TR 38.811: Study on New Radio (NR) to Support Non-Terrestrial Networks (Release 15), [R/OL], 2020-10-08.

[14] Loo, C., "A Statistical Model for a Land Mobile Satellite Link," *IEEE Transactions on Vehicular Technology*, Vol. 34, No. 3, 1985, pp. 122–127.

[15] Corazza, G. E., and F. Vatalaro, "A Statistical Model for Land Mobile Satellite Channels and Its Application to Nongeostationary Orbit Systems," *IEEE Transactions on Vehicular Technology*, Vol. 43, No. 3, pp. 1994, pp. 738–742.

[16] Lutz, E., D. Cygan, M. Dippold, et al., "The Land Mobile Satellite Communication Channel-Recording, Statistics, and Channel Model," *IEEE Transactions on Vehicular Technology*, Vol. 40, No. 2, 1991, pp. 375–386.

CHAPTER 6

Channel Coding for Satellite-Terrestrial Integrated Communication

Channel coding technology is the foundation and core of digital communication systems. It improves the reliability of data transmission by adding redundant information in noisy and interfering channels. Classic channel coding schemes include cyclic, convolutional, turbo, low-density parity-check (LDPC), and polar codes. These coding schemes have been widely utilized in terrestrial and satellite communication systems. This chapter systematically summarized the development and application of channel coding in terrestrial communication systems and satellite communication systems, as well as the new challenges and potential directions faced in responding to the future trend of satellite-terrestrial integration.

6.1 Classical Channel Coding

In wireless communication systems, information transmission must pass through various physical channels. Due to interference, device nonideality, equipment failure, and other factors, the transmitted information symbols will be distorted, resulting in damage to helpful information and misjudgment of the signal at the receiving end. To improve the accuracy of information transmission and enable it to have better resistance to channel noise, interference, and so on, it is necessary to adopt unique error detection and correction methods, namely error control. The task of error control is to detect errors, point out the erroneous signal, or correct the mistake. Error control is mainly achieved by channel coding.

Channel coding refers to detecting and correcting transmission errors by adding redundant information to the original information, thereby improving system reliability. For example, the redundant bit 010 is added to the source bit 1100 according to a specific rule to obtain the code 1100010. The coded bits are transmitted on the wireless channel after signal modulation. Due to the influence of interference, noise, and so on, there are error bits in the demodulated received signal. However, since there are redundant vectors in channel coding, errors can be found and corrected during channel decoding, reducing transmission errors and improving communication reliability. Therefore, designing redundant information and using it to correct mistakes becomes key to channel coding.

In theory, the noisy channel coding theorem (Shannon's second theorem) states that if the capacity of a discrete memoryless stationary channel is C and the length of the input is N, as the information rate to be transmitted is not greater than C, a coding scheme can always be designed. When N is large enough, the decoding error probability is arbitrarily small. In contrast, when the transmission information rate is more significant than C, for codes of any length the decoding error probability must be greater than zero [1]. It should be noted that the channel coding theorem is only an existence theorem, which calculates a limit on the channel capacity. The channel can transmit information almost without distortion when the transmitted information rate does not exceed this limit. In other words, under the premise of ensuring that the transmission information rate is lower than the channel capacity, a coding scheme with an error probability approaching zero exists.

According to how information bits are processed, channel coding can be divided into convolutional and block codes. Specifically, in block codes, the input information bits are divided into several groups, then r supervision bits are generated for each group of information bits according to a specific coding rule, and the supervision bits are only related to the information bits of this group. Furthermore, if the supervision bits are linearly related to the information bits, it is called a linear block code. In convolutional codes, the supervision bits depend not only on the information bits of this group but also on the information bits of the previous L groups, which is expressed explicitly as (N, K, L) code, where N is the code length and L is the constraint length. Since linear block codes and convolutional codes are more commonly utilized, they will be introduced in detail below.

6.1.1 Linear Block Code

Standard linear block codes include RM codes, cyclic codes, LDPC codes, and polar codes. As shown in Figure 6.1, the encoding process of a linear block code (N, K) with information length K and codeword length N can be expressed as

$$\mathbf{c} = \mathbf{mG} \tag{6.1}$$

where $\mathbf{G}$ is a generator matrix with K rows and N columns, $\mathbf{m}$ is the information bit sequence, and $\mathbf{c}$ is the encoded codeword sequence. In the above context, the encoding structure becomes apparent with the determination of the generator matrix $\mathbf{G}$.

6.1.1.1 RM Code

Reed-Muller (RM) code, also known as RM code, was first proposed by Reed in 1954 [2]. RM code is a classic linear block code that can correct multiple errors simultaneously. For positive integers r and m, and satisfying $0 \leq r \leq m$, an RM code of r-order with codeword length $N = 2^m$, denoted by RM (r, m) exists. The information bit length is

$$k(r,m) = \sum_{i=0}^{r} \binom{m}{i} = 1 + \binom{m}{1} + \binom{m}{2} + \cdots + \binom{m}{r} \tag{6.2}$$

The minimum Hamming distance is 2^{m-r}. For example, for the RM code $(5, 10)$, the codeword length $N = 1024$, the information length is 638, and the minimum Hamming distance is 32.

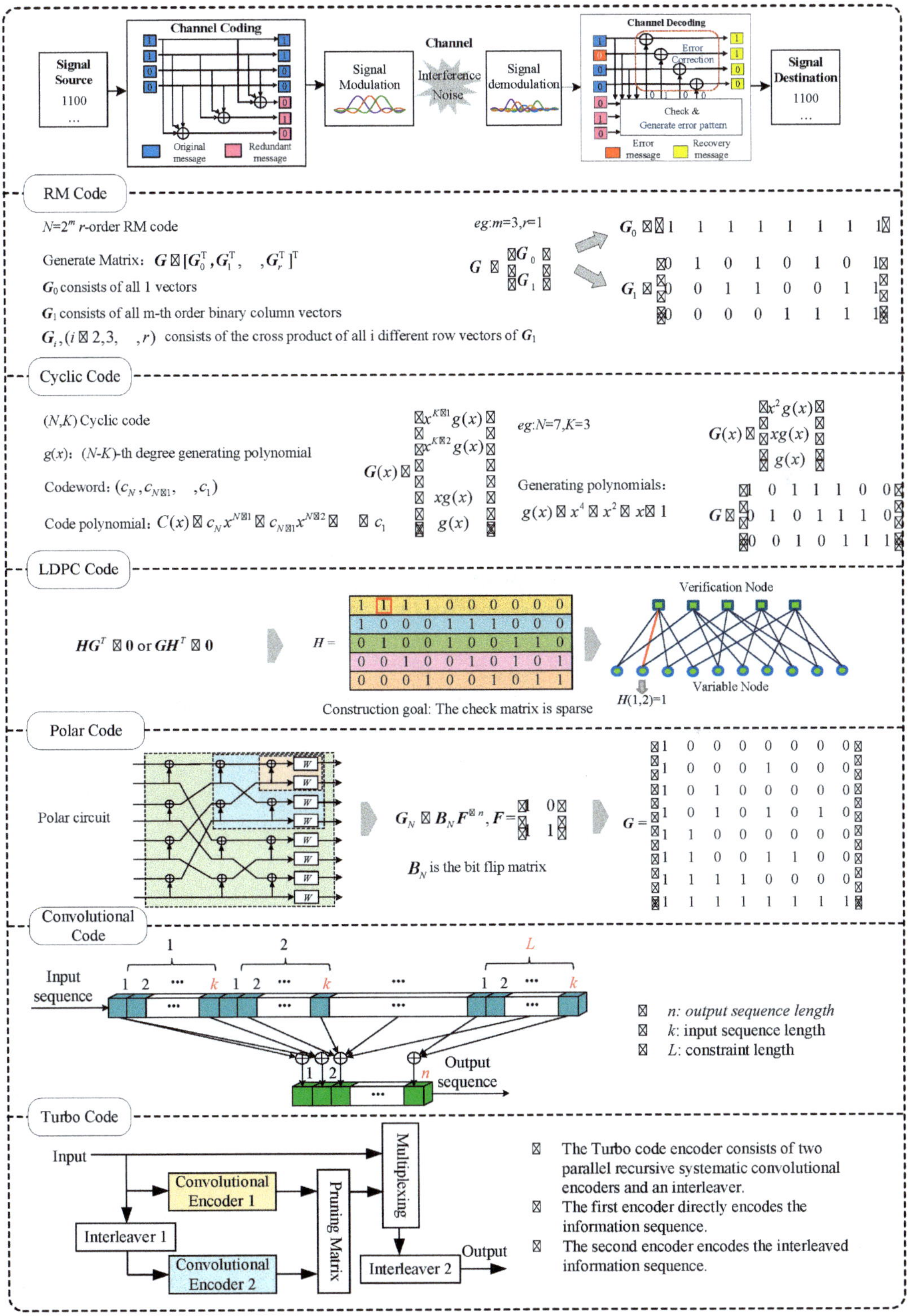

Figure 6.1 Basic principles of classical channel coding.

Like general linear block codes, RM codes can be encoded by generating matrices. The matrix of RM codes can be generated by solving the Hadamard matrix. Specifically, given the matrix with 2×2 kernel

$$F_2 = \begin{bmatrix} 1 & 0 \\ 1 & 1 \end{bmatrix} \tag{6.3}$$

Then the m-order Hadamard matrix is calculated as $F_2^{\otimes m}$, where $\otimes$ represents the Kronecker product. Further, the generator matrix G of the (r, m) RM code consists of all rows in the Hadamard matrix with weight greater than or equal to 2^{m-r}.

RM codes have diverse and straightforward structures with decoding methods, such as hard decisions and soft decisions. Due to its excellent error correction performance, RM was widely used in deep-space communications in the late 1960s and early 1970s.

6.1.1.2 Cyclic Code

Cyclic codes are an important subclass of linear block codes and are a relatively mature coding scheme [3]. In addition to the characteristics of linear block codes, cyclic codes also have a cyclic property: any cyclic shift of a codeword still yields a legal codeword. The generator matrix of a cyclic code can be obtained by cyclic shifting the generator polynomial. For instance, the generator polynomial of a cyclic code is $g(x) = g_{N-K}x^{N-K} + g_{N-K-1}x^{N-K-1} + \cdots + g_1 x + g_0$, where N is the codeword length and K is the information bits length. Then, the generator matrix can be expressed as

$$G_{K\times N} = \begin{bmatrix} g_{N-K} & g_{N-K-1} & \cdots & g_1 & g_0 & 0 & \cdots & 0 \\ 0 & g_{N-K} & g_{N-K-1} & \cdots & g_1 & g_0 & \cdots & 0 \\ \vdots & \vdots & \vdots & \vdots & \vdots & \vdots & \vdots & \vdots \\ 0 & \cdots & 0 & g_{N-K} & g_{N-K-1} & \cdots & g_1 & g_0 \end{bmatrix} \tag{6.4}$$

Bose, Chaudhuri, and Hocquenghem independently discovered BCH codes in 1959 [4], which are crucial cyclic codes. For a BCH code with an error correction capability of t, its generating polynomial contains $2t$ consecutive power roots, which can be efficiently decoded using an iterative decoding algorithm. Specifically, for positive integers m and t, there exists a binary BCH code with a code length of m, whose generating polynomial has roots of t, which can correct t random errors. At the same time, BCH codes also inherit the characteristics of the simple structure of cyclic codes, flexible code length, and code rate design, which also gives them specific applications in early satellite communications such as digital video broadcasting.

The RS code is a multilevel BCH code proposed by S. Reed and G. Solomon of MIT Lincoln Laboratory in 1960 [5]. In the RS code, the input sequence is divided into $k \times m$ groups of k bits. Each group contains k symbols, and each symbol consists of m bits. An RS code can correct t errors has the following parameters: code length is $2^m - 1$, supervision bit length is $2t$, and minimum code distance is $d_{\min} = 2t + 1$.

The encoding process of the RS code is similar to that of the BCH code and is also encoded according to the polynomial $g(x)$. Specifically, it can be implemented by a shift register with feedback. It is worth noting that the RS code can correct burst errors and is widely used in deep-space communications, digital satellite TV systems, and other fields.

6.1.1.3 LDPC Code

Gallager proposed LDPC codes in 1962 [6] but did not attract much attention from experts and scholars then. It was not until the 1990s that MacKay, Spielman, and others discovered that LDPC codes, like turbo codes, could approach the Shannon limit [7], and researchers began to conduct in-depth research on LDPC codes.

LDPC code is a unique linear block code. Its check matrix **H** is sparse and can be represented by the Tanner graph shown in Figure 6.2, where c_i represents the check node and v_j represents the variable node. The line between them represents the element with a value 1 in the check matrix. There are two main methods for constructing the check matrix of LDPC code. The first method is random construction, including Gallager construction, Mackay construction, Davey construction, and PEG construction. This scheme can construct the LDPC code with the best performance when the code length is sufficient, but the complexity is enormous and difficult to implement in hardware. The second method is structured construction, which includes finite geometry construction, combinatorial design, π rotation, and quasi-cyclic construction. Compared with the random construction scheme, the structured LDPC code has a specific structure and cyclic or quasi-cyclic characteristics, and the coding is simple to implement.

Most early LDPC codes were randomly constructed using Gaussian elimination to obtain the generator matrix and complete the encoding. The encoding complexity is $O\left(N^2\right)$, where N is the codeword length. However, as the code length increases, both the storage space and computation amount will dramatically increase. To overcome this defect, Richardson proposed a coding algorithm based on an approximate lower triangular matrix, which reduces the complexity from $O\left(N^2\right)$ to $O\left(N + G^2\right)$, where G is the length of the check sequence. The most common decoding algorithm for LDPC codes is belief propagation (BP) decoding, which completes the decoding by continuously and iteratively updating the information in the variable and checking nodes in parallel.

LDPC code is one of the coding schemes with the best performance at this stage, but its excellent performance can only be fully reflected in the case of long codes. The error correction performance is significantly reduced under short and medium

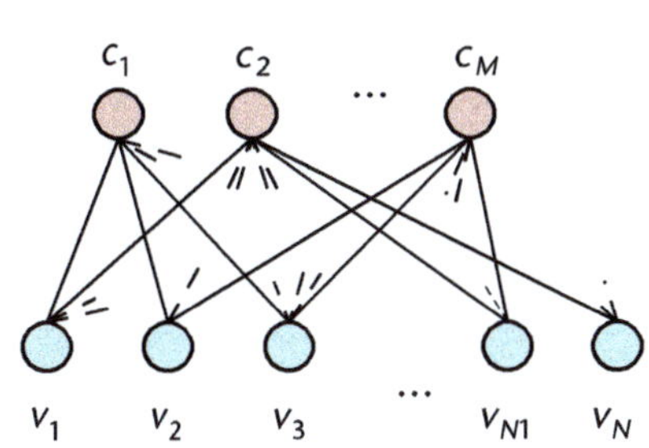

Figure 6.2 LDPC code Tanner graph.

code lengths. In addition, the coding complexity of LDPC codes under the random construction method is high and consumes a great deal of hardware resources. At the same time, LDPC also has false leveling, and it cannot be mathematically proven that the Shannon limit can be reached.

6.1.1.4 Polar Code

Polar codes were created to explore channel coding that approaches the Shannon limit further. Polar codes were proposed by E. Arikan in 2009 [8] and are currently the only coding scheme theoretically proven to reach the Shannon limit. Its creation is a significant breakthrough in channel coding theory and has attracted widespread attention from wireless communication scholars. In recent years, polar codes have significantly progressed in coding construction and decoding.

Polar code encoding is mainly based on the channel polar phenomenon. Channel polarization is combining and splitting N independent channels to obtain N subchannels with certain connections. After the split, the channel capacities of some of the N sub-channels approach 1, and some approach 0. The total channel capacity remains unchanged. As shown in Figure 6.3, channel combination means merging N independent channels W into one channel W_N; channel splitting is to split the merged channel W_N into N subchannels $\{W_N^{(i)} : 1 \leq i \leq N\}$. Since the N subchannels are independent, the total channel capacity of the final separated subchannels is the sum of the capacities of the N independent channels W.

Based on the channel polarization, the bit channel with a capacity close to 1 is selected to transmit valid information, and the bit channel with a capacity close to 0 is assigned to transmit the frozen bits known to the transceiver. Since the two operations of channel joining and splitting are essentially linear calculations, polar codes are also a type of linear block code. Like other block codes, the encoding can be expressed as the multiplication of valid information and the generator matrix. Figure 6.4 shows the polar code encoding process. First, the reliability of each subchannel is calculated based on the known transmission information code length, channel environment, and other factors. Second, the information bit index set is obtained according to sorting the transmission reliability of the polarized subchannels. Then, a mixed mapping of frozen bits and information bits is performed, and the information bits are mapped to subchannels with higher reliability. The remaining subchannels transmit frozen bits to obtain a mixed sequence $u_1^N = \{u_1, u_2, \cdots u_N\}$, and finally, this mixed sequence is multiplied with the generator matrix to obtain the encoded codeword $x_1^N = u_1^N \mathbf{G}_N$. Among them, the polar

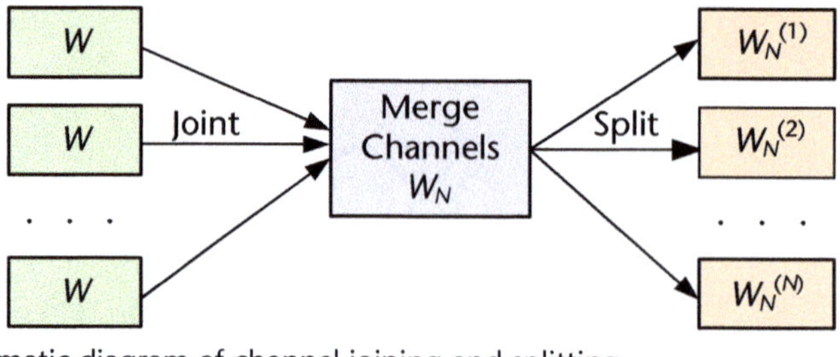

Figure 6.3 Schematic diagram of channel joining and splitting.

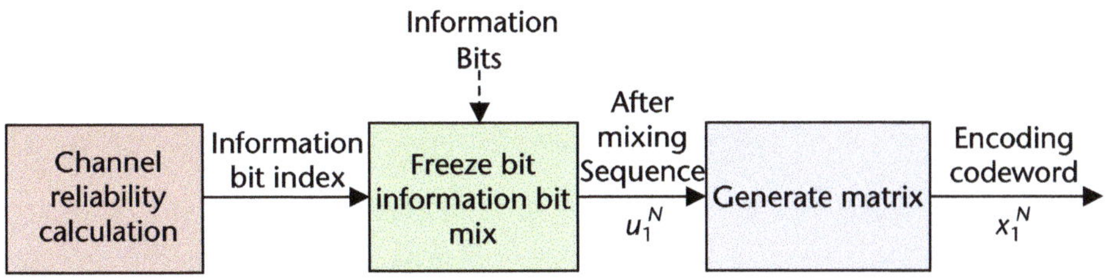

Figure 6.4 Polar code encoding diagram.

code generator matrix $\mathbf{G}_N$ consists of two parts, namely the bit inversion matrix $\mathbf{B}_N$ and the nth Kronecker product of the matrix $\mathbf{F}$.

Regarding polar code decoding, Arkan designed a successive cancellation (SC) decoding algorithm for his polar codes. The SC decoding algorithm adopts a serial decoding mechanism. Specifically, when the SC decoder decodes a bit at a particular position, it needs to use the decoding decision value of the previous bit, which can easily lead to serious error transmission. To reduce the impact of error transmission, the serial cancellation list (SCL) decoding algorithm was proposed [9]. In the SCL decoding process, multiple decoding paths will be retained. When all codewords are decoded, the path metrics of the L most likely decision results are sorted, and the path with the best path metric value is selected as the decoding output. SC and SCL are serial decoding algorithms designed based on the polar principle. In addition, there are parallel decoding algorithms such as belief propagation (BP) and belief propagation list (BPL) for polar codes [10].

As the latest coding scheme, polar code has gradually matured in theory and coding algorithm research after more than 10 years of development. Its evolved version of the PCC polar code was applied to the control channel in the 5G communication eMBB scenario in 2017.

6.1.2 Convolutional Code

P. Elias proposed convolutional codes in 1955 [11]. In contrast to linear block codes, in convolutional codes the parity bits depend on the information bits of the current group and the previous L groups. Specifically, it is represented as (N, K, L) code, where L is the constraint length. Although convolutional codes also encode k bits of information into n bits during the encoding process, the parity bits are related to the previous information and the k bits of information at the current moment. The convolutional code encoder consists of three parts: switch, adder, and register. The example of $(3, 1, 3)$ convolutional code is shown in Figure 6.5.

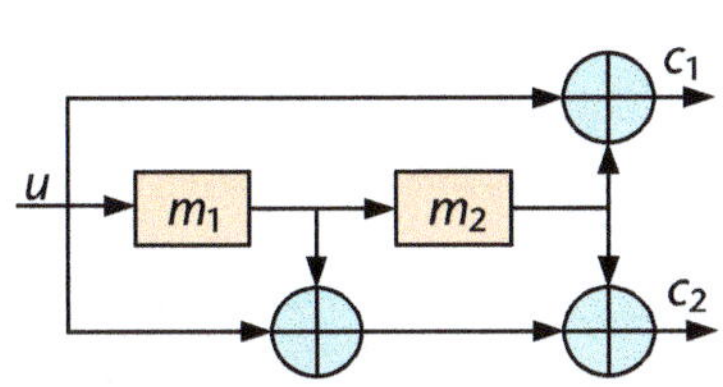

Figure 6.5 (3, 1, 3) Convolutional coding example.

Specifically, if the input sequence is $\cdots b_{i-2}b_{i-1}b_ib_{i+1}\cdots$ and the output of encoder is $c_id_ie_i$, the relationship between input and output is

$$\begin{cases} c_i = b_i \\ d_i = b_i \oplus b_{i-2} \\ e_i = b_i \oplus b_{i-1} \oplus b_{i-2} \end{cases} \tag{6.5}$$

where b_{i-1} and b_{i-2} are the information of the previous time saved in the register.

Convolutional codes can be decoded into two categories: algebraic decoding and probabilistic decoding. Algebraic decoding does not consider the statistical characteristics of the channel but directly uses the algebraic structure of the code itself for decoding. In contrast, the probabilistic decoding of convolutional codes, also known as maximum likelihood decoding, uses the code's characteristics and the channel's statistical characteristics for decoding. Sequential decoding is a probabilistic decoding method proposed for memoryless channels. Another necessary probabilistic decoding of convolutional codes is dimension bit decoding, which has fast speed and high efficiency when the constraint length is short and is currently a widely used decoding method.

Although convolutional codes have the advantages of simple structure and good error correction performance, their performance positively correlates with the constraint length. The longer the constraint length, the higher the hardware storage space requirement. In addition, the number of consecutive bit errors that convolutional codes can correct is limited, and it is challenging to handle burst errors effectively.

Based on the convolutional code, turbo code was proposed by C. Berrou in 1993. Its error correction capability close to the Shannon limit has attracted wide attention from experts and scholars [12]. The parallel concatenated convolutional code (PCCC) is the most widely used structure, as shown in Figure 6.6. This structure is composed of two convolutional encoders in a cascade. The input of convolutional encoder 2 is the interleaved information. The output information of the two encoders is concatenated with the initial input information after passing through the redundancy matrix, and the encoding result can be obtained by interleaving again.

Turbo code uses iterative decoding. As shown in Figure 6.7, the initial input information is divided into three parts: information bit X and check bits Y_1 and Y_2, which are sent to two convolution decoders, respectively. At the same time,

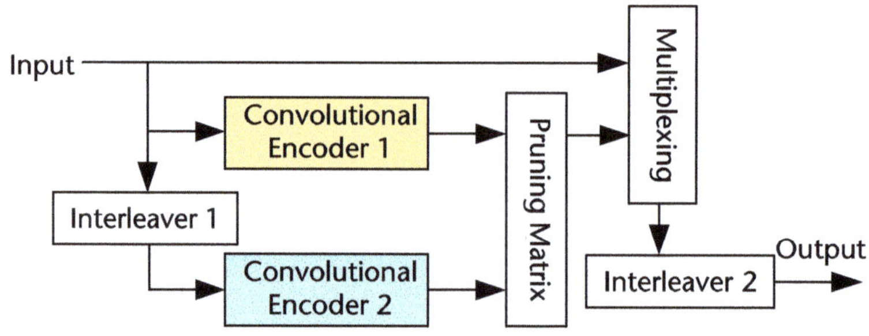

Figure 6.6 PCCC type coding structure.

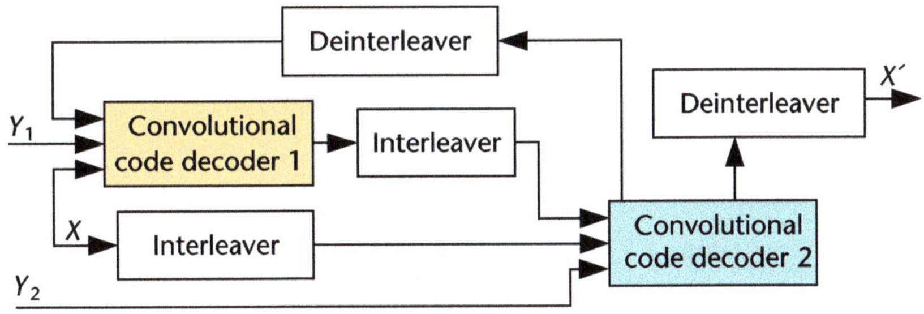

Figure 6.7 PCCC type decoding structure.

the decoding results (external information) of the two convolution decoders are interleaved and deinterleaved and then sent to each other as prior information. The above process is iterated continuously until the external information obtained by the two decoders tends to be stable, and the external information can be output for judgment to complete the decoding process.

Communication standards favor turbo codes due to their error correction capabilities close to the Shannon limit, but their shortcomings are also evident. When the decoding iterations reach a specific number, the bit error rate of the turbo code will change very slowly, and there will be an error floor. In addition, the delay caused by the interleaved and iterative decoding of turbo codes is significant. Turbo codes are challenging to meet the requirements of some communication systems with high real-time requirements.

6.2 Channel Coding for Terrestrial Communication

Channel coding has evolved with the advancement of terrestrial communication technology. Next, we will introduce in detail the application and development of channel coding technology in terrestrial communications such as cellular mobile communications and wireless local area network (LAN) communications.

6.2.1 Cellular Mobile Communication

As shown in Figure 6.8, cellular mobile communications have gone through five generations of development since the 1970s and 1980s, with information rates and reliability continuously increasing. Undoubtedly, channel coding technology plays an indispensable role in each generation of cellular mobile communication systems.

The first-generation (1G) mobile communication system was born in the 1980s and mainly provided analog voice services. The service channel uses analog signals for transmission, and the channel coding scheme used in the control channel is BCH code (n, k, m). The encoded data is sent repeatedly m times to improve the antifading performance, as shown in Table 6.1. For different channel types, the corresponding structure of BCH code is different. Specifically, the forward control channel uses $(40, 28, 5)$ BCH code, and the information transmission rate is 1250–1442 bit/s. The reverse control channel uses $(48, 36, 5)$ BCH code, and the information transmission rate is 1215 bit/s.

The 2G system is a narrowband digital cellular system that expands voice and low-speed data services while increasing system capacity. In addition, 2G

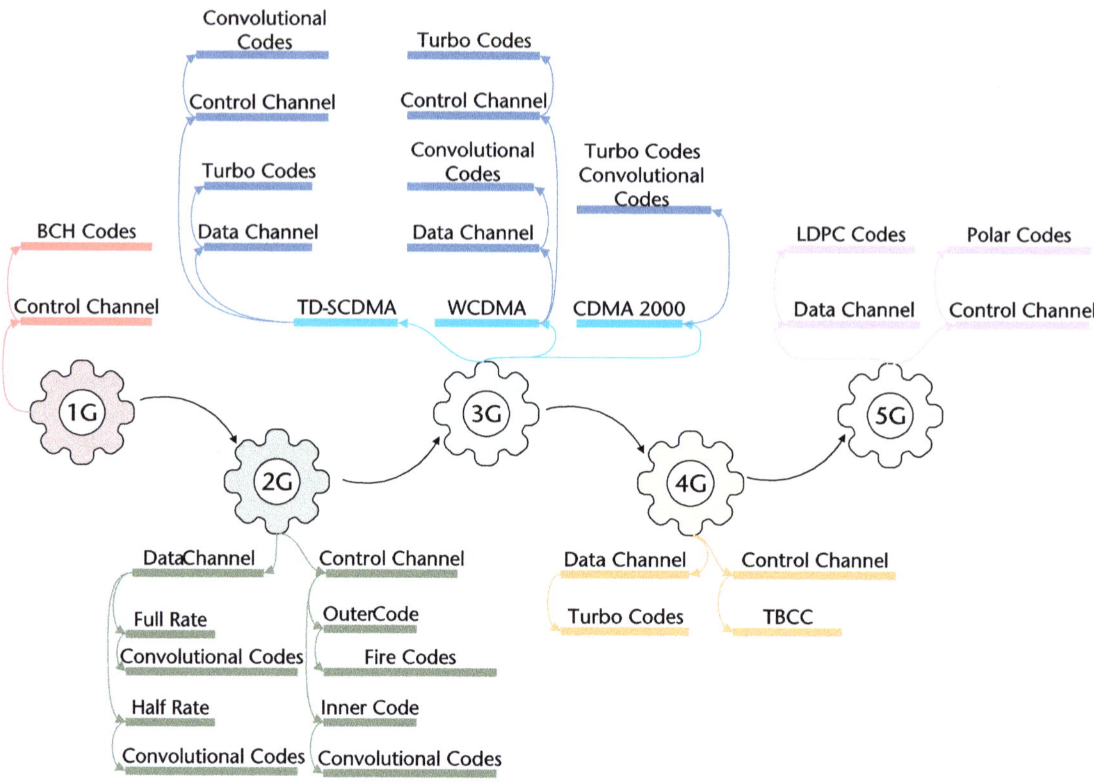

Figure 6.8 Application of channel coding in cellular mobile communication standards.

Table 6.1 Control Channel Coding for 1G Communication

Channel Type	*n*	*k*	*m*	*Information Transmission Rate (bit/s)*
Forward control channel	40	28	5	1215
Reverse control channel	48	36	5	1250–1442

introduced digital modulation technology, mainly using CDMA and time division multiple access (TDMA). Specifically, the US IS-95CDMA uses the CDMA standard, and the main TDMA standards GSM in Europe, D-AMPS in the United States, and PDC in Japan. Currently, GSM has achieved globalization, and different channel coding schemes are used according to its channel type. As shown in Table 6.2, convolutional codes are used in both half-rate data channels and full-rate data channels, but the code length, code rate, and constraint length are different [14]. In the control channel, concatenated coding is used. Specifically, the outer code uses the (228, 184) fire code, and the inner code uses a convolutional code with a code length of 456 and a constraint length of 5.

The 3G system includes three standards: CDMA2000, WCDMA, and TD-SCDMA. The channel coding scheme uses convolutional codes and combines the then-advanced turbo codes according to the channel characteristics [14]. As shown in Table 6.3, CDMA2000 uses convolutional codes with different code rates in the access, reverse public control, and reverse request channels. Turbo codes are used in packet data channels. Similarly, as shown in Table 6.4, TD-SCDMA and WCDMA

Table 6.2 Channel Coding for 2G Communication

Channel Type		*Coding Scheme*
Data channel	Half rate	Convolutional code (211,104,L=7)
	Full rate	Convolutional code (378,189,L=5)
Control channel		Fire (228,184)
		Convolutional code (456,228,L=5)

Table 6.3 Channel Coding for CDMA2000 in 3G Communication

Channel Type	*Coding Scheme*	*Bitrate*
Access channel	Convolutional code	1/3
Enhanced access channel	Convolutional code	1/4
Reverse common control channel	Convolutional code	1/4
Reverse request channel	Convolutional code	1/4
Reverse fundamental channel	Convolutional code	1/2, 1/3, 1/4
Supplemental channel	Convolutional code	1/2, 1/4
	Turbo code Code length 186–20730 bits	1/2, 1/3, 1/4, 1/5
Packet data channel	Turbo code	1/5
Synchronous channel	Convolutional code	1/2

Table 6.4 Channel Coding for TD-SCDMA and WCDMA in 3G Communication

Standard	*Data Channel*	*Control Channel*
TD-SCDMA	Turbo code (1/2, 1/3) Code length 320–5120 bits	Convolutional code (1/2, L=9)
WCDMA	Turbo code (1/2, 1/3) Code length 40–5114 bits	Convolutional code (1/2, 1/3, L=9)

also use two-channel coding schemes, convolutional and turbo codes, according to different channel types. The TD-SCDMA data channel uses turbo codes with a code rate of 1/2 or 1/3, supporting a code length of 320–5120 bits. The control channel uses a convolutional code with a constraint length of 9 and a code rate 1/2. In the WCDMA standard, the data channel uses a turbo code with a code rate 1/3, supporting a code length of 40–5114 bits. The control channel uses a convolutional code with a constraint length of 9 and a code rate of 1/2 or 1/3.

In the 4G system, the 3GPP organization has determined LTE-Advanced as the 4G international standard. In the LTE-Advanced standard, the control and data channels use different coding methods [15]. The data channel uses a turbo code with a code rate of 1,000 Mbps and supports a code length of 40 to 6144 bits. The control channel mainly uses a tail-biting convolutional code (TBCC) with a code rate of 1,000 Mbps and a constraint length of 7, as shown in Table 6.5.

The three major application scenarios of 5G are enhanced mobile broadband, massive machine-type communication, ultrahigh reliability, and low latency communication. Like the 4G mobile communication standard, the 5G mobile

Table 6.5 Channel Coding Scheme in LTE-Advanced

Channel Type	*Coding Scheme*	*Bitrate*
Data channel	Turbo code Code length 40–6144 bits	1/3
Control channel	Tail-biting convolutional code	1/3

communication channel coding scheme is selected according to the channel type. Specifically, as shown in Table 6.6, in the enhanced mobile broadband scenario the data channel uses LDPC coding. The control channel mainly uses cascaded polar code and cyclic redundancy check cascaded polar code [16].

In order to meet the increasing demand for information rate and capacity, mobile communication technology has continuously evolved from the 1G system that only supports voice calls to the 5G system that supports enhanced mobile broadband, massive connections, and ultralow latency reliable transmission. The channel coding scheme is closely related to the actual development needs of mobile communications. There is no doubt that channel coding has made an indelible contribution to ensuring the reliability of each generation of mobile communication systems. In addition, with the development of 5G+ applications such as virtual reality, industrial internet, internet of vehicles, telemedicine, and smart cities, the existing 5G technology will face new challenges. The 6G system, which aims to achieve the interconnection of all things through integrated air, space, and land communications, is in full swing. 6G mobile communications will have significant characteristics such as ubiquity, socialization, and intelligence. The channel propagation environment has the characteristics of ultralarge bandwidth, high-, medium-, and low-frequency bands, and full coverage of air, space, land, and sea. It must support higher throughput transmission at the Tbps level, more complex transmission scenarios, and more heterogeneous and diverse business types, which will bring new challenges to the channel coding scheme.

6.2.2 Wireless Local Area Network Communication

Wireless local area networks (WLAN) are networks that use wireless communication technology to connect computer devices to form a network system that can achieve mutual communication and resource sharing. In a wireless LAN, computers and networks are no longer connected by communication cables. This has the advantages of flexible mobility, simple and convenient installation, low operating cost, strong scalability and easy maintenance, and has therefore developed rapidly.

However, the development of WLAN is also facing challenges such as increasingly crowded frequency bands, channel interference and fading, and therefore improvement in channel coding technology is needed to improve the reliability of information transmission. Since the release of the first WLAN standard IEEE 802.11 in June 1997, it has gradually evolved into IEEE 802.11a/b, then to IEEE 802.11g, and then to the current IEEE 802.11ax, and the channel coding technology used has also been continuously updated and evolved, as shown in Table 6.7.

IEEE 802.11a/b was formulated in 1999, mainly improving the physical layer technology based on IEEE 802.11. Furthermore, in order to improve the transmission performance of IEEE 802.11b, IEEE 802.11g was officially released

Table 6.6 Coding Schemes for Different Channel Types in Enhanced Mobile Broadband Scenarios

Channel Type	*Coding Scheme*
Data channel	LDPC code (N=8848/3840)
Control channel	polar code ($N_{max} = 1024$)

Table 6.7 Channel Coding Schemes in WLAN

WLAN Standard Classification	*Channel Coding Method*	*Time*
IEEE 802.11a	Convolutional code	1999
IEEE 802.11b		1999
IEEE 802.11g		2003
IEEE 802.11n (Wi-Fi 4)	Convolutional code,	2009
IEEE 802.11 ac (Wi-Fi 5)	LDPC code (optional)	2013
IEEE 802.11ax (Wi-Fi 6)	Convolutional code, LDPC code	2019

in 2003, with an operating frequency of 2.4 GHz, a total of 14 frequency bands, and compatible with 802.11b. The channel coding schemes adopted by the above three standards are all convolutional codes with a constraint length of 7. Different code rates are achieved through puncturing and shorting. The specific coding rates include 1/2, 2/3, and 3/4.

To improve the traffic deficiencies of IEEE 802.11a and IEEE 802.11g, IEEE 802.11n was released and implemented in 2009. Compared with the previous standards, IEEE 802.11n increased the bandwidth from 20 to 40 MHz and was backward-compatible with 802.11b, 802.11a, and 802.11g. Subsequently, in 2013, the IEEE 802.11ac standard was released and implemented, operating at a frequency of 5 GHz, with a transmission rate jumping from 600 Mbps of 802.11n to 1 Gbps, reaching the transmission rate of wired cables. Compared with the previous standards, IEEE 802.11n and IEEE 802.11ac have adjusted the channel coding scheme and added a convolutional code with a higher bit rate (5/6). In addition, an optional LDPC code was introduced to replace the original partial convolutional code encoding scheme, with optional code rates including 1/2, 2/3, 3/4, and 5/6, and code lengths including 648, 1296, and 1944 [17]. In 2019, IEEE 802.11ax was released and implemented, which is backward-compatible with 802.11a/b/g/n. Regarding channel coding, unlike the optional LDPC code specified in IEEE 802.11n/ac, LDPC coding is mandatory in IEEE 802.11ax to improve error correction performance at high rates [18].

6.3 Channel Coding for Satellite Communication

Although terrestrial communication technologies such as 5G can provide users with efficient and ultimate services in areas suitable for terrestrial infrastructure construction, deploying terrestrial communication networks in remote areas such as deserts and oceans is unrealistic. Satellite communication has the characteristics of long communication distance and high deployment flexibility, which can effectively make up for the shortcomings of terrestrial communication networks. Satellite communication is the main application scenario of early channel coding technology, promoting continuous upgrading and innovation. As shown in Figure 6.9 [19], this section reviews the development of channel coding in deep-space and near-space communications (based on public information).

6.3.1 Deep-Space Communication

Deep-space communication generally refers to communication between the Earth and the Moon or between the Earth and deep-space spacecraft. Deep-space

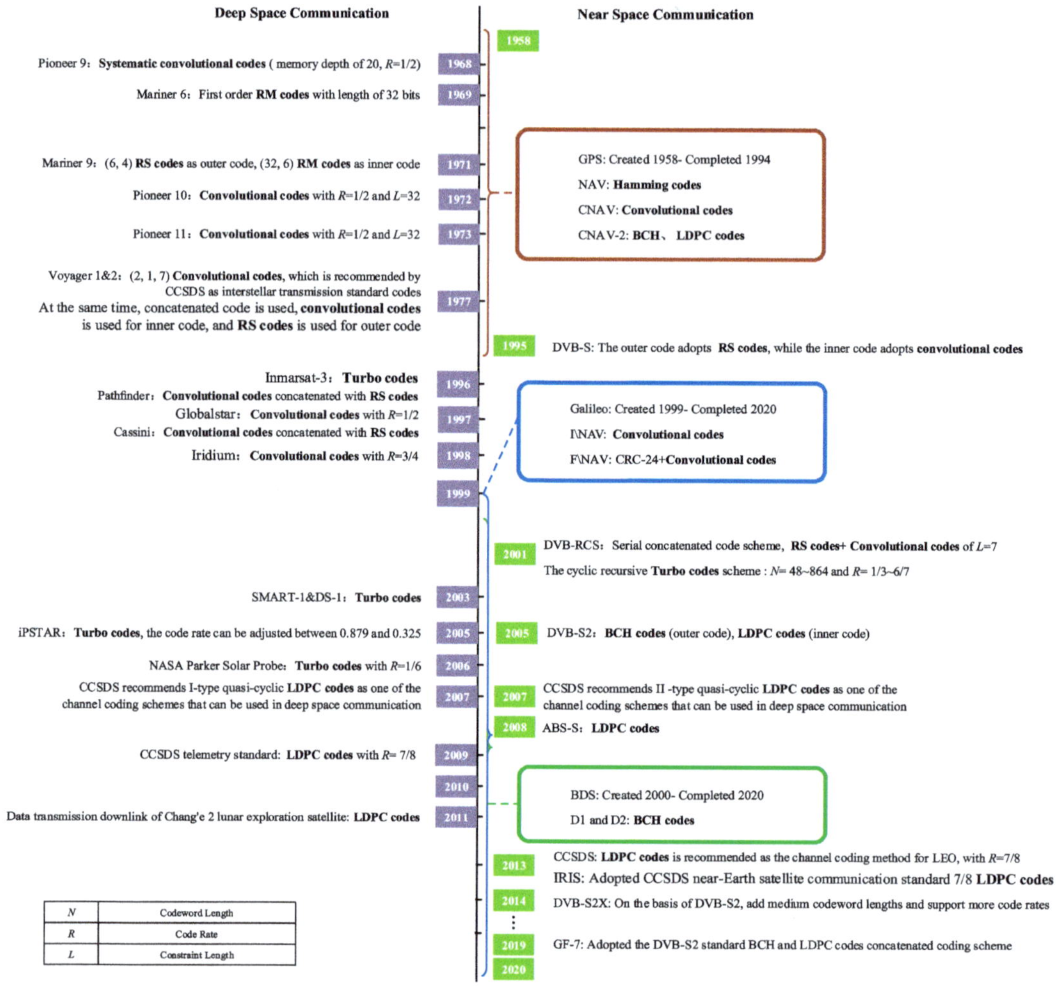

Figure 6.9 Typical applications of channel coding in satellite communications.

communication has a long transmission distance and significant link loss, resulting in an extremely low received signal-to-noise ratio. Therefore, channel coding technology can effectively ensure the reliability of deep-space communication transmission and is indispensable in compensating for the link loss of deep-space communication. Since the 1960s, most deep-space probes have adopted channel coding technology.

In the deep-space communication systems that have been made public, the earliest channel coding technologies used were RM codes and convolutional codes. In 1968, NASA successfully launched the Pioneer 9 probe, which operated in a heliocentric orbit. From 1968 to 1990, it collected data on the electromagnetic and plasma characteristics of interplanetary space and conducted many long-distance communication experiments. The communication system on the probe used a system convolutional code designed by Lin and Lyne with a code rate of 1/2 and a memory depth of 20 [20]. The receiving end used soft iterative decoding based on the Fano algorithm, which could obtain a coding gain of about 6.9 dB. In the early 1970s, NASA successfully launched the Pioneer 10 and 11 probes to explore

Jupiter, Saturn, and the outer solar system, using convolutional codes with a code rate 1/2 and a constraint length of 32.

In 1969, NASA used a 32-bit first-order RM code in the Mariner 6 probe for Mars exploration, which included 6 information bits and 26 check bits. The receiving end used maximum likelihood decoding. Compared with the uncoded system, when the bit error rate (BER) was below 10^{-5}, a coding gain of about 3.2 dB was obtained [21, 22]. In 1977, the famous outer solar system probes Voyager 1 and 2 used the most advanced convolutional code at the time, with a constraint length of 7, and the decoding algorithm used Viterbi decoding. When the signal-to-noise ratio was 2.5 dB, the bit error rate could reach 10^{-5}, and a coding gain of 5.1 dB was obtained. This coding scheme was recommended by the Consultative Committee for Space Data Systems (CCSDS) as the standard coding for interstellar transmission.

RS code was proposed as early as 1960 but was not initially used in satellite communications due to the lack of a simple soft iterative decoding method. In 1967, Forney proposed the concept of a cascade code, which uses RS as the outer code of the cascade code to obtain good error correction capabilities. The Mariner 9 Mars orbiter launched in 1971 used a cascade code structure with RS (6, 4) code as the outer code and RM (32, 6) code as the inner code. Although the coding scheme has a low rate, its good error detection and correction capabilities helped the Mariner 9 Mars orbiter achieve great success. It successfully sent back 7,329 photos and mapped 85% of Mars at a resolution of 1 to 2 km. Compared with the cascade coding structure of the outer code RS code and the inner code RM code, the inner code uses a convolutional code to achieve better error correction performance and has since been widely used in various satellite communications. Voyager 1 and 2 used this concatenated code, which effectively ensured the transmission reliability of compressed data when passing through the outer solar system, such as Uranus and Neptune. Specifically, the inner code uses (2, 1, 7) convolutional code, and the outer code uses (255, 223) RS code to obtain a 2 dB coding gain. Similar concatenated codes are also included in the DVB-S standard, in which the outer code uses (204, 188) RS code and the inner code uses (2, 1, 7) convolutional code. After puncturing, different code rates, such as 2/3, 3/4, 5/6, and 7/8, can be obtained.

The invention of turbo codes in 1993 brought new changes to channel coding in deep-space communications. Due to its performance being close to the Shannon limit, research and standardization work on turbo codes in deep-space communications was carried out rapidly. The CCSDS even recommended it as one of the channel coding schemes that can be used for deep-space communications. Specifically, a parallel cascaded turbo code structure, which includes two 16-state recursive convolutional codes with precisely the same structure, can use four code rates of 1/2, 1/3, 1/4, and 1/6. The information block lengths are 1784 bits, 3568 bits, 7136 bits, and 8920 bits, respectively. The first commercial system that supports turbo codes in satellite communications is Inmarsat-3, which allows users to achieve a communication rate of 64 kbps with satellites [23]. In addition, turbo codes are also used in satellite systems such as Anik F2, DirecTV, and DISH Network, as well as various deep-space exploration missions. In 2003, the ESA launched the Lunar Exploration Mission Wisdom 1, successfully applying turbo codes. NASA's DS-1 asteroid probe adopted the CCSDS standard (8920, 1/6) turbo code, achieving a data rate of 250 kbps for deep-space exploration missions. In 2006, NASA applied turbo codes with a code rate 1/6 in its solar probe, achieving a data transmission rate of 960 kps.

LDPC code has the characteristics of flexible code length and code rate. It was approved by CCSDS as one of the deep-space communication channel coding schemes in 2007. CCSDS defines four code rate LDPC code check matrix structures of 1/2, 2/3, 4/5, and 223/255 (approximately 7/8) for deep-space exploration, remote control, and other tasks. In 2009, the CCSDS telemetry standard adopted an LDPC (8160, 7136) coding scheme with a code rate 7/8. When the bit error rate is of the order of magnitude, the required signal-to-noise ratio is about 3.8 dB, and the coding gain is 7 dB.

In summary, channel coding has always played an essential role in developing deep-space communication. Since 1968, convolutional codes, RM codes, RS codes, turbo codes, and LDPC codes have been successively applied to deep-space communication, promoting the development of deep-space communication.

6.3.2 Near-Space Communication

Near-space communication refers to the communication between communication entities on the earth and aircraft in space less than 2×10^6 km from the earth. These aircraft include various artificial satellites, manned spacecraft, and space shuttles. Next, we will introduce the application of channel coding in two near-space communication scenarios: digital video broadcasting (DVB) and global satellite navigation.

6.3.2.1 Digital Video Broadcasting

The DVB standard is considered one of the most important sets of standards for television broadcasting and is widely utilized around the world. Table 6.8 summarizes the channel coding methods in various DVB standards.

DVB-S is the standard for digital satellite TV systems. The digital satellite TV system transmits digitally encoded and compressed TV signals to the user via geosynchronous satellites. There are two primary forms of transmission. The first is to communicate the digital TV signal to the cable TV front end, which is converted into an analog signal by the cable TV station and transmitted to the user's home. The second is directly transmitting the digital signal to the user's home. Due

Table 6.8 Channel Coding Schemes in Different DVB Standards

DVB Standard Classification	*Channel Coding Methods*	*Time*
DVB-S	External code RS code Inner code convolutional code	1994
DVB-S2	External code BCH code Inner code LDPC code	2004
DVB-C	RS code	1994
DVB-T	External code RS code Inner code convolutional code	2004
DVB-RCS	External code RS code Inner code convolutional code; Turbo code	2000

to the complex satellite channel conditions, severe interference, and fading, the DVB-S standard needs to adopt channel coding technology to improve the reliability of information transmission. Specifically, the DVB-S standard adopts a cascade coding scheme of outer code RS code and inner code convolutional code.

With the continuous development of channel coding technology and the continuous improvement of human demand for satellite communication transmission, the core technologies, such as coding and modulation in the original DVB-S, can no longer meet the demand. Therefore, the second-generation satellite DVB standard DVB-S2 has made improvements in coding and modulation: (a) the inner code uses LDPC code, and the outer code uses BCH code, which has better error correction performance than the cascade coding scheme in the DVB-S standard, (b) supports multiple code rates such as $1/4 \sim 9/10$, (c) supports multiple modulation modes such as QPSK, 8PSK, 16APSK, and 32APSK, (d) adaptive coding and modulation technology; according to different signal transmission environments, the coding and modulation scheme is adaptively selected, significantly improving the system's reliability. Compared with the DVB-S standard, the channel coding scheme adopted by the DVB-S2 standard significantly improves the error correction performance, which is only $0.7 \sim 1$ dB away from the Shannon limit.

The DVB-RCS standard is an industry standard defined for interactive digital television applications based on satellite channels. The standard provides two independent channel coding schemes: (1) serial concatenation coding scheme, in which the outer code is RS code and the inner code is convolutional code with a constraint length of 7, and (2) multibase cyclic recursive turbo code scheme, with a codeword length of 48 to 864 and code rate of 1/3 to 6/7.

6.3.2.2 Global Navigation Satellite System

GNSS refers to all satellite navigation systems, including the Global Positioning System (GPS), the Galileo Satellite Navigation System, and the BeiDou Navigation Satellite System (BDS). GNSS can provide real-time navigation and positioning and is widely used in communications, engineering, and military. In the GNSS satellite navigation and positioning process, the satellite completes the positioning and measurement tasks by sending navigation messages to the terrestrial unit. Therefore, correctly sending and receiving navigation messages is key to the entire navigation and positioning process. However, the satellite-to-terrestrial link has a long transmission distance and poor channel conditions. Therefore, adding channel coding technology to the navigation message can effectively improve the reliability of the system [24].

GPS navigation messages include NAV, CNAV (Civil Navigation Message), and CNAV-2 [25]. Specifically, the channel coding in the NAV message uses extended Hamming code $(32, 26)$, which can detect and correct individual errors. CNAV uses CRC cascade convolutional code $(600, 300)$, in which the 24-bit cyclic redundancy check bits are located at the end of the codeword. CNAV-2 navigation messages use coding methods such as BCH and LDPC codes, and different subframes correspond to different coding methods. For example, for more critical subframe data, BCH $(51, 8)$ coding and maximum correlation probability decoding are used to maximize the decoding accuracy and ensure high-reliability transmission of essential data.

There are two types of message structures disclosed by the Galileo satellite navigation system: free navigation (F\NAV) message and integrity navigation (I\NAV) message. Compared with GPS navigation messages, Galileo satellite navigation messages have strong error correction capabilities and short positioning time. In the I\NAV message, the $(171, 133)$ convolutional code is used. The encoding method in the F\NAV message uses the CRC-24 cascaded convolutional code $(488, 244)$.

With the successful launch of the last satellite of the BeiDou-3 network, my country of China has fully built its global navigation system. BDS has also surpassed Galileo and can stand shoulder-to-shoulder with GPS. There are two main types of BDS navigation messages [26]: D1 and D2. Both messages use the $(15, 11, 1)$ BCH coding scheme, where the code length is 15, and the information bit length is 11, which can correct 1-bit random errors.

In summary, channel coding also plays a vital role in near-space communications. Specifically, in digital video broadcasting systems, channel coding can be used to encode and decode video, audio, and other signals to improve transmission quality and stability and reduce data loss and errors during transmission. In global satellite navigation systems, channel coding is used to encode and decode navigation signals transmitted by satellites to improve signal reliability and accuracy.

6.4 Integrated Satellite-Terrestrial Channel Coding

With the continuous innovation of communication technology, satellite communication, with its unique advantages, will gradually achieve deep integration with terrestrial mobile communication. At present, terrestrial mobile communication technology is developing rapidly, and 4G communication is already very mature. 5G communication has also been officially put into commercial use to support the communication needs of the information society. Research on the sixth generation of mobile communication related technologies is also in full swing and is expected to be launched for the first time around 2030. As the terrestrial network develops and replaces rapidly, achieving space-based innovation, breaking the separation of the sky network and the terrestrial network, and promoting the integration of satellite communication and terrestrial networks have also become the focus of discussion in all walks of life. As early as 2016, the International Telecommunication Union announced preliminary research results such as the system architecture and deployment scenarios of satellite-terrestrial integration. 3GPP has been conducting research on satellite-terrestrial integration since Release 14. At present, the satellite-terrestrial integrated network has become one of the 6G candidate technologies recognized by many research institutions at home and abroad, and is the key to realizing the 6G vision of "full coverage and intelligent connection of all things."

Although satellite-terrestrial integrated network technology is an inevitable trend of development, satellite communications and terrestrial communications are quite different in terms of signal propagation environment, data processing capabilities, computing/storage resources, or business service quality assurance. To achieve true satellite-terrestrial integrated, there are still many key technical issues that need to be studied, overcome and solved one by one. In this section, we focus on channel coding technology in satellite-terrestrial integrated communications. First, we summarize the current status of channel coding research in satellite-terrestrial integrated. Secondly, we analyze the potential advantages and disadvantages of different coding schemes and propose candidate coding schemes. Finally, we present the possible challenges and solutions in future integrated satellite-terrestrial.

6.4.1 The Current Research Status

Looking back at the development of terrestrial and satellite communication, channel coding plays a pivotal role in both communication networks. With the advent of the new era of 6G, satellite-terrestrial integrated communication has also become a hot topic. The channel coding scheme for satellite-terrestrial integration has also become a research focus. Currently, most studies will design corresponding channel coding schemes based on certain specific characteristics of satellite-terrestrial communication (see Figure 6.10). To improve the spectrum efficiency of satellite-terrestrial integrated communication under power limitation, a dual binary turbo code was proposed, which can match high-order modulation well and significantly improve spectrum efficiency. However, when a synchronization is offset,

Years	Main issues solved	Coding scheme	Advantage	Insufficient
2015	Spectral efficiency of satellite-terrestrial integrated communication under power limitation	Duobinary Turbo code	Very good match for high-order modulation, improving spectral effect	When there is synchronization offset, there are still huge challenges in improving spectral efficiency
2014	There is a huge transmission delay in satellite-terrestrial integration, and the ARQ mechanism is not applicable	CCDS proposes the use of rateless coding scheme	\	\
2018	There is a huge transmission delay in satellite-terrestrial integration, and the ARQ mechanism is not applicable	Proposed a rateless simulation Fountain code	Solve the problem of large transmission delay caused by ARQ	CSI needs to be obtained, and it is difficult to balance error correction and complexity
2017	Will 5G air interface technology continue to be used for satellite-terrestrial integrated communications? (Standard integration)	3GPP discussed this issue and discussed in some agreements the key role that 5G NR may play in satellites	\	\
2019	Will 5G air interface technology continue to be used for satellite-terrestrial integrated communications? (Standard integration)	Comparative analysis of the characteristics of satellite-terrestrial integrated communications shows that 5G air interface channel coding can still be used for satellite-terrestrial integration	5G air interface technology is better than satellite communication standards, such as DVB-S2X, which will bring performance gains	5G air interface technology cannot be directly copied, and adaptive design becomes a problem
2021	Potential channel coding schemes for future satellite-earth integration	Proposes a variety of coding schemes that may be used for satellite-terrestrial integrated communications, such as polar codes, turbo codes, and concatenated codes	Analyze the shortcomings of different coding schemes in future integrated communications	No specific adaptive design solution in integrated communications is given
2024	Channel coding schemes for 6G satellite-terrestrial integrated communications	The adaptive construction, decoder hardware architecture, and sync-free transmission of PCC polar codes are proposed for satellite-terrestrial integrated communications	High spectral efficiency, strong error correction, and high throughput	\

Figure 6.10 Current status of channel coding research for satellite-terrestrial integrated communications.

the scheme still faces significant challenges in enhancing the spectrum efficiency. In addition, there is a vast transmission delay in satellite-terrestrial integrated communication, which makes the traditional automatic repeat-request (ARQ) mechanism not applicable. To solve this problem, CCSDS recommends using a rateless coding scheme. Based on this, a rateless analog fountain coding was proposed, effectively solving the considerable transmission delay caused by ARQ. However, this coding scheme requires obtaining channel state information (CSI) in advance, and it is challenging to balance error correction performance and complexity.

On the other hand, many studies are devoted to discussing the feasibility of integrating 5G air interface technology with satellite communication standards. Among them, 3GPP also attempts to specify unified terrestrial and satellite communications standards. For example, the critical role of satellites in 5G NR is discussed in TR38.913 and TS22.261, the evaluation methods and critical factors are discussed in TR38.811, and the potential critical technologies of 5G NR under satellite communication conditions are mentioned in TR22.822 and TR38.821. Further, some researchers have deeply analyzed and compared the characteristics of terrestrial and satellite communications. Experimental analysis shows that 5G air interface technology will bring more significant performance gains than DVB-S2X. Therefore, it can be preliminarily concluded that the 5G NR channel coding scheme has broad application prospects in future satellite-terrestrial integrated communications. However, the difficulty lies in the adaptive design, and the 5G air interface technology cannot be directly copied. Recently, some studies have focused on analyzing the currently popular polar codes, LDPC codes, turbo codes, and cascade codes and proposed their possible advantages and disadvantages in integrated communications, which provides guidance for the coding design in future satellite-terrestrial integrated communications. The only drawback here is the lack of corresponding coding adaptability design.

According to the current status of channel coding research in satellite-terrestrial integrated communication mentioned above, turbo code, LDPC code, and polar code, which have the most extensive influence in academia and industry, are the mainstream coding schemes in satellite-terrestrial integrated communication research. Among them, turbo code is favored by various communication standards because its error correction ability is close to the Shannon limit. However, turbo codes have two obvious disadvantages: (1) error flattening, and (2) not suitable for occasions with high real-time requirements. LDPC code is also a coding paradigm that approaches the Shannon limit. Its excellent performance can only be fully reflected when the codeword length is long. However, the high complexity of LDPC code brings catastrophic challenges to satellite-terrestrial integrated communication with limited hardware resources. Is there a coding scheme that is more compatible with satellite-terrestrial integration scenarios than turbo code and LDPC code? The polar code that has emerged in recent years provides an answer to this puzzle. Polar code adopts the concept of channel polar and can theoretically reach Shannon capacity with infinite codeword length. In addition, polar codes have the characteristics of strong error correction capability, low encoding and decoding complexity, and synchronization-free capability, which can meet the requirements of future satellite-terrestrial integrated communications for transmission reliability, rate, and resource utilization.

6.4.2 Possible Challenges

Satellite-terrestrial integrated communications have broad development prospects. Channel coding, as an indispensable technology, faces the following challenges.

6.4.2.1 Low Signal-to-Noise Ratio

Under a large time and space span, information transmission must pass through channels with diverse transmission characteristics, including free space, atmosphere, and near the ground, and suffer the overlapping influence of different physical processes. One of the biggest difficulties faced by satellite-terrestrial integrated communication transmission is the huge link attenuation, a large part of which is the free-space attenuation caused by the long communication distance. The free-space attenuation is related to the frequency and will increase with the increase of the communication frequency band. For example, the free-space attenuation of the communication link from the ground to the geosynchronous orbit satellite is about 190 dB in the S-band and about 210 dB in the Ka-band. In addition, rain attenuation is also an important factor affecting information transmission. Under normal circumstances, light to moderate rain will cause about 3–5 dB of attenuation to the Ka-band electromagnetic waves, and heavy rain will cause even more than 10 dB of attenuation. For Q- and V-band electromagnetic waves, due to the continued increase in frequency, the impact of rain attenuation is more serious, reaching more than 30 dB. In addition, factors such as multipath effect, Doppler effect, and channel susceptibility to interference make the information have a very high bit error rate during transmission. Therefore, channel coding with strong error correction performance is urgently needed to ensure high-reliability information transmission of satellite-terrestrial integrated communication.

6.4.2.2 High-Mobility Communication Scenarios

Due to different orbital altitudes, the operating speed of satellites varies greatly. For example, LEO orbit satellites travel very fast relative to the ground, and it takes about 2 hours to orbit the earth. For terrestrial users, the single pass time of each LEO satellite is only about a few minutes to more than 10 minutes, and the Doppler frequency shift is very obvious, about 600–700 kHz, which brings great challenges to communication synchronization. Coupled with factors as rain attenuation, and human interference, the traditional synchronous transmission schemes based on pilots, training sequences, and so on, are prone to failure of the receiver synchronization module, resulting in a high probability of communication link interruption.

6.4.2.3 Limited Hardware Resources

There are great differences between satellites and ground in terms of available resources and node capabilities. The power supply, computing, storage and other resources of terrestrial network nodes will not be greatly restricted, but for satellites, due to their limited total weight, the available power, computing power and storage capacity are very limited. With the continuous deepening of satellite-terrestrial integrated network, the sharp increase in the number of terrestrial users and the widespread application of intersatellite links have enriched the scenarios of satellite-terrestrial integrated communication. However, satellites are constrained by available resources and node capabilities, making it difficult to accurately guarantee the complex and diverse services and scenarios of satellite-terrestrial integrated communication.

6.4.3 Possible Channel Coding Schemes

In order to solve the above challenges, research can be carried out from the following technologies.

6.4.3.1 Adaptive Construction

With the continuous integration of satellite-terrestrial networks in the future, the convergence of technical systems will continue to deepen, but there are fundamental differences between the characteristics of space and ground, and the adaptive design of coding technology has become very important. Specifically, satellite-terrestrial integrated communication intends to realize information transmission in a more generalized, liberalized, and complex time and space. After passing through various channels with different transmission characteristics such as free space, the earth's atmosphere, and the ground, the degree of attenuation suffered by the signal is bound to be different, which in turn leads to different error probabilities of different information bits. Therefore, the channel coding scheme in satellite-terrestrial integration needs to have both powerful error correction capability and flexible coding structure. PCC polar code cascades polar code with parity check code, and exhibits error correction performance close to the capacity limit under limited code length. In addition, the parity bit position distribution and value of the PCC polar code are flexible, and it has the characteristics of naturally matching the channel. It can provide special protection for specific information bits, and is expected to solve the problem of high bit errors under low signal-to-noise ratio in satellite-terrestrial integrated communications.

To this end, a universal adaptive coding construction scheme based on PCC polar code is proposed. As shown in Figure 6.11, the architecture is composed of a constructor and an evaluator. Specifically, the evaluator decodes the demodulated information. When the decoding result fails to pass the frame check, the frame error rate value under the current channel state and construction mode is counted and fed back to the constructor. The constructor updates and optimizes the check bit position and the information bit checked by it according to the feedback result until a construction scheme that can meet the given frame error rate condition is obtained.

6.4.3.2 Low-Complexity Decoder Design

At present, the bent-pipe system adopted by communication satellites does not require high onboard coding and decoding capabilities. The onboard coding and decoding of only a small amount of control instructions, status monitoring and other information is involved. However, with the continuous development of aerospace networks and the continuous integration of ground-ground networks, the demand for onboard processing capabilities has greatly increased. In the future,

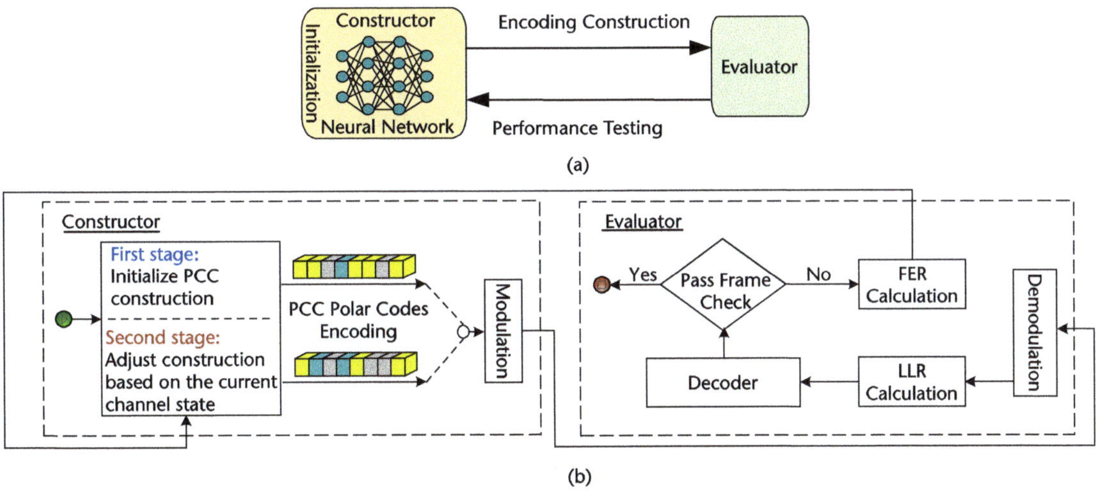

Figure 6.11 Adaptive construction based on PCC polar code: (a) construct-evaluate architecture, and (b) adaptive PCC polar codes universal architecture.

aerospace channel coding technology should provide excellent gain while occupying less resource overhead. Compared with base stations in terrestrial networks, satellites are severely limited in terms of factors such as volume, weight, power, and computing/storage capabilities, which in turn restricts the improvement of the transmission rate and data processing capabilities of satellite communication systems. The mainstream high-performance decoding methods of polar codes are mostly designed based on successive cancellation list (SCL) decoding. However, the existing SCL decoder hardware implementation mostly uses a list parallel structure. Due to the limitation of continuous deletion decoding rules, the calculation modules need to wait for each other for a long time. There is a lot of idle time in the circuit, and the resource utilization efficiency is low.

To this end, a list serial SCL decoding architecture is proposed to adapt to the satellite-terrestrial integrated communication scenario with limited onboard hardware resources. Its hardware architecture is shown in Figure 6.12, which includes three parts: storage unit, calculation unit, and control unit. According to the calculation rules of serial pipeline, after the log-likelihood ratio (LLR) of a path is calculated, the path management operation is performed in the form of a pointer, during which part and calculation are performed simultaneously. The time consumption of the path management process can be covered by other operations, and the main calculation unit is in a full-load state, which effectively reduces the consumption of hardware resources. In the calculation of a single path, a semiparallel structure is adopted to reduce the excessive decoding delay caused by serial calculation by increasing the number of processing units. While maintaining the same error correction performance, the throughput is significantly improved and the utilization efficiency of hardware resources is improved.

6.4.3.3 Synchronous-Free Transmission

Traditional satellite communications use synchronous transmission schemes such as pilots and training sequences. When subject to rain attenuation and human

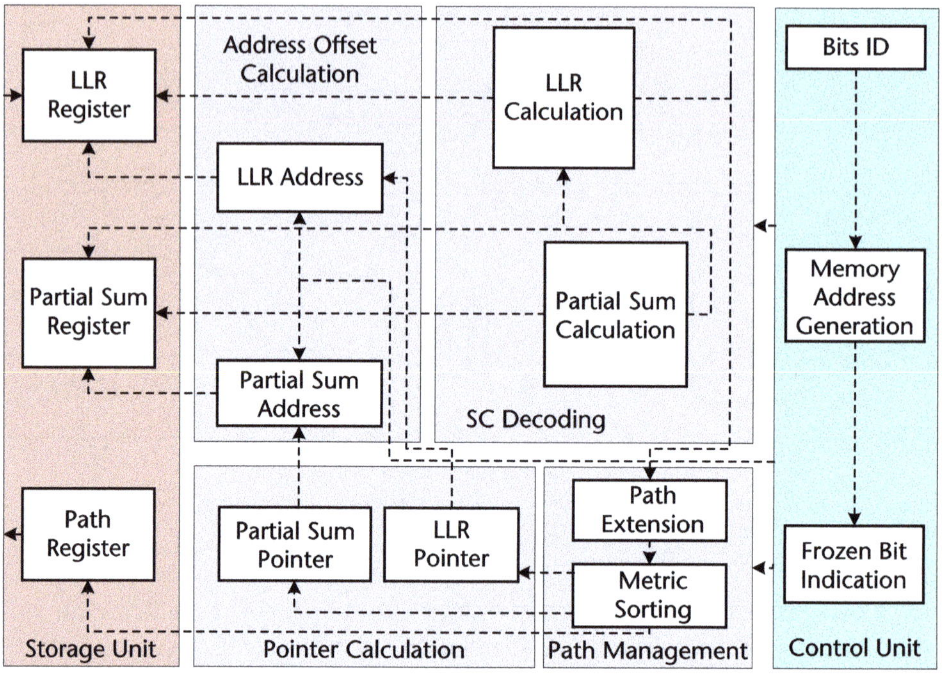

Figure 6.12 SCL decoder design architecture with low hardware resource overhead.

interference, the receiver synchronization module is prone to failure and the communication link interruption rate is high. Since the pilot content used for synchronization is fixed and sent repeatedly, the communication signal is easily intercepted by monitoring, and it is urgent to develop new high-reliability transmission technologies. Synchronous-free transmission can effectively improve the link interruption caused by the failure of the synchronization module by adopting encoding and decoding with self-synchronization capabilities. At the same time, the saved channel resources can be further used to improve the coding error correction performance and improve communication reliability. In addition, synchronization-free transmission does not transmit repeated fixed synchronization signals, which greatly increases the difficulty of signal detection. Therefore, synchronization-free transmission with anti-interception performance will play a vital role in the future satellite-terrestrial integrated communication.

Figure 6.13 shows the synchronization-free transmission architecture based on PCC polar code in satellite-terrestrial integrated communication. The signal is transmitted through the satellite channel after PCC polar code coding and modulation. The receiver includes the design of signal preprocessing and self-synchronization decoding module. The preprocessing operation is to process the received signal and demodulate and generate multiple sets of candidate sequences of synchronization parameters according to correction parameters such as sampling point, carrier frequency, and phase offset. Subsequently, the codeword receiving sequence is input into the PCC polar code decoder with self-synchronization capability for decoding, and the codeword with the best metric value is output as the decoding result, where the metric value is used to characterize the log-likelihood information of the decoding result corresponding to the current decoding bit. This

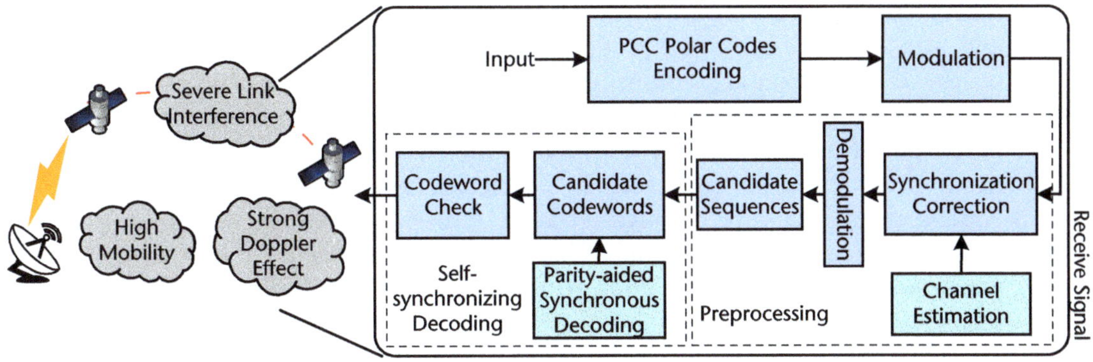

Figure 6.13 Plane synchronization transmission architecture based on PCC polar code.

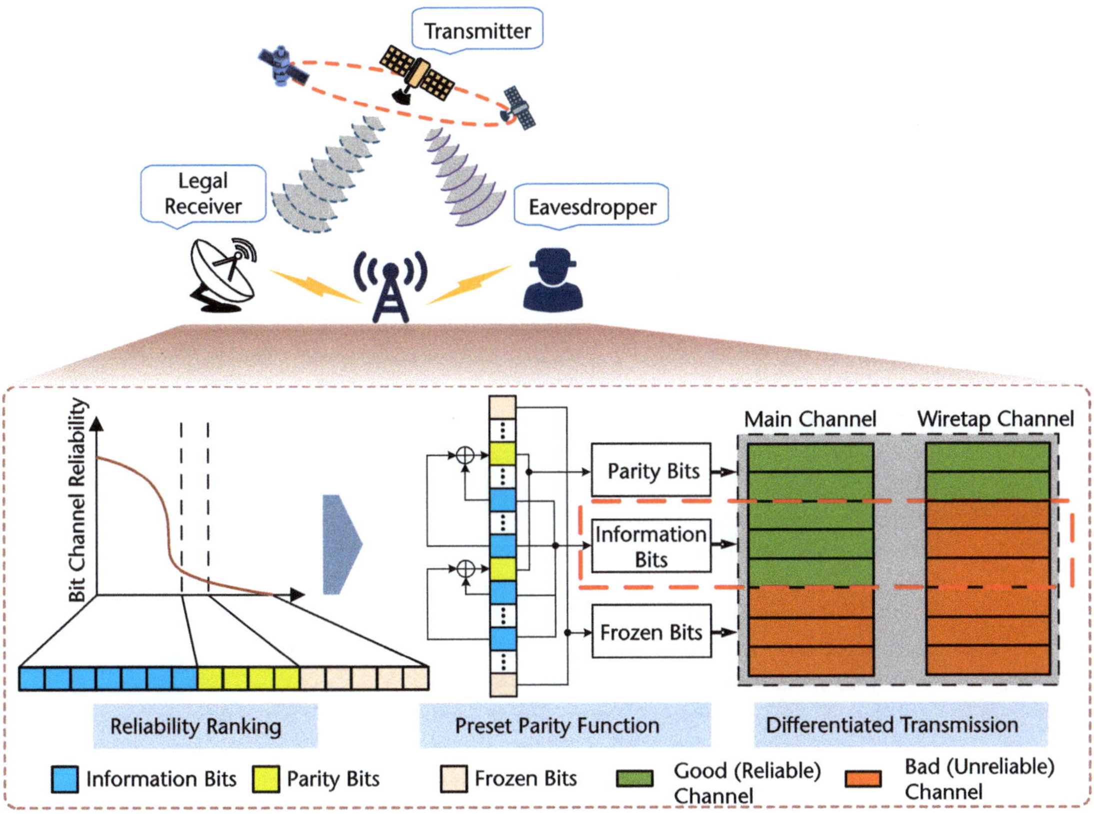

Figure 6.14 Physical layer security coding architecture based on PCC polar code.

method eliminates the pilot structure without reducing the communication rate and communication frame error rate, and can complete accurate synchronization according to the credibility index, simplify the receiver, and improve the probability of successful communication of satellite-terrestrial integrated communication.

6.4.3.4 Physical Layer Security Coding

It is foreseeable that in the future, satellite-terrestrial integrated communication channels will be open, broadcast, and complex, and the information transmitted through them will be extremely vulnerable to security threats such as eavesdropping and attacks. Traditional cryptography-based security mechanisms rely on high-overhead encryption algorithms and are difficult to resist the powerful computing power of terminals. In contrast, physical layer security transmission technology is based on information theory and makes full use of wireless channel characteristics to provide a lightweight security protection idea for wireless communications. Among them, the core of physical layer security coding lies in the joint design of coding structure and channel characteristics, relying on the error correction performance of the mother code to stimulate the dual effects of reliability and security.

To this end, it is important to investigate the physical layer security and reliable coding architecture based on PCC polar code, as shown in Figure 6.14. For degraded eavesdropping channels, differential mapping is achieved based on the channel polar phenomenon, and a verification relationship is constructed to assist legitimate users in decoding. Since private information is only transmitted on the "bad" channel of the eavesdropper, the probability of transmission errors will be greatly increased; legitimate users can further use the preset verification relationship between the verification bit and the private information bit to correct the private information.

References

[1] Shannon, C. E., "*A Mathematical Theory of Communication*," The Bell System Technical Journal, Vol. 27, No. 3, 1948, pp. 379–423.

[2] Muller, D. E., "Application of Boolean Algebra to Switching Circuit Design and to Error Detection," *Transactions of the IRE Professional Group on Electronic Computers*, Vol. 3, 1954, pp. 6–12.

[3] Prange, E., Cyclic Error-Correcting Codes in Two Symbols, AFCRC-TN-57, Cambridge, MA: Air Force Cambridge Research Center, 1957.

[4] Bose, R. C., and D. K. Raychaudhuri, "On a Class of Error Correcting Binary Group Codes," *Information and Control*, Vol. 3, No. 1, 1960, pp. 68–79.

[5] Reed, I. S., and G. Solomon, "Polynomial Codes Over Certain Finite Fields," *Journal of the Society for Industrial and Applied Mathematics*, Vol. 8, No. 2, 1960, pp. 300–304.

[6] Gallager, R., "Low-Density Parity-Check Codes," *IRE Transactions on Information Theory*, Vol. 8, No. 1, 1962, pp. 21–28.

[7] Davey, M. C., and D. J. C. MacKay, "Low Density Parity Check Codes Over GF(q)," in *Information Theory Workshop*, Killarney, 1998, pp. 70–71.

[8] Arikan, E., "Channel Polarization: A Method for Constructing Capacity-Achieving Codes for Symmetric Binary-Input Memoryless Channels," *IEEE Transactions on Information Theory*, Vol. 55, No. 7, 32009, pp. 3051–3073.

[9] Tal, I., and A. Vardy, "List Decoding of Polar Codes," *IEEE Transactions on Information Theory*, Vol. 61, No. 5, 2015, pp. 2213–2226.

[10] Elkelesh, A., M. Ebada, S. Cammerer, et al., "Belief Propagation List Decoding of Polar Codes," *IEEE Communications Letters*, Vol. 22, No. 8, pp. 2018, pp. 1536–1539.

[11] Elias, P., "Coding for Noisy Channels," *IRE Convention Record*, Vol. 3, 1955, pp. 37–46.

[12] Berrou, C., and A. Glavieux, "Near Optimum Error Correcting Coding and Decoding: Turbo-Codes," *IEEE Transactions on Communications*, Vol. 44, No. 10, 1996, pp. 1261–1271.

[13] UNE-EN 300909 V7.3.1-2006, *Digital Cellular Telecommunications System (Phase 2+) (GSM); Channel Coding* (GSM 05.03 Version 7.3.1 Release 1998, Madrid: The Spanish Association for Standardization and Certification, 2006.

[14] ITU-R V1-2006, Migration to IMT-2000 Systems-Supplement 1 of the Handbook on Deployment of IMT-2000 Systems, Geneva, Switzerland: ITU Publications, 2006.

[15] 3GPP TS 36.212 V13.2.0-2016, Multiplexing and Channel Coding, Sophia Antipolis, France: European Telecommunications Standards Institute, 2016.

[16] 3GPP TS 38.212 V16.4.0-2020, Multiplexing and Channel Coding, Sophia Antipolis, France: European Telecommunications Standards Institute, 2020.

[17] Perahia, E., and R. Stacey, *Next Generation Wireless LANs: 802.11n and 802.11ac*, Cambridge, United Kingdom: Cambridge University Press, 2013.

[18] Hoefel, R. P. F., "IEEE 802.11ax: On Performance of Multi-Antenna Technologies with LDPC Codes," in *2018 IEEE Seventh International Conference on Communications and Electronics (ICCE)*, IEEE, 2018, pp. 159–164.

[19] Liu, Y., L. Xiao, W. Liu, et al., "Channel Coding for Satellite-Terrestrial Integrated Communication: Classic Applications, Key Technologies, and Challenges," *IEEE Wireless Communications*, Vol. 31, No. 3, 2024, pp. 348–354.

[20] Lin, S., and H. Lyne, "Some Results on binary Convolutional Code Generators (Corresp.)," *IEEE Transactions on Information Theory*, Vol. 13, No. 1, 1967, pp. 134–139.

[21] Cooke, B., "Reed Muller Error Correcting Codes," *MIT Undergraduate Journal of Mathematics*, Vol. 1, No. 06, 1999, pp. 21–26.

[22] Berlekamp, E., and L. Welch, "Weight Distributions of the Cosets of the (32, 6) Reed-Muller Code," *IEEE Transactions on Information Theory*, Vol. 18, No. 1, 1972, pp. 203–207.

[23] Franchi, A., A. Howell, and J. Sengupta,"Broadband Mobile Via Satellite: INMARSAT BGAN, in *IEE Seminar on Broadband Satellite: The Critical Success Factors-Technology, Services and Markets* (Ref. No. 2000/067), IET, 2002.

[24] Huang, J., Y. Su, W., Liu, et al. "Adaptive Modulation and Coding Techniques for Global Navigation Satellite System Inter-Satellite Communication Based on the Channel Condition," *IET Communications*, Vol. 10, No. 16, 2016, pp. 2091–2095.

[25] Navstar GPS Space Segment/User Segment L1C Interfaces: IS-GPS-800, Revision, 2013.

[26] BeiDou Satellite Navigation System Space Signal Interface Control File–Public Service Signal B3I (Version 1.0), China Satellite Navigation System Management Office, 2018.

CHAPTER 7

Signal Modulation for Satellite-Terrestrial Integrated Communication

Signal modulation is the cornerstone of achieving high-speed and high-reliability satellite-terrestrial integrated communications. It maps the digital signal to a specific waveform suitable for wireless channel transmission. Specifically, modulation technology can be divided into baseband modulation, single-carrier modulation, and multicarrier (MC) modulation according to the carrier mapping method.

For the case of integrated communication, especially for LEO satellite communication, the critical challenges mainly lie in its limited spaceborne resources and high-mobility channel environment, which requires the designed waveform to have a lower peak-to-average power ratio (PAPR), stronger resistance to Doppler shift, and a reduced complexity on signal processing. Conventional OFDM cannot meet all these requirements and a range of novel waveforms have been investigated in the current literature.

This chapter aims to analyze the advantages and tradeoffs of potential waveforms, thereby providing enhanced design guidelines for future integrated communications.

7.1 Classic Modulation Waveforms

Modulation technology is an important part of wireless communication, which can be divided into analog modulation and digital modulation. Since 2G communication, digital modulation has gained popularity and can be categorized into three distinct types: baseband modulation, SC modulation, and MC modulation.

7.1.1 Baseband Modulation

This section introduces multiple phase shift keying (MPSK) and quadrature amplitude modulation (QAM).

7.1.1.1 MPSK Scheme

MPSK is a constant envelope digital modulation, which uses phase to carry modulation information. The carrier phase in its modulation has different values, and the constellation diagram consists of one ring [2]. The MPSK signal is represented as

$$x(t) = A(t)\cos\left[\omega_c t + \phi(t)\right] \tag{7.1}$$

while $A(t)$ is the amplitude of $x(t)$, ω_c is the carrier angular frequency, $\phi(t)$ is the phase component of $x(t)$, containing modulation information. Within one cycle, $\phi(t)$ is a constant, and therefore (7.1) is rewritten as

$$x\left(t\right)=A\left(t\right)\cos\left[\omega_c t+\phi_k\right], kT\le t\le\left(k+1\right)T \tag{7.2}$$

Furthermore, (7.2) can be obtained by

$$\begin{aligned}x\left(t\right)&=A\left(t\right)\cos\phi_k\cos\left(\omega_c t\right)-A\left(t\right)\sin\phi_k\sin\left(\omega_c t\right)\\&=I_k\cos\left(\omega_c t\right)-Q_k\sin\left(\omega_c t\right)\end{aligned} \tag{7.3}$$

where $I_k=A\left(t\right)\cos\phi_k$ and $Q_k=A\left(t\right)\sin\phi_k$ are the amplitude values of the kth in-phase component and the orthogonal component. The phase ϕ_k of the kth symbol is represented as

$$\phi_k=\phi_{k-1}+\Delta\phi \tag{7.4}$$

where ϕ_{k-1} is the phase of the $k-1$th symbol and $\Delta\phi$ is the change in phase of the kth symbol. Figure 7.1 presents the constellation diagrams of BPSK, QPSK, and 8PSK.

Explicitly, Figure 7.2 characterizes the modulation principle block diagram of the 8PSK signal. As shown in the figure, the input binary information sequence undergoes serial parallel conversion, generating a 3-bit code set $(b_3b_2b_1)$ each time, resulting in a symbol rate of 1/3 in the bit stream. After differential phase encoding, code group $(b_3b_2b_1)$ corresponds the bit information with the position information of differential 8PSK constellation points to obtain signal (I_k, Q_k). The signal after phase mapping is a pulse signal with a large number of high-frequency components, which is not suitable for transmission on the channel and requires shaping filtering.

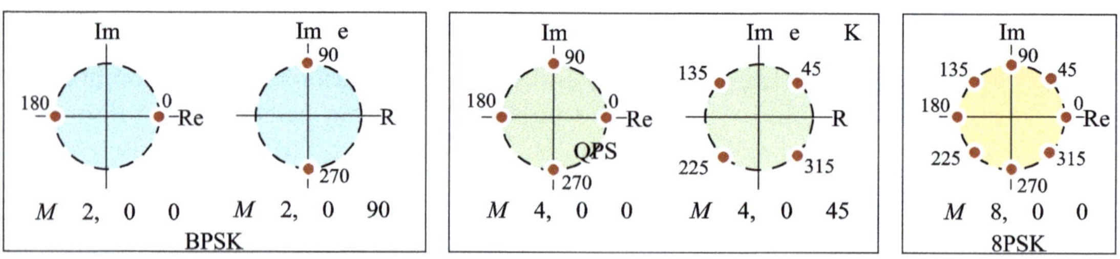

Figure 7.1 Multibase PSK constellation diagram (BPSK, QPSK, 8PSK).

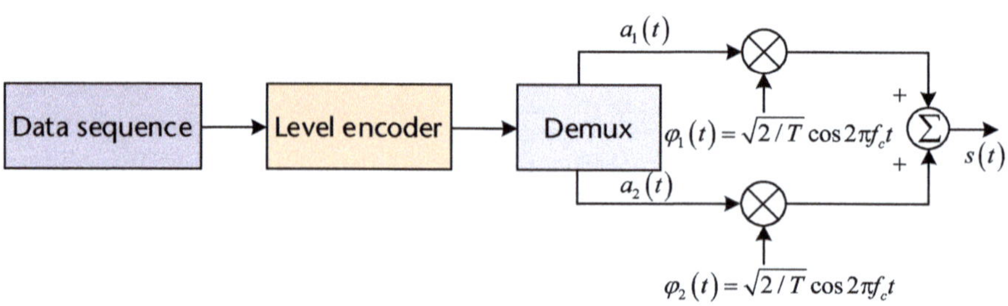

Figure 7.2 8PSK modulation principle block diagram.

Forming a filter to suppress the high-frequency part of the signal can avoid signal crosstalk and reduce the error rate. The filter coefficients used should comply with Nyquist's first law, generally using low-pass filters with linear phase characteristics and square root raised cosine characteristics. After filtering, the discrete baseband sampling values are multiplied by the local output discrete orthogonal carriers to complete spectrum shifting. The baseband signal is modulated into an intermediate frequency signal and sent to the subsequent circuit for processing.

Furthermore, Figure 7.3 shows the demodulation process of MPSK. As shown in the figure, the received MPSK modulated signal is orthogonally downconverted, multiplied by two mutually orthogonal local carriers, and then filtered out by a low-pass filter to remove high-order frequency components. The two zero-IF signals are represented as

$$\begin{cases} x = G_x \left[a_k \cos(\phi - \theta) - b_k \sin(\phi - \theta) + N_x\right] \\ y = G_y \left[a_k \sin(\phi + \theta) + b_k \cos(\phi + \theta) + N_y\right] \end{cases} \tag{7.5}$$

where G_x and G_y are the gains of the in-phase and quadrature paths, which determine the input amplitude of the two signals of the intermediate frequency digital processor, ϕ is the in-phase phase error of the recovered carrier, θ is the quadrature phase error of the recovered carrier, a_k and b_k are the modulated code streams at the transmitter, which take several characteristic discrete values with a certain probability at the sampling moment of each code element, and N_x and N_y are the projection components of the channel additive noise on the in-phase and quadrature reference axes.

7.1.1.2 QAM Scheme

As shown in Figure 7.4, as M increases, the distance between adjacent phases gradually decreases, resulting in a decrease in noise tolerance and an increase in bit error rate. To improve the noise tolerance at higher levels of M, QAM is proposed [3]. The amplitude and phase of the QAM signal are modulated as two independent parameters simultaneously, using two orthogonal carriers $\cos\omega_c t$ and $\sin\omega_c t$ as the basis functions. For QAM scheme with a symbol interval of T_s, the modulated

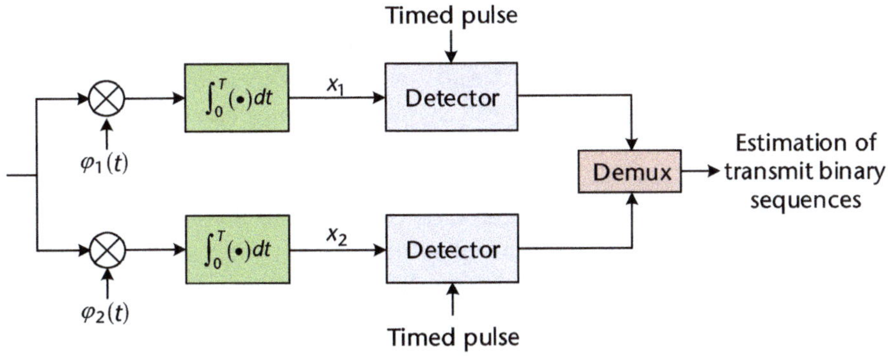

Figure 7.3 MPSK signal demodulation block diagram.

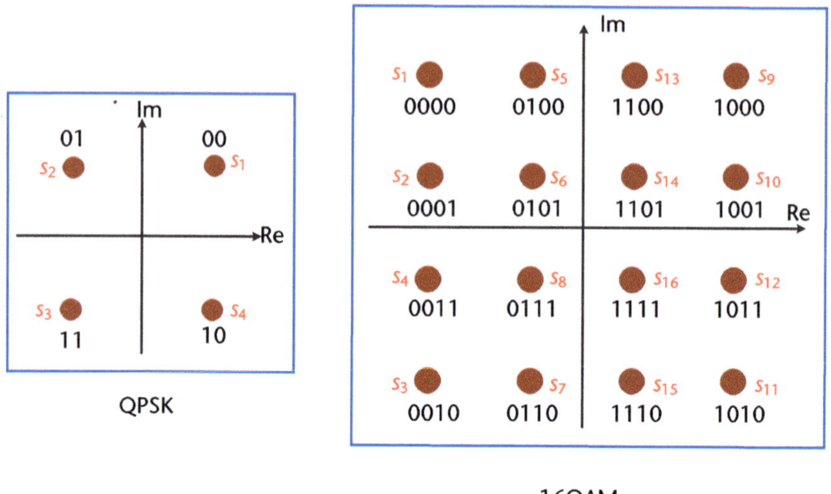

Figure 7.4 (a) 4QAM, and (b) 16QAM constellation diagram.

waveform of the transmitted signal is

$$s(t) = \sum_n A_n g(t - nT_s)\cos(\omega_c t + \theta_n) \tag{7.6}$$

while A_n is the amplitude of the baseband signal and $g(t - nT_s)$ is the waveform of a single baseband signal with a width of T_s. The orthogonal form of (7.6) is represented as

$$s(t) = \left[\sum_n A_n g(t - nT_s)\cos\theta_n\right]\cos\omega_c t - \left[\sum_n A_n g(t - nT_s)\sin\theta_n\right]\sin\omega_c t \tag{7.7}$$

Let $X_n = A_n g(t - nT_s)\cos\theta_n$, $Y_n = -A_n g(t - nT_s)\sin\theta_n$. Equation (7.7) can be rewritten as

$$s(t) = \sum_n X_n \cos\omega_c t + \sum_n Y_n \sin\omega_c t \tag{7.8}$$

According to (7.8), $s(t)$ can be seen as the sum of two orthogonal amplitude keying signals. In (7.6), if the θ_n value can only be taken as $\pi/4$ and $-\pi/4$, then the value of A_n can only be taken as $+A$ and $-A$, and at this point, the QAM signal becomes a QPSK signal. Therefore, QPSK signal is the simplest QAM signal.

In constellation mapping, a constellation diagram is usually used to represent a two-dimensional pattern of QAM modulated signals. The horizontal axis of a constellation is the real number axis (I-axis), and the vertical axis is the imaginary number axis (Q-axis). The number of bits that each symbol can carry on a constellation diagram is called the modulation order. If the number of constellation points on the constellation diagram is L, then the modulation order is $\log_2 L$. The smaller the modulation order, the smaller the number of information bits carried by a constellation symbol, and the lower the requirement for signal-to-noise ratio.

In contrast, the higher the modulation order, the more information bits a constellation symbol carries, and the higher the requirement for signal-to-noise ratio. At present, BPSK (including phase rotation $\pi/2$BPSK), 4QAM, 16QAM, 64QAM, 256QAM have all been adopted by the 5G standard. For ease of understanding, Figure 7.4 presents the constellations of 4QAM and 16QAM. Tables 7.1 and 7.2 present 4QAM and 16QAM mapping table, respectively.

The initial satellite communication standards mainly used PSK modulation. Due to the increasing scarcity of satellite communication spectrum resources, QAM is gradually being applied in satellite communication due to its high frequency band utilization, which will be introduced in detail in the following sections.

7.1.2 OFDM-Based MC Modulation

With the rapid development of large-scale integrated circuits, the OFDM technique becomes the most popular multicarrier modulation scheme. In the OFDM scheme, high-speed data streams are transformed into parallel data streams through inverse discrete Fourier transform. With the assistance of cyclic prefixes, channel linear convolution is transformed into circular convolution. Based on this feature, the receiving end can use a single tap equalizer to recover the signal after undergoing discrete Fourier transform (DFT). Among them, inverse discrete Fourier transform (IDFT) and DFT can be achieved through fast Fourier transform (FFT) and inverse fast Fourier transform (IFFT). Therefore, OFDM technology can effectively combat multipath channel fading with low implementation complexity.

Figure 7.5 presents the system diagram of OFDM. The time-domain signal of the nth OFDM symbol can be expressed as

$$s(k) = \frac{1}{\sqrt{M}} \sum_{m=0}^{M-1} x_{m,n} e^{j2\pi mk/M}, k = 0, 1, \cdots, M-1 \tag{7.9}$$

where $1/\sqrt{M}$ is the power normalization factor.

Table 7.1 4QAM Constellation Mapping Table

Input Signal	*Output Signal*
00	$(1/\sqrt{2}, 1/\sqrt{2}j)$
01	$(-1/\sqrt{2}, 1/\sqrt{2}j)$
11	$(-1/\sqrt{2}, -1/\sqrt{2}j)$
10	$(1/\sqrt{2}, -1/\sqrt{2}j)$

Table 7.2 16QAM Constellation Mapping Table

Input Signal	*Output Signal*	*Input Signal*	*Output Signal*
0000	$(-3/\sqrt{10}, 3/\sqrt{10}j)$	1000	$(3/\sqrt{10}, 3/\sqrt{10}j)$
0001	$(-3/\sqrt{10}, 1/\sqrt{10}j)$	1001	$(3/\sqrt{10}, 1/\sqrt{10}j)$
0010	$(-3/\sqrt{10}, -3/\sqrt{10}j)$	1010	$(3/\sqrt{10}, -3/\sqrt{10}j)$
0011	$(-3/\sqrt{10}, -1/\sqrt{10}j)$	1011	$(3/\sqrt{10}, -1/\sqrt{10}j)$
0100	$(-1/\sqrt{10}, 3/\sqrt{10}j)$	1100	$(1/\sqrt{10}, 3/\sqrt{10}j)$
0101	$(-1/\sqrt{10}, 1/\sqrt{10}j)$	1101	$(1/\sqrt{10}, 1/\sqrt{10}j)$
0110	$(-1/\sqrt{10}, -3/\sqrt{10}j)$	1110	$(1/\sqrt{10}, -3/\sqrt{10}j)$
0111	$(-1/\sqrt{10}, 1/\sqrt{10}j)$	1111	$(1/\sqrt{10}, -1/\sqrt{10}j)$

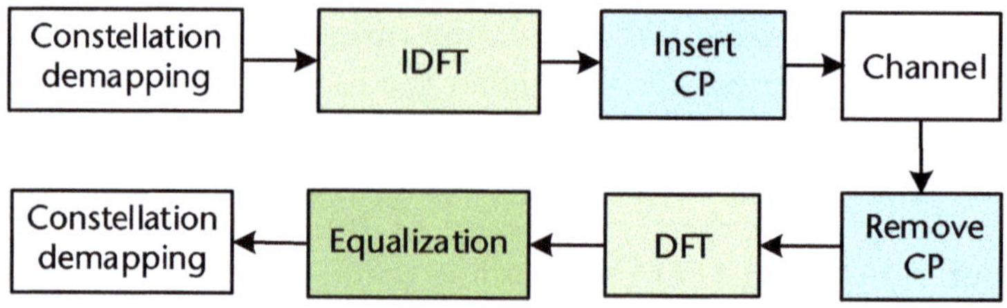

Figure 7.5 OFDM system block diagram.

Assuming the wireless channel is an ideal one, the received signal $y(k)$ is equal to the transmitted signal $s(k)$. At this point, OFDM signals do not need to insert cyclic prefixes to maintain orthogonality between subcarriers and the demodulate waveform for each subcarrier is $e^{-j2\pi mk/M}$. The transmission symbols of each subcarrier are restored to

$$\begin{aligned}\hat{x}_{m,n} &= \frac{1}{\sqrt{M}}\sum_{k=0}^{M-1} y(k)e^{-j2\pi mk/M} \\ &= \frac{1}{\sqrt{M}}\sum_{k=0}^{M-1}\left[\frac{1}{\sqrt{M}}\sum_{m_0=0}^{M-1} x_{m_0,n}e^{j2\pi m_0 k/M}\right]e^{-j2\pi mk/M} \\ &= \frac{1}{M}\sum_{k=0}^{M-1}\sum_{m_0=0}^{M-1} x_{m_0,n}e^{j2\pi(m_0-m)k/M}\end{aligned} \tag{7.10}$$

According to orthogonality, (7.10) has the following characteristics:

$$\hat{x}_{m,n} = \frac{1}{M}\sum_{k=0}^{M-1}\sum_{m_0=0}^{M-1} x_{m_0,n}e^{j2\pi(m_0-m)k/M} = \begin{cases} x_{m,n}, m = m_0 \\ 0, m \neq m_0 \end{cases} \tag{7.11}$$

If $m = m_0$ is satisfied, we have $\hat{x}_{m,n} = x_{m,n}$. If $m \neq m_0$ holds true, we have $\hat{x}_{m,n} = 0$. As a result, the transmission symbols of each subcarrier in the OFDM system can be restored without loss.

Figure 7.6 shows the orthogonal schematic diagram of each subcarrier in the OFDM system. As shown in the figure, OFDM modulates parallel low-speed data onto several parallel high-speed orthogonal subcarriers through IDFT operation. Figure 7.7 shows the cyclic prefix design of OFDM. As shown in the figure, $s(k)$ is the signal obtained by IDFT conversion of the transmission symbols of each subcarrier. Copy and insert some end signals of $s(k)$ into the front end to obtain a cyclic prefix signal. The time length of OFDM signals containing cyclic prefixes is $\bar{T} = T + T_{cp}$, where T and T_{cp} represent the length of $s(k)$ and the cyclic prefix, respectively.

After inserting a cyclic prefix, the transmission signal of the OFDM system is

$$s_{cp}(k) = \frac{1}{\sqrt{M}}\sum_{m=0}^{M-1} x_{m,n}e^{j2\pi mk/M}, k = -L_{cp}, \cdots, -1, 0, 1, \cdots, M-1 \tag{7.12}$$

where $L_{cp} = T_{cp}/T_s$ is the length of the sampling point for the cyclic prefix.

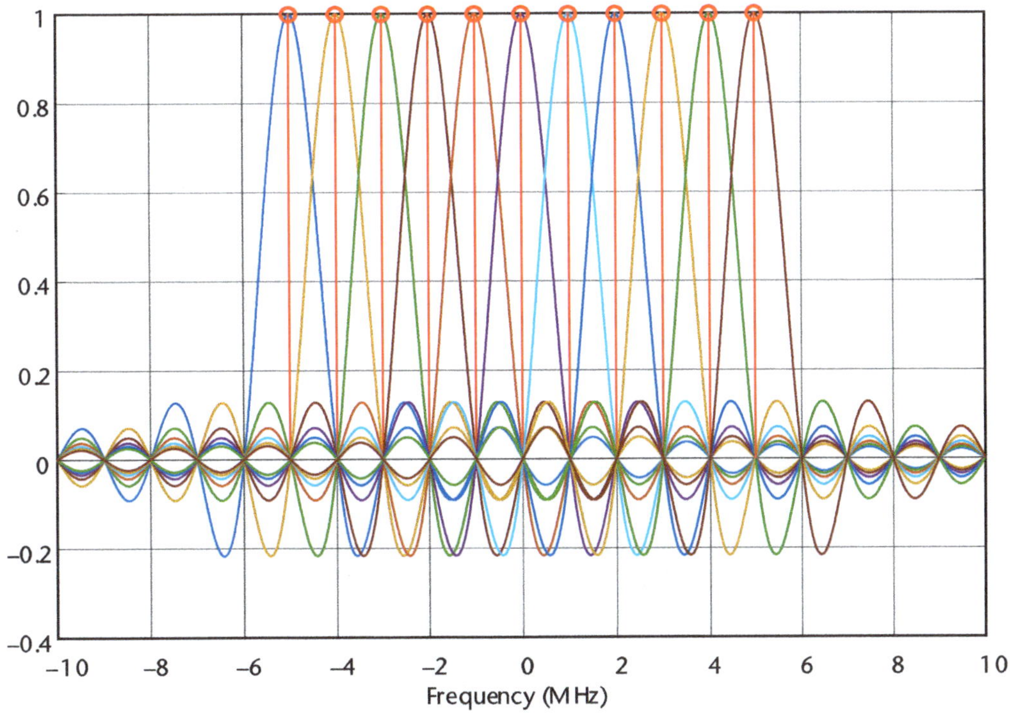

Figure 7.6 OFDM subcarrier orthogonality diagram.

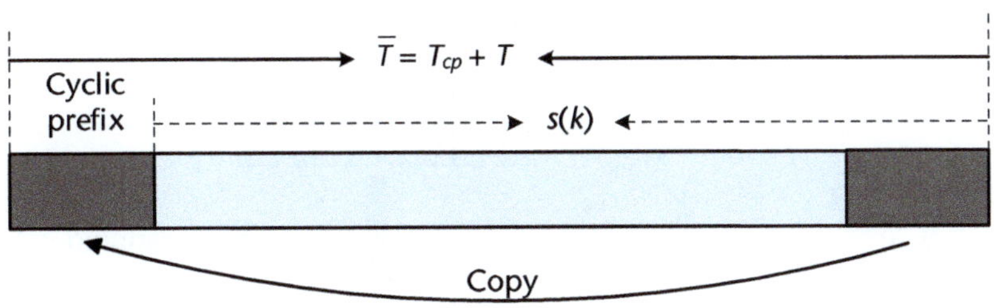

Figure 7.7 OFDM cyclic prefix.

OFDM is a low-complexity yet high-spectral-efficiency technique that has been widely used in both fourth-generation and fifth-generation wireless communication systems. Owing to its powerful capability of combating the multipath effect, it has also been adopted in the satellite communication standards DVB-SH and ATSC 3.0 since 2007, which will be introduced in the following section.

7.1.3 DFT-s-OFDM-Based SC Modulation

Single-carrier modulation modulates the data stream to a single carrier for transmission. Generally, it exhibits lower PAPR and is insensitive to carrier synchronization and timing deviation. Discrete Fourier transform spread orthogonal frequency division multiplexing (DFT-S-OFDM) [5, 6] is the popular scheme and has been applied in 4G and 5G communication.

Figure 7.8 presents the system diagram of DFT-S-OFDM. The transmitting end of DFT-S-OFDM includes a N point DFT module and a M point IDFT module.

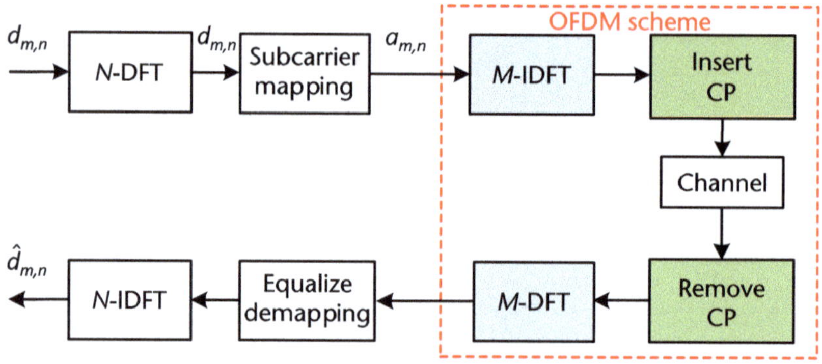

Figure 7.8 DFT-S-OFDM system block diagram.

The receiving end consists of a N point IDFT module and a M point DFT module, where we have $N < M$. The mth transmission symbol $d_{m,n}$ on the nth subcarrier first undergoes DFT transformation at the Mth point. After subcarrier mapping, N point IDFT transformation and cyclic prefix insertion are employed. Explicitly, the signal with N point DFT module can be expressed as

$$\tilde{d}_{m,n} = \frac{1}{\sqrt{N}} \sum_{k=0}^{N-1} d_{m,n} \mathrm{e}^{-\mathrm{j}2\pi mk/N} \tag{7.13}$$

Due to $N < M$, after subcarrier mapping, only a portion of the subcarriers are occupied by $\tilde{d}_{m,n}$. Figure 7.9 illustrates the subcarrier mapping methods

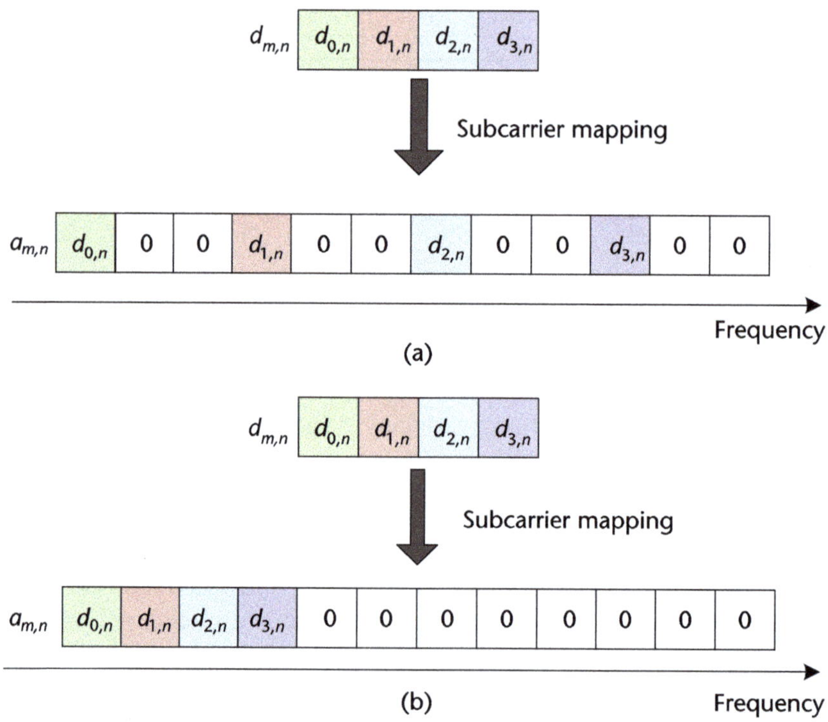

Figure 7.9 DFT-S-OFDM subcarrier mapping methods: (a) distributed mapping, and (b) centralized mapping.

of DFT-S-OFDM, including distributed mapping and centralized mapping. Distributed mapping is the equal interval mapping of DFT output signals onto OFDM subcarriers. When the interval is equal to $\lambda = M/N$, this special distributed mapping is also known as interleaved mapping. Centralized mapping is the mapping of DFT output signals onto a continuous segment of OFDM subcarriers. More specifically, the distributed mapping can be expressed as

$$a_{m_0+\lambda m,n} = \tilde{d}_{m,n}, m = 0, 1, \cdots, N-1, n \in \tag{7.14}$$

where, m_0 is the starting position of the subcarrier mapping, and λ is a positive integer greater than 1, representing the subcarrier spacing of the mapping. Centralized mapping can be formulated as

$$a_{m_0+m,n} = \tilde{d}_{m,n}, m = 0, 1, \cdots, N-1, n \in \tag{7.15}$$

Due to the special structure of DFT-S-OFDM, it exhibits lower PAPR over OFDM counterpart maintaining the capability of combating multipath fading. Similar to OFDM scheme, it has also been applied to 4G and 5G communications.

7.2 Modulation Standard for Cellular Mobile Communication

7.2.1 Modulation for 1G Communication

1G was born in the 1980s. It mainly uses cellular mobile communication mechanism technology based on analog signals to provide voice services. Frequency modulation (FM) modulates the user's analog voice signal to the carrier frequency of the subcarrier assigned to the user. Specifically, FM modulation converts the voice between 300 and 3,400 Hz to a high-frequency carrier frequency and transmits it through a wireless channel. The receiver uses carrier technology to restore the voice information. The transmission channels in each direction include voice channels and control channels. The voice channel transmits analog voice using FM modulation and the control channel transmits data control signaling using frequency shift keying modulation.

Typical analog mobile communication systems include Bell Labs' Advanced Mobile Phone System (AMPS), Nordic Mobile Telephony (NMT), and the UK's Total Access Communication System (TACS). Due to the use of analog signal transmission, 1G can generally only transmit voice signals with the challenges of low spectrum efficiency, low data rate, limited capacity, narrow coverage, poor confidentiality, unstable signals, and low voice quality. In the 1G era, communication standards in different countries were not unified, which hindered the development of wireless communication technology.

7.2.2 Modulation for 2G Communication

The mainstream technical standards of the 2G system include GSM in Europe and IS-95 in the United States. The digital modulation technology used in the GSM system includes Gaussian minimum frequency-shift keying (GMSK). GMSK modulation is to insert a Gaussian low-pass premodulation filter before the MSK modulator for premodulation filtering to reduce the jump energy when switching between two carriers of different frequencies, so that the modulated signal spectrum is compact and the bit error rate is good. IS-95 uses quadrature phase shift keying (QPSK) for the forward link and offset-QPSK (OQPSK) for the reverse link. OQPSK overcomes the 180-phase jump of QPSK, the phase does not exceed the zero point, so that the peak-to-average power ratio is small. Moreover, the signal has a small envelope fluctuation after passing through the bandpass filter and the power amplifier can work in an efficient nonlinear state, resulting in better performance.

2G communications have a milestone significance in the development and promotion of mobile communications, completing the transition from analog technology to digital technology and opening the era of digital mobile communications. It overcomes the weaknesses of analog mobile communication systems, greatly improves voice quality and confidentiality, and can perform automatic roaming within and between areas. However, the 2G standard is not unified, and it can only roam in the same standard coverage area, and cannot roam globally. Additionally, the bandwidth is limited and cannot provide high-speed data transmission. Due to the weak anti-interference and antifading capabilities and low frequency utilization, 2G communication cannot meet the growing communication needs of multimedia services such as pictures and videos.

7.2.3 Modulation for 3G Communication

In order to overcome the shortcomings of the 2G system and meet the demand for mobile multimedia services, 3G communications have gradually developed. IMT-2000 refers to the International Mobile Telecom System that operates in the 2,000-MHz frequency band and was put into commercial use around 2000. It includes both ground communication systems and satellite communication systems. The broadband mobile communication system based on IMT-2000 is called the third generation mobile communication system.

The main features of 3G are:

1. A system with global penetration and global seamless roaming. 3G is a system that is covered and used worldwide. It uses a common frequency band and global unified standards.
2. It has the ability to support multimedia services, especially internet services. 3G can support voice packet data and multimedia services and can provide the required bandwidth as needed.
3. It has a variable high-speed data rate and can meet the requirements of three environments: fast mobile environment, outdoor to indoor or walking environment, and indoor environment.
4. It is easy to transition and evolve. 3G can be flexibly evolved on the basis of 2G network and is compatible with fixed network.
5. It has high spectrum efficiency, high service quality, low cost, and high capacity. With the development of smart terminals, people can effectively process a variety of media services such as images, music, and video streams. Humanity has officially entered the era of mobile multimedia.

3G supports services with a rate of up to 2 Mbps, and the types of services will involve voice, data, images, and multimedia services. The 3G system still uses digital modulation technology, including BPSK, QPSK, 8PSK, and other modulation methods. The mainstream standards of 3G are wideband code division multiple access (WCDMA), CDMA2000, and TD-SCDMA. Both WCDMA and CDMA2000 use BPSK for uplink and QPSK for downlink. TD-SCDMA uses QPSK and 8PSK at high rates.

7.2.4 Modulation for 4G Communication

The 4G communications are further developed and optimized on the basis of 3G, using more advanced communication technology and communication protocols, and further broadening the spectrum. Although 3G can be used to transmit video data, if you want to transmit high-definition video data, you need to use 4G. The 4G system supports QPSK, 16QAM, 64QAM, and 256QAM modulation methods, and also adopts a new multicarrier modulation—OFDM. OFDM converts serial data streams into high-speed parallel carrier transmission, which can not only significantly improve spectrum efficiency, but also use its structural characteristics to effectively combat multipath fading and improve transmission reliability. LTE uses OFDM in the downlink, while DFT-S-OFDM is used in the uplink and direct links because DFT-S-OFDM has a lower PAPR than OFDM.

Usually, international 4G standards include Long Term Evolution-Advanced (LTE-A) led by the 3GPP and the Institute of Electrical and Electronics Engineers (IEEE)-led Wireless Metropolitan Area Network-Advanced (Wireless MAN-Advanced) (802.16m). 4G can provide users with a wider range of services, such as internet access, image transmission, video on demand, data mutual transmission, and even real-time viewing of TV programs and other data or multimedia services.

7.2.5 Modulation for 5G Communication

4G networks have greatly promoted the rapid development of the internet. With the rapid growth of smart terminals, new services continue to emerge, such as augmented reality, virtual reality, and smart industrial Internet of Things, which put forward higher requirements for communication speed and communication latency, further promoting the development of 5G communications. In the face of diversified communication services, 5G communications define three major application scenarios: enhanced mobile broadband (eMBB), ultra-reliable low latency communication (URLLC), and massive machine-type communication (mMTC). In terms of technology, 5G communications use OFDM multicarrier modulation with different parameters and large-scale MIMO technology. In addition, 5G further expands spectrum resources, including low-frequency and millimeter-wave high-frequency communications.

Like LTE, 5G communications support QPSK, 16QAM, 64QAM, and 256QAM modulation modes for both uplink and downlink. In addition, the uplink supports $\pi/2$-BPSK to further reduce the peak-to-average ratio, thereby improving the efficiency of the power amplifier at low data rates, which is very important for large-scale machine-type communication services. Because 5G communication supports a wide range of application scenarios, it is likely that the supported modulation methods will need to be further expanded in the future. For example, 1024QAM may also be included for specific scenarios.

In the frequency range up to at least 52.6 GHz, NR uses Cyclic Prefix OFDM (CP-OFDM) for both uplink and downlink. Compared with LTE, CP-OFDM is only used for LTE downlink transmission, while DFT-S-OFDM is used for uplink transmission. Using the same waveform for uplink and downlink simplifies the overall design, especially for wireless backhaul and device-to-device (D2D) communication. In addition, for scenarios with limited uplink coverage, the option of using DFT-S-OFDM is provided through single-stream transmission. In the implementation, the gNB can choose the uplink waveform (CP-OFDM or DFT-S-OFDM), and the UE should support both OFDM and DFT-S-OFDM modes. Any operation that is transparent to the receiver (such as windowing/filtering) can be carried out based on the NR waveform to improve the spectrum restrictions.

Obviously, from 1G to 5G, the communication spectrum is constantly expanding, the communication air interface technology is constantly being innovated, and the communication rate and communication quality have made a qualitative leap. Among them, 1G is based on analog modulation, 2G and 3G mainly use low-frequency communication and SC modulation technology, where data symbols are transmitted through one signal frequency. As the bandwidth increases, the sampling period of the system becomes smaller, the multipath effect of the wireless channel is further enhanced, and the SC modulation technology is no longer applicable. OFDM multicarrier modulation technology divides the system bandwidth into multiple orthogonal subcarriers, which significantly increases the symbol period of each subcarrier, can effectively combat the channel multipath effect, simplifying the system design. Moreover, OFDM can effectively combine with channel coding, multiantenna, and other technologies to improve the system's spectrum efficiency and has become a key modulation technology for 4G and 5G communications.

7.3 Modulation Standard for Satellite Communication

Since the available bandwidth of satellite communication systems is limited, in order to achieve high-speed information transmission within the limited bandwidth, high-order modulation methods with high bandwidth utilization must be used. At present, the research on satellite communication system modulation technology mainly focuses on three aspects:

1. Based on the power-limited characteristics of satellite communication systems, research on modulation methods have to consider how to improve power efficiency;
2. Based on the bandwidth-limited characteristics of frequency resources, research on modulation methods have to consider how to improve spectrum efficiency;
3. Modulation technology is necessary suitable for nonlinear channels [19].

Table 7.3 lists the modulation methods of common satellite communication standards.

Table 7.3 Modulation Methods of Common Satellite Transmission Standards

Number	*Standard Name*	*Standard Proposer*	*Modulation Mode*
1	DVB-S		QPSK
2	DVB-S2		QPSK,8PSK, 16APSK,32APSK
3	DVB-S2X	European Telecommunications Standards Institute (ETSI)	4APSK,8APSK,16APSK, 32APSK,64APSK, 128APSK,256APSK
4	DVB-SH		QPSK,8PSK,16APSK, 16QAM,CP-OFDM
5	ATSC	Advanced Television Systems Committee of the United States (ATSC)	QPSK,8PSK,16PSK, 16QAM,64QAM, 256QAM,CP-OFDM
6	ISDB-S	Japanese Digital Broadcasting Expert Group (DiBEG)	BPSK,QPSK, 8PSK,TC8PSK

7.3.1 Modulation for DVB-S Communication

In 1994, the European Telecommunications Standards Institute (ETSI) released the Digital Video Broadcasting-Satellite (DVB-S) standard [20]. The DVB-S standard has been widely used around the world and has become the mainstream transmission standard in the field of satellite broadcasting and television. The DVB-S system standard describes the frame structure, channel coding, and modulation method for satellite digital multiplexing. The standard stipulates the use of the MPEG-2 codec to transmit digital video via satellite in fixed satellite and broadcast satellite service bands. DVB-S uses QPSK as the modulation mode.

7.3.2 Modulation for DVB-S2 Communication

After entering the twenty-first century, with the rapid changes in communication needs, the DVB-S standard has defects such as low transmission efficiency, inability to transmit high-definition television HDTV, and inability to perform IP networking, making it difficult to meet business needs. To solve these problems, in March 2005, ETSI released the second generation of digital satellite video broadcasting (Digital Video Broadcasting-Satellite-Second Generation (DVB-S2)) standard [21]. DVB-S2 is an improvement on DVB-S, which reduces the design cost and provides better performance and more flexible services than the DVB-S standard.

Compared with the DVB-S standard, the technical features of the DVB-S2 standard include:

1. *Flexible and efficient multicoding rate and multimodulation transmission technology*: The DVB-S2 standard supports 11 coding rates from 1/4 to 9/10. It supports high-order modulation methods such as QPSK, 8PSK, 16APSK and 32APSK, and can be flexibly selected according to the link conditions. The reason for choosing APSK is that it is highly robust to the high power amplifier (HPA) distortion inherent in satellite transponders. The use of 32APSK makes the spectrum efficiency of the DVB-S2 standard (32APSK 9/10) 257% higher than that of the DVB-S standard (QPSK 7/8);
2. *Adaptive coding and modulation (ACM)*: DVB-S2 equipment provides real-time variable adaptive coding and modulation methods according to the

different signal transmission environments of the terminal and the gateway. ACM can automatically optimize frame-by-frame coding and modulation. The terminal with a poor signal uses low-order modulation, and the terminal with a strong signal uses high-order modulation, thereby enhancing the system's ability to resist interference such as rain attenuation and improving the reliability of the system's RF signal transmission;

3. *A variety of selectable spectrum roll-off coefficients*: When setting transmission parameters, users can independently select three roll-off coefficients of 0.20, 0.25, and 0.35 for square root raised cosine filtering to meet users' audio, video, and data transmission needs.

By adopting the above advanced coding and modulation technologies, the DVB-S2 system has the following advantages: it can support more transmission service types and source formats, better channel coding gain, higher channel spectrum utilization and parameter efficiency, and backward compatibility with the DVB-S standard.

7.3.3 Modulation for DVB-S2X Communication

Between 2005 and 2014, the digital satellite broadcast television sector experienced substantial transformations. On one hand, user demands have continually evolved, with UHD TV satellite broadcasting to homes and high-speed IP access based on TV broadcast satellites leading to an increase in users within VSAT application scenarios, thereby generating additional revenue. On the other hand, nonstandardized proprietary technologies such as NS3 significantly surpass the existing DVB-S2 standard in terms of spectral efficiency. In this context, satellite television broadcasting technology must also keep pace with contemporary advancements. In February 2014, the DVB-S2X standard was officially introduced. Technological innovations in DVB-S2X encompass reduced roll-down coefficients, advanced filtering techniques, and higher-order modulation.

The roll-down coefficients utilized by the DVB-S2 system are 0.35, 0.25, and 0.20, respectively; conversely, those employed by the DVB-S2X system have been decreased to 0.15, 0.10, and 0.05, respectively. The smaller roll-down coefficient of DVB-S2X combined with its advanced filtering technology effectively mitigates sidelobes on both sides of the spectrum, thereby conserving actual physical bandwidth occupied by a channel and allows for adjacent physical channels to be spaced as closely as just 1.05 times the symbol rate apart. Consequently, spectral efficiency for the DVB-S2X system can be enhanced by up to 15% compared to that of the DVB-S2 system.

While DVB-S2 employs four PSK modulation modes: 4APSK, 8APSK, 16APSK, and 32APSK, DVBS-2X adopts a higher-order PSK modulation scheme—up to 256APSK—facilitating easier compensation for nonlinearities present in satellite transponders while achieving superior spectrum utilization rates. Furthermore, larger satellite-borne antennas can be deployed alongside increased transmission power from satellites, which is crucial for enabling point-beam regional broadcasting within Ka-band frequencies. Due to its higher-order modulation capabilities, DVB-S2X enhances spectral efficiency by an impressive 51% when compared against DVB-S2, approaching closer toward Shannon's limit.

7.3.4 Modulation for DVB-SH Communication

Digital Video Broadcasting-Satellite Services to Handheld (DVB-SH) is a new generation of digital multimedia broadcasting standard proposed by the DVB organization. It mainly supports mobile TV services, as well as various types of information services such as video on demand and data transmission [22]. The main features of this standard are that it can be used in a satellite/terrestrial hybrid network, and can achieve signal coverage of a large area through GEO satellites. For densely populated buildings or indoor environments where satellite direct transmission signals cannot reach, the signal can be relayed through ground auxiliary base stations to ensure safe and reliable transmission of information. Because the DVB-SH protocol combines the satellite end and the ground relay end to achieve signal coverage, it is universal; that is, it can be used in any indoor or outdoor environment. When the transmission environment between the satellite and the target receiving user is good, the signal can be directly transmitted without going through the ground auxiliary base station; when the link between the satellite and the target receiving user is blocked, the signal can be forwarded through the ground auxiliary base station [23].

Figure 7.10 shows the network module of DVB-SH. As shown in the figure, DVB-SH has two architectures: SH-A and SH-B. In SH-A, both the satellite and the ground system use OFDM multicarrier modulation, which can effectively solve the multipath problem. In SH-B, the satellite link uses a time division multiplexing (TDM) structure to transmit the signal, while the ground link uses OFDM to transmit the signal. Compared to single-carrier systems, multicarrier systems have a larger peak power to average power ratio and higher requirements for power amplifiers. Therefore, the SH-A method is used when bandwidth resources are limited, while SH-B can be used in power-limited satellite systems.

In DVB-SH, QPSK, 8PSK, and 16APSK can be selected as constellation modulation options when TDM is used. When the OFDM multicarrier method is used, QPSK, 16QAM, and nonuniform 16QAM (supporting hierarchical modulation) can be selected as constellation modulation options. The roll-off factor can be selected as 0.15, 0.25, and 0.35. In addition to the usual 2K-FFT, 4K-FFT, and 8K-FFT modes,

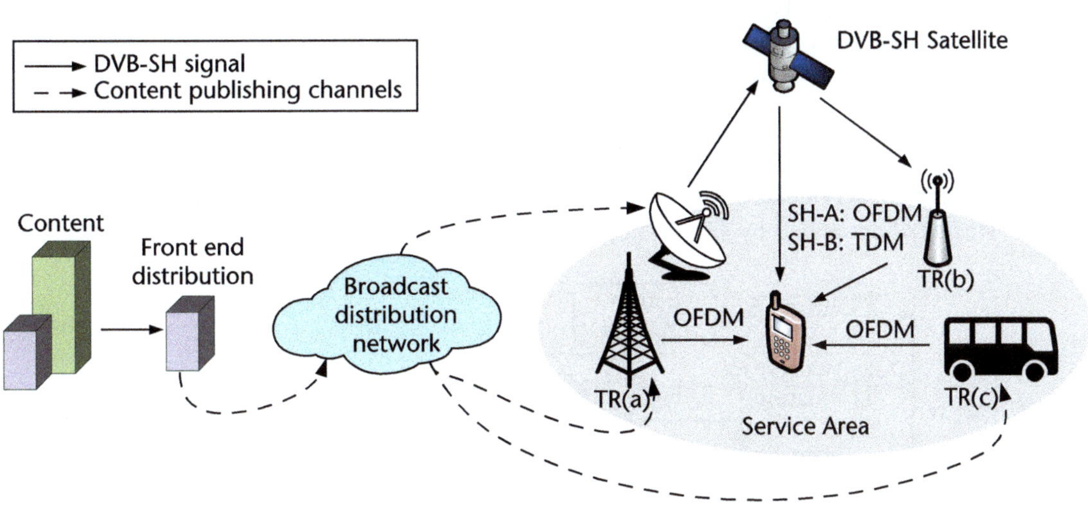

Figure 7.10 DVB-SH network module.

the DVB-SH standard adds a 1K-FFT mode, which is mainly used in the L-band, with a planned channel width of 1.7 MHz. The DVB-SH system can select L- and S-band frequencies between 1 and 3 GHz. The DVB-SH system is applicable when the terminal is in a mobile state of up to 3,200 km/h or in a static state indoors.

7.3.5 Modulation for ATSC Communication

The ATSC standard is a set of American standards for transmitting digital television through terrestrial, cable, and satellite networks. The United States officially adopted ATSC as the national digital television standard on December 24, 1996. The ATSC standard has the following major technical advantages: low noise threshold (close to the theoretical value of 14.9 dB), large transmission capacity (6-MHz bandwidth transmits 19.3 Mbps), long transmission distance, wide coverage, and easy implementation of reception scheme [24].

The latest ATSC version is ATSC3.0. In addition to the traditional downlink, the physical layer of ATSC3.0 also includes an uplink. Among them, the uplink uses the return channel to further expand the interactive function of digital television; the downlink achieves the function of supporting future ultra-high-definition television transmission such as 4K and 8K through the reasonable use of input formatting and signaling. ATSC3.0 adopts cyclic prefix orthogonal frequency division multiplexing technology, inserts a cyclic prefix as a protection interval at the beginning of each OFDM symbol, and needs to insert a pilot for channel estimation.

Figure 7.11 shows the physical layer system diagram of ATSC3.0. As shown in the figure, the physical layer of ATSC3.0 is connected to the link layer through two links, uplink and downlink, and the return channel transmits the uplink data to the link layer. The downlink transmits control information and signaling so that the data is formatted, coded, and modulated to generate waveforms to achieve ultra-high-definition video transmission. The ATSC3.0 system can support both broadcast services and interactive services without relying on other network infrastructure.

ATSC3.0 defines 6 QAMs with different modulation orders, including irregular 16QAM, 64QAM, 256QAM, 1024QAM, 4096QAM, and regular QPSK. Compared with regular constellation modulation, the use of irregular constellation modulation can achieve a performance gain of more than 1 dB.

7.3.6 Modulation for ISDB-S Communication

Integrated Services Digital Broadcasting-Satellite (ISDB-S) is a satellite broadcasting standard proposed by Japan in October 1999. ISDB-S operates in the Broadcasting

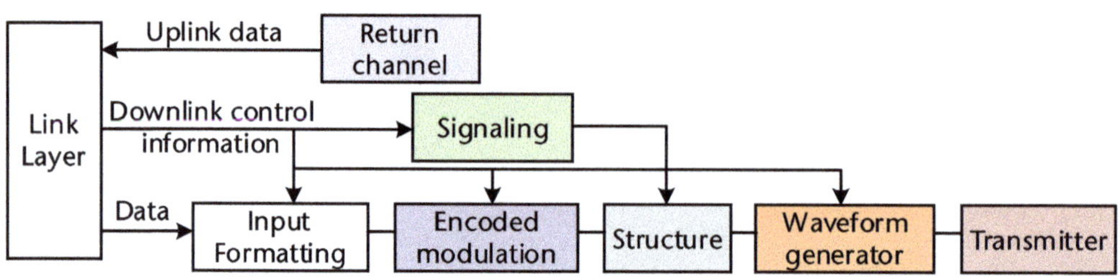

Figure 7.11 ATSC3.0 physical layer system diagram.

Satellite Service (BSS) band between 11.7 and 12.2 GHz [25]. Table 7.4 summarizes the ISDB-S standard features.

Compared with DVB-S, the ISDB-S system uses a hierarchical modulation scheme. The optional modulation schemes include trellis-coded eight PSK (TC8PSK), QPSK, and BPSK. In a satellite channel, the selection of TC8PSK can achieve a maximum information bit rate of 52.2 Mbit/s, while the occupied bandwidth is expanded from the traditional 27 to 34.5 MHz.

ISDB-S is highly flexible. ISDB-S can use multiple modulation schemes at the same time to meet the requirements of integrating multiple services on a satellite channel. For example, ISDB-S uses BPSK to transmit audio data, which can effectively reduce the impact of rain attenuation. TC8PSK can provide the maximum transmission capacity, so it is used to transmit video data. ISDB-S uses the transmission and multiplexing configuration control (TMCC) signal to send a modulation scheme selection command according to the transmission signal format. TMCC consists of 384 bits of fixed-length data. The receiver selects the corresponding demodulation scheme to process the received signal according to the pre-TMCC signal. The TMCC used for carrier recovery is provided by the fixed modulation scheme BPSK.

7.4 Potential Modulation for Integrated Communication

In the future integrated communication, QAM modulation may be sensitive to nonlinear distortion. Irregular constellation modulation, such as golden amplitude modulation (GAM) [4] has the potential for performance enhancement with a lower PAPR. Moreover, OFDM and DFT-s-OFDM waveforms tend to be impaired by severe intercarrier interference in the case of high-mobility wireless channel, which may suffer from significant performance loss in the context of LEO satellite communication. Furthermore, the time-domain rectangular window gives rise to severe out-of-band leakage in the frequency domain, which readily exceeds the radiation limit of the LEO satellite communication system. Accordingly, novel waveforms are necessary for development.

To provide reliable communication in high-mobility environment, Doppler-resilience waveforms (i.e., OTFS, orthogonal chirp division multiplexing (OCDM), and affine frequency division multiplexing (AFDM)), have been investigated in [5–7]. However, the signal detection complexity of these waveforms needs to be further reduced to accommodate onboard resource constraints. Moreover, synchronization and sidelobe suppression represent pivotal challenges in waveform design for satellite communications. The waveforms, such as filter bank multicarrier

Table 7.4 ISDB-S Standard Characteristics

Modulation Scheme	*TC8PSK, QPSK, BPSK*
Rising cosine rolling factor	0.35
Transmission symbol rate	28.86 Mbaud
Internal code rate	BPSK(1/2) QPSK(1/2,2/3,3/4,5/6,7/8) TC8PSK(2/3)
Packet size	188 bytes

(FBMC) [8], universal filtered multicarrier (UFMC) [9], and generalized frequency division multiplexing (GFDM) [10] can provide a strong immunity to these issues at the cost of performance degradation in the presence of severe Doppler spreads. The quest for an eminent waveform has sparked a flurry of research activities. However, the vast majority of current research mainly focuses on a single waveform in the context of terrestrial communication; the specific waveform design guidelines for LEO satellite communication remain in their infancy, warranting further exploration. Explicitly, Figure 7.12 presents the potential baseband and carrier waveforms.

7.4.1 Irregular Baseband Modulation

Irregular modulation has always been a hot research topic in wireless communication research due to its flexible design. Typical irregular baseband waveforms include triangular-QAM (TQAM) [11] and GAM.

To be more specific, the signal points of TQAM are distributed at the vertices of adjacent equilateral triangles, consistently maintaining the minimal distance from the decision boundary. It was demonstrated in [11] that the TQAM scheme was capable of providing higher power gain than QAM by virtue of denser constellation points. To further improve the performance, GAM aided irregular modulation was proposed, where the angle and amplitude of the constellation points can be flexibly designed to have lower PAPR or higher mutual information.

The performance of both regular and irregular modulation is associated with bit-to-symbol mapping method. Gray mapping performs excellently in the context of regular modulation, where the Hamming distance between neighboring constellation points is one. However, due to the different Euclidean distances of neighboring constellation points in the irregular modulation, Gray mapping is extremely challenging.

To address this issue, graph theory assisted bit-to-symbol mapping optimization approach was proposed in [4]. It is worth noting that the GAM scheme with the mapping method of [4] is capable of achieving comparable BER performance to its QAM counterpart with lower PAPR. It is expected that the irregular modulation with an advanced configuration is capable of outperforming its QAM counterpart in terms of BER and PAPR. Moreover, the constellation size of irregular modulation can be designed as an arbitrary number, which facilitates rate matching and rate adaptation, thereby simplifying the design of satellite transmitters. As a result, irregular modulation constitutes an appealing baseband waveform for LEO satellite communications.

7.4.2 FBMC Multicarrier Modulation

OFDM has the advantages of simple implementation and low complexity. However, due to the use of rectangular windows with high spectral sidelobes for each subcarrier, the transmitted signal of OFDM suffer from serious out-of-band leakage characteristics. Compared to the main lobe, the spectral sidelobes only decrease by more than 10 dB and decrease slowly. Therefore, OFDM requires a longer frequency protection interval, resulting in low spectrum utilization. In addition, OFDM users need to maintain strict synchronization and orthogonality, resulting in a huge waste of communication resources and an increase in access delay as the

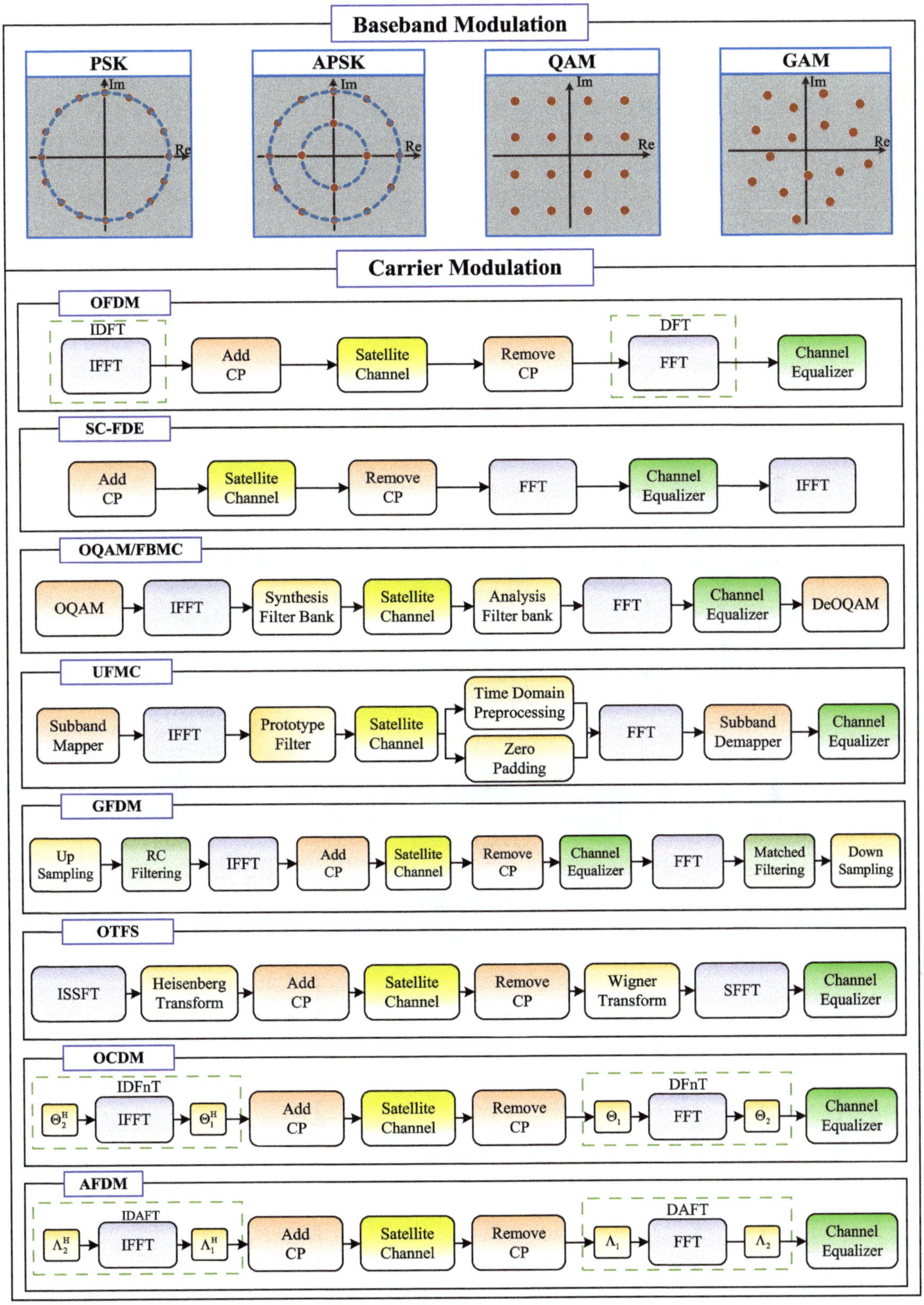

Figure 7.12 Potential baseband and carrier waveforms.

number of users increases. In this context, FBMC technology with low-frequency spectral sidelobe characteristics has received significant attention [9].

Figure 7.13 illustrates the system diagram of FBMC modulation. As shown in the figure, FBMC consists of M subcarriers, which are shaped by a prototype filter at the transmitting terminal carrier signal and demodulated by a matched filter at the receiving end. Usually, the coefficients of FBMC prototype filters are real and symmetric. Therefore, the coefficients of the shaping filter at the transmitting end and the matching filter at the receiving end are both $g(k)$. Similar to OFDM, the modulation waveform of FBMC subcarriers is $\mathrm{e}^{\mathrm{j}2\pi mk/M}$, and the demodulation waveform is $\mathrm{e}^{-\mathrm{j}2\pi mk/M}$. m is the subcarrier sequence number, and k is the time sampling sequence number. Due to the fact that the real and imaginary parts of complex QAM symbols need to be separated by half a symbol period, $\frac{M}{2}$ sampling points are selected to correspond to half a symbol period; that is, $\frac{T}{2}$. The symbols $\uparrow \frac{M}{2}$ and $\downarrow \frac{M}{2}$ represent upsampling and downsampling with intervals of $\frac{M}{2}$ sampling points, respectively. In addition, a phase factor $\mathrm{e}^{\mathrm{j}\pi(m+n)/2}$ and

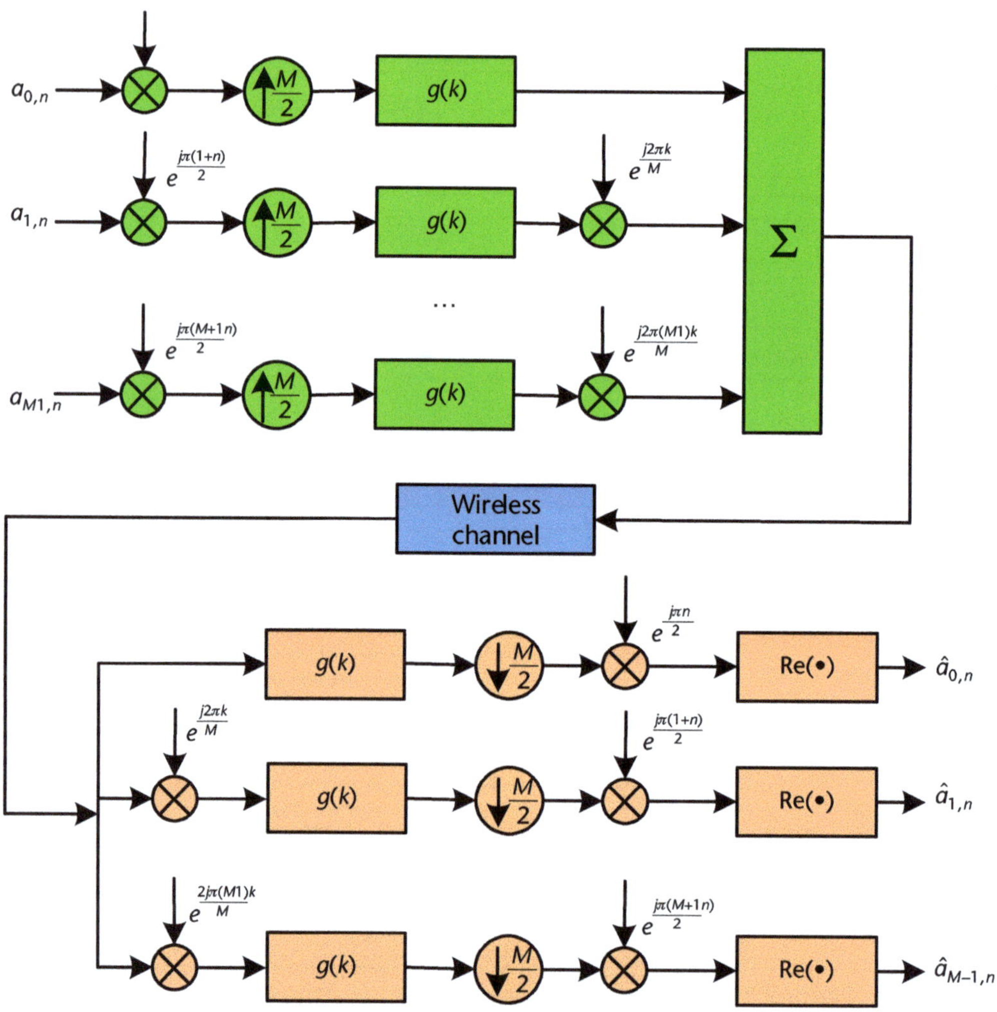

Figure 7.13 System block diagram of FBMC multicarrier modulation.

$e^{-j\pi(m+n)/2}$ is designed for each subcarrier at the sending and receiving ends to meet the orthogonality conditions of FBMC.

The real number transmission symbol $a_{m,n}$ of FBMC is modulated by the transmitter, and the baseband equivalent transmission signal $s(k)$ is

$$s(k) = \sum_{m=0}^{M-1} \sum_{n\in \mathbf{Z}} a_{m,n} g\left(k - n\frac{M}{2}\right) e^{j2\pi mk/M} e^{j\pi(m+n)/2} \tag{7.16}$$

where $\mathbf{Z}$ is the set of natural numbers. Assuming that $s(k)$ passes through an ideal distortion free channel, the received signal is $r(k) = s(k)$. At the receiving end, after demodulation of $r(k)$, the demodulation symbol at the time-frequency position (m, n) is

$$\hat{a}_{m,n} = \sum_{k=-\infty}^{\infty} r(k) g\left(k - n\frac{M}{2}\right) e^{-j2\pi mk/M} e^{-j\pi(m+n)/2} \tag{7.17}$$

Obtaining the transmission symbol by taking the real part of the demodulation symbol

$$\mathrm{Re}\{\hat{a}_{m,n}\} = a_{m,n} \tag{7.18}$$

In order to achieve maximum transmission efficiency in OFDM, the subcarrier interval Δf and symbol interval T must meet the following requirements;

$$\Delta f = \frac{1}{T} \tag{7.19}$$

At this point, the symbols of OFDM are orthogonal to each other and do not interfere with each other, achieving maximum transmission efficiency. However, due to the fact that the orthogonality condition of FBMC only holds in the real field, FBMC cannot directly transmit complex symbols. It is necessary to take the real and imaginary parts of the symbols separately, and transmit them as real part symbols at intervals of half a symbol cycle. Therefore, in FBMC, the subcarrier spacing and symbol spacing satisfy

$$\Delta f = \frac{2}{T} \tag{7.20}$$

Figure 7.14 illustrates the time-frequency data blocks of OFDM and FBMC, respectively. As shown in the figure, OFDM multicarrier modulation sends complex symbols with a symbol interval of T, where FBMC multicarrier modulation sends real symbols with a symbol interval of $T/2$. In theory, FBMC and OFDM transmit the same amount of data in the same amount of time.

Figure 7.15 illustrates the process of obtaining FBMC real number symbols. As shown in the figure, let the complex symbol be $x_{m,n}$, where m is the subcarrier sequence number and n is the time sequence number. Extract the real and imaginary parts of the complex symbols of FBMC to obtain the sending symbol $a_{m,n}$. The real number symbol sent by FBMC is

$$a_{m,2n} = \mathrm{Re}\{x_{m,n}\}, a_{m,2n+1} = \mathrm{Im}\{x_{m,n}\} \tag{7.21}$$

where $\mathrm{Re}\{\cdot\}$ and $\mathrm{Im}\{\cdot\}$ are the operations of taking the real part and taking the imaginary part, respectively.

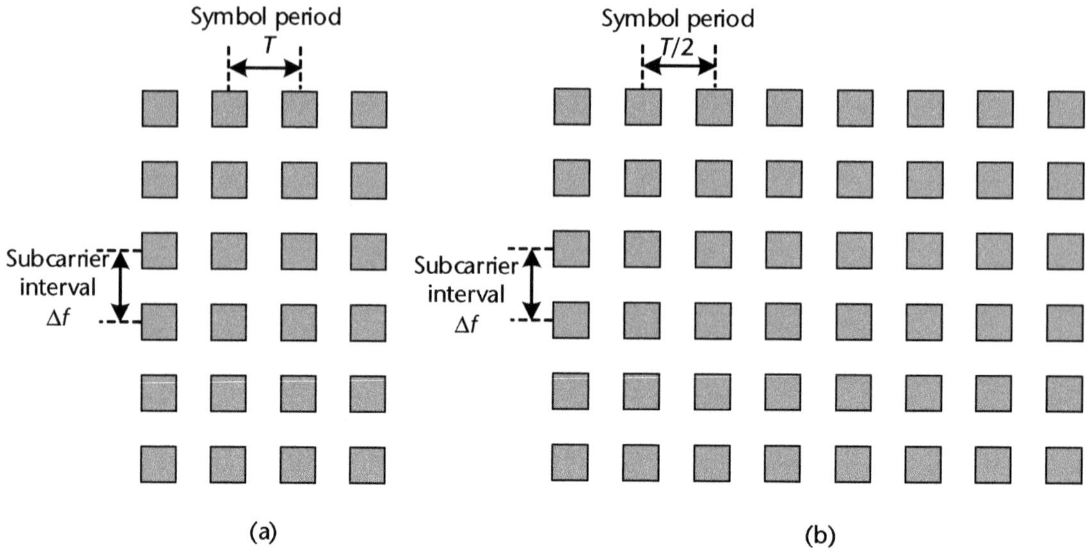

Figure 7.14 Time-frequency data blocks of the OFDM system and FBMC system. (a) Time frequency resource data blocks for classical OFDM systems (imaginary number symbols) and (b) time frequency resource data block of FBMC system (real number symbols).

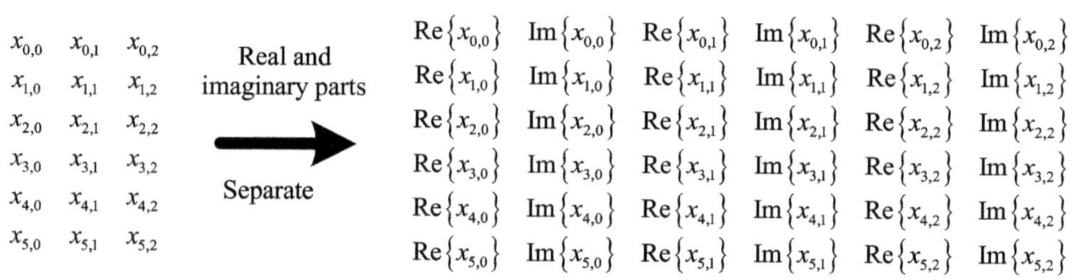

Figure 7.15 The process of obtaining the real number symbol of FBMC.

The advantages of FBMC are as follows:

1. FBMC uses a prototype filter with very low spectral sidelobes for signal shaping, which makes the out-of-band leakage of FBMC signals very weak and almost does not cause interference to other users. Users can use smaller frequency band protection intervals to effectively use fragmented spectra.
2. There are many narrow-spectrum interference signals in wireless communication environments, which can cause significant damage to OFDM signals with high and slow spectral sidelobes. However, the impact on FBMC signals with fast spectral downscaling is very weak.
3. FBMC is an asynchronous transmission technology that uses FBMC modulation to eliminate the need for strict synchronization and orthogonality between user signals, greatly reducing the synchronization overhead and access delay of the system, making system design more flexible and efficient.

The disadvantages of FBMC are as follows:

1. There is no orthogonality between subcarriers in the FBMC system, resulting in interference between subcarriers.

2. Due to the use of nonrectangular waveforms in the FBMC system, there is intersymbol interference in the time domain, and corresponding interference cancellation techniques need to be adopted.

7.4.3 UFMC Multicarrier Modulation

In response to the carrier interference problem caused by OFDM frequency offset, Vakilian proposed the UFMC technology in 2013 [17]. UFMC filters subband signals, which not only reduces out-of-band sidelobe levels but also minimizes carrier interference between adjacent users during asynchronous transmission. UFMC filters a group of continuous subcarriers, and can configure the number of subcarriers according to actual needs, thereby increasing or shortening the length of the filter. It can flexibly support different frame structures and reduce system complexity. Figure 7.16 presents the different filtering methods of three technologies.

Figure 7.17 shows the system diagram of UFMC. As shown in the figure, first divide the subcarriers into B subbands, and overlay a FIR Chebyshev filter on each subband, with a filter length of L_f and an FFT length of N. The frequency domain signal of the lth transmission subband is $X_l = [X_l(0), X_l(1), \cdots, X_l(M-1)]$, and the independent random variable bitstream X_l is transformed by IDFT to obtain the time-domain signal $x_l = [x_l(0), x_l(1), \cdots, x_l(N-1)]$ of the subband, where $x_l(n)$ is

$$x_l(n) = \frac{1}{N} \sum_{m \in B} X_l(m) e^{j2\pi mn/N} n = 0, \cdots, N-1 \tag{7.22}$$

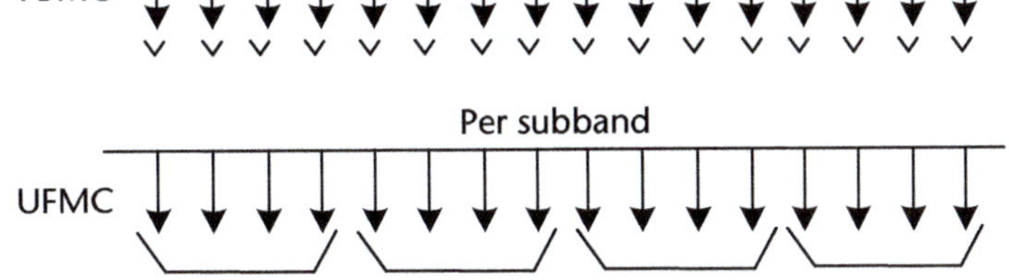

Figure 7.16 Comparison of filtering methods among OFDM, FBMC, and UFMC.

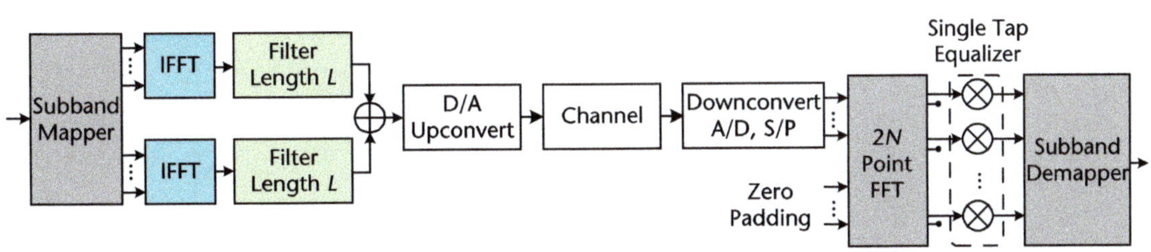

Figure 7.17 UFMC system block diagram.

The time-domain signal X_l of the subband is accumulated through the FIR Chebyshev filter, and the resulting baseband time-domain signal is

$$X = \sum_{l=1}^{B} x_l * f \tag{7.23}$$

Assuming that the time-varying pulse response of the channel is $\mathbf{h}$ and the additive white Gaussian noise is η, the time-domain signal $\mathbf{y}$ at the receiving end is

$$y = x * h + \eta = \sum_{l=1}^{B} x_l * f * h + \eta \tag{7.24}$$

At the receiving end, based on the $2N$ point FFT detection, the time-domain signal $\mathbf{y}$ is transformed into a frequency-domain signal $\mathbf{Y}$. The kth subcarrier $Y(k)$ obtained after signal demodulation is represented as

$$\begin{aligned} Y(k) &= \sum_{l=1}^{B} \tilde{X}_l(k) F(k) H(k) + W(k) \\ &= \tilde{X}_l(k) F(k) H(k) + \sum_{\substack{a=1 \\ a \neq l}}^{B} \tilde{X}_a(k) F(k) H(k) + W(k) \end{aligned} \tag{7.25}$$

where $W(k)$ is additive white Gaussian noise. To distinguish the signal and interference of $Y(k)$ in (7.25), the expression of $\tilde{X}_l(k)$ is first derived as

$$\begin{aligned} \tilde{X}_{lodd}(k) &= \sum_{n=0}^{2N-1} x_l(n) e^{-j2\pi kn/2N} \\ &= \sum_{n=0}^{N-1} x_l(n) e^{-j2\pi kn/4N} = X_l\left(\frac{k}{2}\right) \end{aligned} \tag{7.26}$$

where k is an even number.

$$\begin{aligned} \tilde{X}_{leven}(k) &= \sum_{n=0}^{2N-1} x_l(n) e^{-j2\pi kn/2N} = \sum_{n=0}^{N-1} x_l(n) e^{-j\pi kn/N} \\ &= \sum_{n=0}^{N-1} \frac{1}{N} \sum_{m \in B_l} X_l(m) e^{j2\pi mn/N} \cdot e^{-j\pi kn/N} \\ &= \sum_{m \in B_l} X_l(m) \cdot e^{j\frac{\pi}{2}(2m-k)\left(1-\frac{1}{N}\right)} \cdot \frac{\sin\left[\frac{\pi}{2}(2m-k)\right]}{N \sin\left[\frac{\pi(2m-k)}{2N}\right]} \end{aligned} \tag{7.27}$$

where k is an odd number and B_l is the set of frequency domain subcarrier index numbers for the lth subband. Therefore, according to (7.26) and (7.27), $\tilde{X}_l(k)$ is represented as

$$\tilde{X}_l(k) = \begin{cases} \sum\limits_{m \in B_l} X_l(m) e^{j\frac{\pi}{2}(2m-k)\left(1-\frac{1}{N}\right)} \frac{\sin\left[\frac{\pi}{2}(2m-k)\right]}{N \sin\left[\frac{\pi(2m-k)}{2N}\right]}, & k \text{ is odd} \\ X_l(k/2), & k \text{ is even} \end{cases} \tag{7.28}$$

where $X_l(k)$ is the frequency domain signal of the kth subcarrier in the lth subband. Furthermore, it can be represented as

$$X_l(k) = \begin{cases} 0, k \notin B_l \\ X_l(k), k \in B_l \end{cases} \tag{7.29}$$

According to (7.29), due to the FFT detection based on $2N$ points at the receiving end, all odd-numbered subcarriers contain two parts: ICI and signal. Therefore, (7.28) can be rewritten as the sum of signal and interference

$$\tilde{X}_l(k) = \tilde{X}_{l,s_m}(k) + \tilde{X}_{l,ICI_m}(k) \tag{7.30}$$

where $\tilde{X}_{l,s_m}(k)$ is the mth subcarrier signal obtained by FFT detection transformation based on the $2N$th point of the subcarrier k in the mth subband of the transmitting end. $\tilde{X}_{l,ICI_m}(k)$ is the interference from the kth subcarrier received by the mth subcarrier signal obtained through FFT detection transformation based on the $2N$ point. Due to the fact that the receiver is based on $2N$ point FFT demodulation, all odd-numbered subcarriers are additional signals. Therefore, only the subcarriers of even-numbered points with $m = 0, 2, \ldots, 2N - 2$ need to be considered. When performing linear convolution, it is necessary to consider the subcarriers of odd points and take $k = 0, 1, 2, \ldots, 2N - 1$. $\tilde{X}_{l,s_m}(k)$ and $\tilde{X}_{l,ICI_m}(k)$ are respectively represented as

$$\tilde{X}_{l,s_m}(k) = \begin{cases} X_l\left(\frac{m}{2}\right), k \\ X_l\left(\frac{m}{2}\right) e^{j\frac{\pi}{2}(m-k)\left(1-\frac{1}{N}\right)} \frac{\sin\left[\frac{\pi}{2}(m-k)\right]}{N \sin\left[\frac{\pi(m-k)}{2N}\right]}, k \end{cases} \tag{7.31}$$

$$\tilde{X}_{l,ICI_m}(k) = \begin{cases} 0, k \\ \sum\limits_{i \in B, i \neq m/2} S_l(i) e^{j\frac{\pi}{2}(2i-k)\left(1-\frac{1}{N}\right)} \frac{\sin\left[\frac{\pi}{2}(2i-k)\right]}{N \sin\left[\frac{\pi(2i-k)}{2N}\right]}, k \end{cases} \tag{7.32}$$

Substituting (7.31) and (7.32) into (7.25) yields

$$\begin{aligned} Y(k) &= \tilde{X}_{l,s_m}(k) F(k) H(k) \cdots \text{signal} \\ &+ \tilde{X}_{l,ICI_m}(k) F(k) H(k) \cdots \text{ICI} \\ &+ \sum_{\substack{a=1 \\ a \neq l}}^{B} \tilde{X}_a(k) F(k) H(k) + W(k) \quad \cdots \text{IBI} \end{aligned} \tag{7.33}$$

According to (7.33), subcarrier $Y(k)$ is not only affected by interference between subcarriers within the same subband, but also by interband interference (IBI) between subcarriers.

The signal received by the receiving end is first processed by analog-to-digital conversion, filtering, and so on, to obtain two digital baseband signals of IQ. Then, time-frequency synchronization is performed to determine the position of the frame and FFT window, and frequency offset is estimated and compensated. The transmission symbol length of UFMC is $N + L - 1$. Under the influence of the subband filter, the receiver needs to perform $2N$ point FFT to fully convert it into a frequency domain signal, remove odd subcarriers to obtain a N point frequency domain symbol, and then compensate for residual frequency offset and phase difference. Finally, perform unmapping and decoding operations to obtain the original data stream.

Due to the special structure, the advantages of UFMC are as follows:

1. There is no CP in UFMC, the spectrum utilization is high, and the number of subcarriers in the subband can be flexibly configured according to actual needs.
2. UFMC filters each subband, effectively reducing interband interference and supporting asynchronous transmission mode.
3. UFMC uses shorter waveguides with low implementation difficulty, and supports short structures suitable for burst communication.
4. Due to the implementation of QAM constellation mapping in UFMC, it is compatible with MIMO technology.

The disadvantages of UFMC are as follows:

1. UFMC does not use CP and has poor resistance to multipath interference, making it difficult to meet the application scenarios of loose time synchronization to save power.
2. UFMC amplifies the noise power in the system and is sensitive to time bias.
3. Because UFMC needs to perform waves in the subband at the transmitting end, the symmetry between the transmitting and receiving ends of the FFT at the $2N$ point is poor, resulting in higher system implementation complexity.

7.4.4 GFDM Multicarrier Modulation

GFDM proposed by Fettweis et al. is a multicarrier modulation scheme based on symbol blocks [18]. GFDM modulates data symbols onto two-dimensional time-frequency blocks for transmission by dividing time-domain time slots and frequency-domain subcarriers, and uses parameter adjustable shaping filters. Unlike FBMC, GFDM uses cyclic convolution during subcarrier filtering to avoid trailing of the prototype filter, exhibits good out-of-band leakage characteristics, and is suitable for dynamic spectrum access.

GFDM divides time slots into M subslots (subsymbols) at equal intervals, while also dividing the bandwidth of time slot M into K subcarriers at equal intervals. GFDM data symbols are transmitted through two-dimensional time-frequency domain structural blocks. To reduce out-of-band radiation, an adjustable shaping filter can be selected. GFDM technology can effectively utilize fragmented spectrum without strict requirements for system synchronization performance, and has strong flexibility.

Figure 7.18 illustrates a typical GFDM wireless communication system. As shown in the figure, the binary information sent by the source is first channel encoded and mapped into complex data symbols through constellation diagrams. The mapping rule depends on the modulation type and modulation order. The complex data symbol forms a baseband signal after undergoing serial parallel conversion, GFDM modulation, and adding CP. After receiving the signal, the receiving end eliminates the effects of channel delay, frequency offset, and time-domain window function through synchronization and channel estimation, window function removal, CP removal, channel equalization, and other operations. Then, complex data symbols are obtained through GFDM demodulation module demodulation and parallel serial conversion. Finally, the transmission binary stream is obtained through constellation symbol judgment and channel decoding.

Unlike OFDM, GFDM uses a more flexible data block structure, where data symbols are modulated in a block format. Let d represent a symbol vector containing N sampling points, where $N = K \times M$, K and M are positive integers. By serial-to-parallel conversion, the $N \times 1$-dimensional symbol vector d is divided into K vectors of length M. The block matrix form of d is

$$D = \begin{pmatrix} d_0 & d_1 & \cdots & d_{K-1} \end{pmatrix}^T = \begin{pmatrix} d_{0,0} & \cdots & d_{0,M-1} \\ \vdots & \ddots & \vdots \\ d_{K-1,0} & \cdots & d_{K-1,M-1} \end{pmatrix}, KM = N \tag{7.34}$$

where $d_k = [d_{k,0}, d_{k,1}, \ldots, d_{k,M-1}]^T, k = 0, 1, \cdots, K-1$. Define K as the number of GFDM subcarriers and M as the number of GFDM subsymbols, then each GFDM subcarrier contains M subsymbols.

Figure 7.19 shows the modulation principle of GFDM. As shown in the figure, the transmitter first divides the high-speed transmission data symbol sequence $d = [d_0, \cdots, d_{N-1}]$ into K low-speed parallel subdata symbol sequences. Each path carries M data symbols; that is, $d_k = [d_{k,0}, \cdots, d_{k,M-1}]$, and then each data symbol of each path is upsampled by N times, and the signal is dispersed into M different

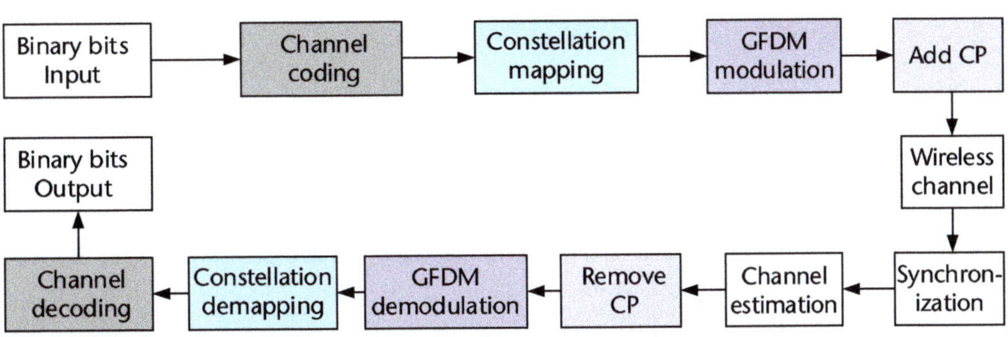

Figure 7.18 GFDM system block diagram.

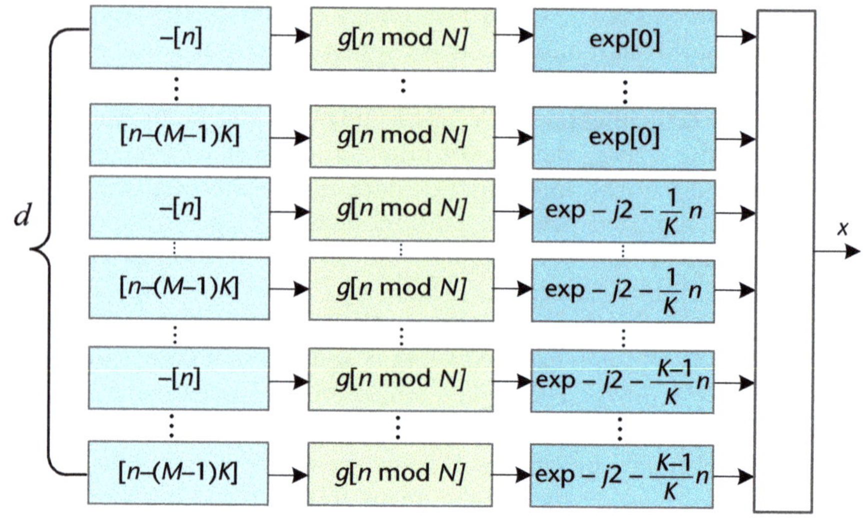

Figure 7.19 GFDM modulation principle.

time slots through a circular shift filter. Finally, each signal is modulated to the corresponding center frequency to obtain a GFDM symbol.

Therefore, the GFDM modulation process is expressed as

$$x\left(n\right)=\sum_{k=0}^{K-1}\sum_{m=0}^{M-1}g_{k,m}\left(n\right)d_{k,m},n=0,\ldots,N-1 \tag{7.35}$$

where $g_{k,m}\left(n\right)$ represents the coefficient of the shaping filter, specifically:

$$g_{k,m}\left(n\right)=g\left(\left(n-mk\right)\bmod N\right)e^{-j2\pi\frac{k}{K}n} \tag{7.36}$$

At the receiver, after time-frequency synchronization and CP removal, the GFDM received signal is expressed as

$$y=Hx+w \tag{7.37}$$

where H is the channel response matrix with dimension $N\times N$. This matrix is a circulant matrix, the first column of which is the time domain channel impulse response, and the other columns are cyclic shifts of the first column. w represents a Gaussian white noise sampling sequence with a mean of zero and a variance of σ_w^2.

The advantages of GFDM are as follows:

1. GFDM has CP, which can effectively combat multipath interference.
2. It can be implemented using IFFT/FFT technology, and the system implementation is easy.
3. The use of cyclic filters greatly increases the out-of-band attenuation rate of the main lobe, and has lower out-of-band radiation than OFDM, and higher spectrum utilization.
4. It can perform filtering according to the required waveform characteristics, and the structural design is flexible.

The disadvantages of GFDM are as follows:

1. GFDM as a multicarrier system faces the problem of high PAPR. In order to prevent nonlinear distortion of GFDM signals after power amplification, the power consumption of the HPA must be increased and a certain back-off power must be retained.
2. The subcarriers do not need to be orthogonal, so the inherent ICI and ISI in the system make the receiver processing relatively complex.

7.4.5 OTFS Multicarrier Modulation

Future wireless communication systems need to provide reliable communication for a wide range of communication scenarios and meet the needs of Internet of Vehicles communication, UAV communication, high-speed train communication, low-orbit satellite communication, and underwater acoustic communication. Therefore, future wireless communication channel scenarios will be more complex. Classic multicarrier modulation OFDM technology suffers serious performance loss under fast time-varying channels. Multicarrier modulation technology that can adapt to fast time-varying fading channels has become a key factor for reliable transmission of future broadband mobile communications.

In 2016, Hadani et al. proposed orthogonal time frequency space (OTFS) modulation [19]. The OTFS system is a new multicarrier modulation technology that introduces the delay domain and Doppler domain. It converts the dual dispersion channel into an approximately nonfading delay-Doppler domain channel through a series of two-dimensional transformations. Each symbol in the data frame of the delay-Doppler domain channel experiences almost the same fading, achieving a more significant performance gain than existing modulation schemes. At the same time, the OTFS system can design preprocessing and postprocessing modules based on the OFDM system to achieve compatibility with the 5G architecture. Compared with OFDM modulation, OTFS modulation has stronger antifrequency offset performance and lower peak-to-average ratio and frequency band leakage.

Figure 7.20 describes the basic principle of OTFS modulation. As shown in the figure, first, the information bits are modulated by amplitude and phase to generate $M \times N$ constellation symbols. Among them, M is the number of subcarriers; N is the number of time slots. Then, the constellation symbols undergo the inverse symplectic finite Fourier transform (ISFFT) to convert the delay Doppler domain signal

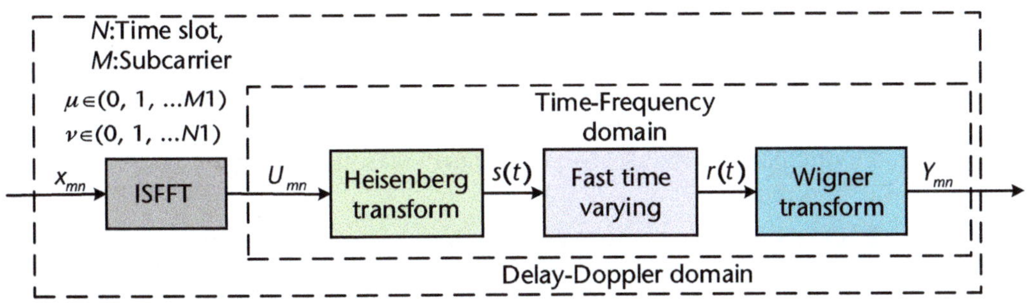

Figure 7.20 OTFS modulation principle.

into a time-frequency domain signal. The signal then undergoes the Heisenberg transform and is transmitted over the wireless channel. At the receiving end, the signal undergoes a Wigner transform and then a symplectic finite Fourier transform (SFFT) for signal recovery.

Specifically, the OTFS modulation process is described by taking the information bit of length $MN\log_2(L)$ as an example. The information bit is subjected to amplitude and phase modulation to obtain the following MN constellation symbols:

$$\mathbf{X} = \begin{bmatrix} x_{11} & x_{12} & \cdots & x_{1N} \\ x_{21} & x_{22} & \cdots & x_{2N} \\ \vdots & \vdots & \ddots & \vdots \\ x_{M1} & x_{M2} & \cdots & x_{MN} \end{bmatrix} \tag{7.38}$$

The constellation symbol is converted into a two-dimensional time-frequency symbol U_{mn} by ISFFT:

$$U_{mn} = \frac{1}{\sqrt{MN}} \sum_{k=0}^{N-1} \sum_{l=0}^{M-1} x_{kl} e^{j2\pi\left(\frac{nk}{N} - \frac{ml}{M}\right)}, m = 1, 2, \cdots, M, n = 1, 2, \cdots, N \tag{7.39}$$

Next, the time-frequency symbol U_{mn} is processed by the Heisenberg transform and two-dimensional windowing to obtain the transmission signal $s(t)$

$$s(t) = \sum_{n=0}^{N-1} \sum_{m=0}^{M-1} U_{mn} g_{\text{tx}}(t - nT) e^{j2\pi m\Delta f(t-nT)} \tag{7.40}$$

At the receiving, the received signal is subjected to Wigner transformation and sampling to obtain

$$Y_{mn} = \left[\int g_{\text{rx}}^{*}(t-\tau) r(t) e^{-j2\pi\upsilon(t-\tau)} d_t \right] \Big|_{\tau=nT,\upsilon=m\Delta f} \tag{7.41}$$

Corresponding to the transmitting end, the received signal will be obtained through SFFT transformation;

$$y_{kl} = \frac{1}{\sqrt{NM}} \sum_{n=0}^{N-1} \sum_{m=0}^{M-1} Y_{mn} e^{-j2\pi\left(\frac{nk}{N} - \frac{ml}{M}\right)} \tag{7.42}$$

Finally, a low-complexity signal detection algorithm is designed to further recover the signal.

The advantages of OTFS are as follows:

1. Compared with OFDM, OTFS provides high data rate and better flexibility and reliability;
2. OTFS exhibits better performance than OFDM in the context of high mobility channels;
3. Compared with OFDM and GFDM, OTFS exhibits lower PAPR [20].

The main disadvantage of OTFS is its high complexity in signal processing. Specifically, the minimum computational complexity of the current OTFS detector is $O\left(MN\log\left(N\right)\right)$, which is much higher than the OFDM detector complexity. At the same time, the current OTFS system mainly considers integer Doppler shift. If the fractional Doppler shift assumption that is closer to reality is adopted, the complexity will be further increased.

7.4.6 OCDM Multicarrier Modulation

Chirp signals have the ability to compress pulses and spread spectrum, and play an important role in radar and communication systems. Based on orthogonal chirp signals, orthogonal chirp division multiplexing (OCDM) technology was proposed for high-speed communication [21]. The OCDM system can achieve the maximum spectrum utilization of chirp spread spectrum.

OCDM uses a set of mutually orthogonal chirp signals as its basic waveform:

$$\psi_n(t) = e^{j\frac{\pi}{4}}e^{-j\pi\frac{N}{T^2}\left(t-n\frac{T}{N}\right)^2}, \quad 0 \le t \le T \tag{7.43}$$

where n is the index of the linear frequency modulation waveform, N is the number of linear frequency modulation waves, and T is the duration of the linear frequency modulation waveform. The linear frequency modulation waveform used by OCDM is also orthogonal to each other, just like the sine wave used by OFDM:

$$\int_0^T \psi_m^*(t)\psi_k(t)dt = \int_0^T e^{j\pi\frac{N}{T^2}\left(t-m\frac{T}{N}\right)^2}e^{-j\pi\frac{N}{T^2}\left(t-k\frac{T}{N}\right)^2}dt = \delta(m-k) \tag{7.44}$$

At the transmitting end, the OCDM waveform is expressed as

$$s\left(t\right) = \sum_{k=0}^{N-1} x\left(k\right)\psi_k\left(t\right), 0 \le t \le T \tag{7.45}$$

At the receiving end, the matched filter is used to extract the mth OCDM signal, and the result can be expressed as

$$\begin{aligned} x'\left(m\right) &= \int_0^T s\left(t\right)\psi_m^*\left(t\right) \\ &= \sum_{k=0}^{N-1} x\left(k\right)\delta(m-k) = x\left(m\right) \end{aligned} \tag{7.46}$$

Figure 7.21 shows the block diagram of the transmitter and receiver of the OCDM system. As shown in the figure, the difference between the receiving and transmitting process of the OCDM system and the OFDM system is that OCDM uses the inverse discrete Fresnel transform (IDFnT) at the transmitter and the discrete Fresnel transform (DFnT) at the receiver. The discrete Fresnel (inverse) transform can be obtained by simply multiplying the coefficients of the discrete Fourier (inverse) transform. Therefore, the OCDM system can be obtained on the

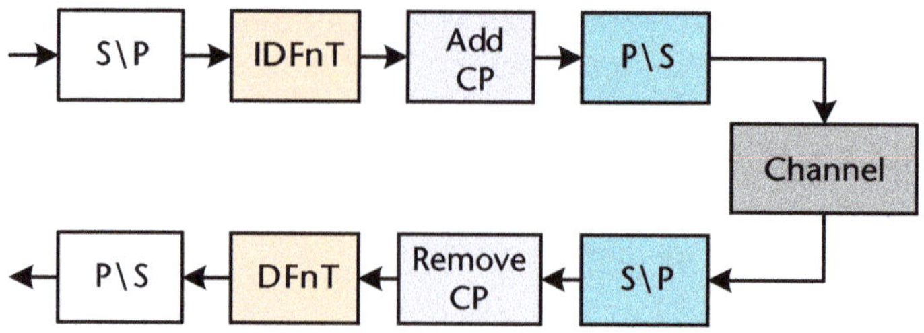

Figure 7.21 OCDM receiving and transmitting process block diagram.

basis of the OFDM system, and the complexity of the two is the same, which greatly reduces the difficulty of implementing the OCDM system.

The OCDM system can obtain the transmitted signal $s(n)$ by sampling the analog signal $s(t)$:

$$\begin{aligned} s(n) &= F_{\psi}^{-1}\{x(k)\} \\ &= \begin{cases} \sum_{k=0}^{N-1} x(k)\psi_k\left(n\frac{T}{N}\right) & N \equiv 0(\bmod 2) \\ \sum_{k=0}^{N-1} x(k)\psi_k\left(n\frac{T}{N} + \frac{T}{2N}\right) & N \equiv 1(\bmod 2) \end{cases} \\ &= e^{j\frac{\pi}{4}} \sum_{k=0}^{N-1} x(k) \times \begin{cases} e^{-j\frac{\pi}{N}(n-k)^2} & N \equiv 0(\bmod 2) \\ e^{-j\frac{\pi}{N}\left(n-k+\frac{1}{2}\right)^2} & N \equiv 1(\bmod 2) \end{cases} \end{aligned} \tag{7.47}$$

where the expression of IDFnT is slightly different due to the difference in the parity of N. The matrix form of OCDM modulation is as follows:

$$s = \phi^H x \tag{7.48}$$

where $s = [s(0), s(1), \ldots, s(N-1)]^T$ is the discrete time domain signal after OCDM modulation; $x = [x(0), x(1), \ldots, x(N-1)]^T$ is the symbol vector after mapping and ϕ is the IDFnT matrix. The matrix ϕ is expressed as

$$\phi(m,n) = \frac{1}{\sqrt{N}} e^{-j\frac{\pi}{4}} \times \begin{cases} e^{j\frac{\pi}{N}(m-n)^2} & N \equiv 0\,(\bmod 2) \\ e^{j\frac{\pi}{N}\left(m+\frac{1}{2}-n\right)^2} & N \equiv 1\,(\bmod 2) \end{cases} \tag{7.49}$$

Since the DFnT matrix is a unitary matrix, at the receiver the DFnT matrix can be inverted to recover the transmitted symbols:

$$x' = \phi s = x \tag{7.50}$$

The advantages of OCDM are as follows: (1) OCDM is more robust than OFDM when the guard interval is insufficient, and (2) OCDM is proven to have significantly better performance than OFDM in the presence of NBI and TBI. The disadvantages of OCDM is that its performance can be further improved compared with the OTFS scheme.

7.4.7 AFDM Multicarrier Modulation

AFDM is also a new multicarrier modulation scheme with orthogonal chirp signals as subcarriers. By setting the discrete affine Fourier transform (DAFT) parameters of AFDM, the overlap of time domain channel paths with significant delay or Doppler shift in the DAFT domain can be avoided, effectively overcoming high Doppler shift [22]. The DAFT domain impulse response conveys the complete delay-Doppler representation of the channel, so AFDM can achieve full diversity of linear time-varying channels.

AFDM is a new type of multichirp waveform generated and demodulated using DAFT. Figure 7.22 shows the modulation and demodulation architecture of the AFDM system. As shown in the figure, the AFDM system uses the inverse discrete affine Fourier transform (IDAFT) at the transmitter and DAFT at the receiver. DAFT can be obtained by multiplying DFT with a matrix, and using an OFDM modulator/demodulator as the core, AFDM modulation/demodulation can be effectively implemented.

Let **x** represent N QAM modulated symbol vectors, mapped to the time domain by IDAFT:

$$
\begin{aligned}
s_n &= \sum_{m=0}^{N-1} x_m \phi_n(m), n = 0, \cdots, N-1 \\
&= \sum_{m=0}^{N-1} x_m \frac{1}{\sqrt{N}} e^{j2\pi\left(c_1 n^2 + c_2 m^2 + nm/N\right)}
\end{aligned}
\tag{7.51}
$$

where c_1 and c_2 are the parameters of AFDM. When the parameters c_1 and c_2 are set reasonably, AFDM can achieve full diversity of dual-dispersion channels. OFDM and OCDM are special cases of $c_1 = 0, c_2 = 0$ and $c_1 = \frac{1}{2M}, c_2 = \frac{1}{2M}$, respectively. Its matrix form can be expressed as

$$
s = \Lambda_{c_1}^H F^H \Lambda_{c_2}^H x \tag{7.52}
$$

where F is the DFT matrix item, $\Lambda_c = diag(e^{-j2\pi cn^2}, n = 0, 1, \cdots, N-1)$.

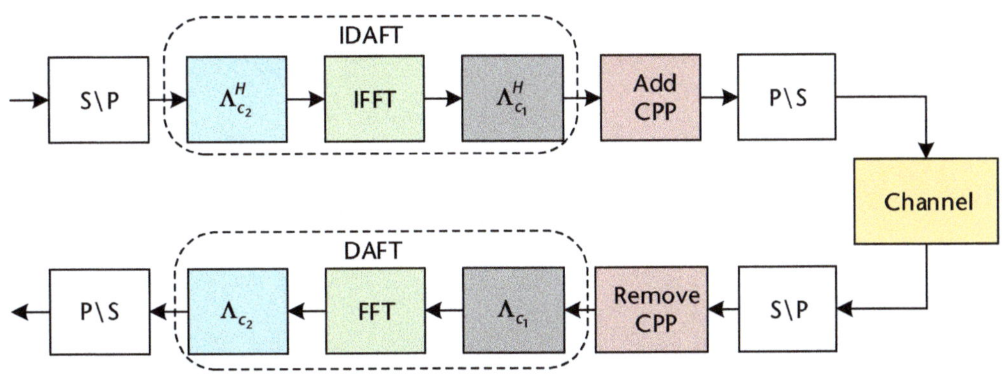

Figure 7.22 AFDM modem architecture.

To combat the multipath effect, AFDM needs to design a prefix to put the signal in a periodic domain. However, the AFDM signal period is different from that of OFDM, so a chirp-periodic prefix (CPP) is used. The prefix is

$$s_n = s_{N+n}e^{-j2\pi c_1\left(N^2+2Nn\right)}, n = -L, \cdots, -1 \tag{7.53}$$

When $2Nc_1$ is an integer and N is an even number. After parallel-to-serial conversion, the signal reaches the receiving end through the time-varying channel. The received signal is

$$r_n = \sum_{l=0}^{\infty} s_{n-l}g_n\left(l\right) + w_n \tag{7.54}$$

where w_n is additive white Gaussian noise, $g_n\left(l\right)$ is the channel impulse response at time n and delay l, and the channel impulse response is

$$g_n\left(l\right) = \sum_{i=1}^{P} h_i e^{-j2\pi f_i n}\delta\left(l - l_i\right) \tag{7.55}$$

Each delay division can have a Doppler frequency spread, as long as different paths are allowed to have the same delay but different Doppler shifts. $\upsilon_i \triangleq Nf_i = \alpha_i + a_i \in \left[-\upsilon_{\max}, \upsilon_{\max}\right]$ is the Doppler shift normalized to the subcarrier spacing, $\alpha_i \in \left[-\alpha_{\max}, \alpha_{\max}\right]$ is the integer part, and a_i is the fractional part and satisfies $\frac{-1}{2} < a_i \leq \frac{1}{2}$.

After removing CPP, the matrix form of the received signal is expressed as

$$r = Hs + w \tag{7.56}$$

where $H = \sum\limits_{i=1}^{P} h_i\Gamma_{CPP_i}\Delta_{f_i}\Pi^{l_i}$, $\Delta_{f_i} \triangleq diag\left(e^{-j2\pi f_i n}, n = 0, \ldots, N-1\right)$, Π is the forward cyclic shift matrix, and Γ_{CPP_i} is

$$\Gamma_{CPP_i} \triangleq diag\left(\begin{cases} e^{-j2\pi c_1\left(N^2-2N\left(l_i-n\right)\right)}, n < l_i \\ 1, n \geq l_i \end{cases}, n = 0, \ldots N-1\right) \tag{7.57}$$

At the receiver, the DAFT domain output symbol is represented as

$$y_m = \sum_{n=0}^{N-1} r_n\phi_n{}^*\left(m\right) \tag{7.58}$$

The matrix form of the output signal is expressed as

$$y = \Lambda_{c_2}F\Lambda_{c_1}r = H_{eff}x + \tilde{w} \tag{7.59}$$

where $H_{eff} = \Lambda_{c_2}F\Lambda_{c_1}H\Lambda_{c_1}^H F^H\Lambda_{c_2}^H$, $\tilde{w} = \Lambda_{c_2}F\Lambda_{c_1}w$ and $w \sim CN\left(0, N_0 I\right)$. $\Lambda_{c_2}F\Lambda_{c_1}$ is a unitary matrix, then $\tilde{w}$ and w have the same covariance.

Define $H_i \triangleq \Lambda_{c_2} F \Lambda_{c_1} \Gamma_{CPP_i} \Delta_{f_i} \Pi^{l_i} \Lambda_{c_1}^H F^H \Lambda_{c_2}^H$, then the output can be rewritten as

$$y = \sum_{i=1}^{P} h_i H_i x + \tilde{w} \tag{7.60}$$

In order to make the DAFT domain impulse response constitute a complete delay-Doppler representation of the channel, the corresponding parameters c_1 and c_2 should be set. For the only nonzero item in each row of path i, it should not coincide with the position of the only nonzero item in the same row of H_j, and the loc_i trajectory range is $-\alpha_{\max} + 2Nc_1 l_i < loc_i \leq \alpha_{\max} + 2Nc_1 l_i$. Therefore, in order to make the nonzero items of H_i and H_j not overlap, the intersection of the corresponding ranges of loc_i and loc_j is empty; that is,

$$\{-\alpha_{\max} + 2Nc_1 l_i, \ldots, \alpha_{\max} + 2Nc_1 l_i\} \cap \{-\alpha_{\max} + 2Nc_1 l_j, \ldots, \alpha_{\max} + 2Nc_1 l_j\} = \emptyset \tag{7.61}$$

If two paths have the same time delay but different Doppler shifts, they always occupy two different positions in the DAFT domain. For paths with different time delays $(l_i \neq l_j)$, assuming $l_j > l_i$, the above formula is equivalent to

$$2Nc_1 > \frac{2\alpha_{\max}}{l_j - l_i} \tag{7.62}$$

If the time domain impulse response of the channel does not have sparsity, the minimum value of $l_j - l_i$ is 1, and c_1 should satisfy

$$c_1 = \frac{2\alpha_{\max} + 1}{2N} \tag{7.63}$$

When $2\alpha_{\max} l_{\max} + 2\alpha_{\max} + l_{\max} < N$ is satisfied in the case of integer Doppler shift, the channel paths with different delays or different Doppler shifts are separated in the DAFT domain, so that H_{eff} has the delay-Doppler representation of the channel in the DAFT domain. When $c_1 = \frac{2\alpha_{\max}+1}{2N}$, and c_2 is set to any irrational number or a rational number sufficiently smaller than $\frac{1}{2N}$, AFDM can achieve full diversity of the LTV channel [22]. This parameter setting results in subcarrier $\phi_n(m) = \frac{1}{\sqrt{N}} e^{j2\pi\left(c_1 n^2 + c_2 m^2 + nm/N\right)}$ having a time-frequency content that is different from all existing waveforms to date.

The advantages of AFDM are as follows:

1. Under the same time and frequency resource occupation, the complexity of AFDM will be far lower than that of OTFS.
2. When considering channel estimation, due to the two-dimensional structure of the underlying transform of OTFS, its pilot overhead is twice that of AFDM. This difference translates into a significant advantage of AFDM over OTFS in terms of spectral efficiency.
3. In time-varying channels, AFDM and OTFS have almost the same bit error rate performance because both AFDM and OTFS implement full-delay

Doppler representation of the channel. Each path in the path of the effective channel (in the DAFT domain of AFDM, in the delayed Doppler domain of OTFS) corresponds to a delay-tap Doppler bin pair of the wireless propagation channel, and each path in AFDM and OTFS have the same path gain (at least strictly equal in the case of integer-valued Doppler shifts). This feature enables AFDM to achieve the optimal diversity order of the LTV channel.

The disadvantages of AFDM are as follows:

1. AFDM and OFDM have the same high PAPR.
2. Research on MIMO technology is a conventional method to improve the system diversity order. Transmit diversity of MIMO-AFDM system [19] is challenging.
3. Channel estimation and signal detection for AFDM systems is challenging. How to design low-complexity channel estimation and signal detection technology deserves further research.

7.4.8 Performance Analysis

Figure 7.23 compares the performance of GAM with the proposed mapping method to that of PSK, APSK, and QAM with Gray mapping for $M = 64$ and $M = 256$. Obviously, the GAM scheme with the proposed mapping principle is capable of providing significant performance gain over the PSK, APSK schemes and offers comparable performance to QAM with Gray mapping. It should be noted that the GAM constellation is not yet optimized, and we would expect better performance after performing constellation shaping design.

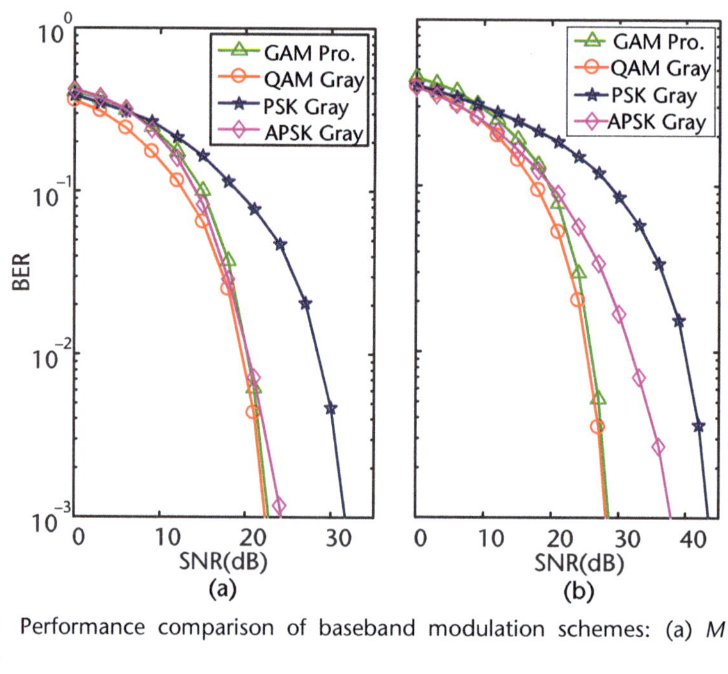

Figure 7.23 Performance comparison of baseband modulation schemes: (a) $M = 64$, and (b) $M = 256$.

Figure 7.24 compares the PAPR of potential modulation waveforms. It is obvious that the PAPR performance of SC modulation is significantly better than that of MC modulation schemes. In the MC schemes, OTFS scheme exhibits better PAPR performance over other MC alternatives. To provide further insights, Figure 7.25

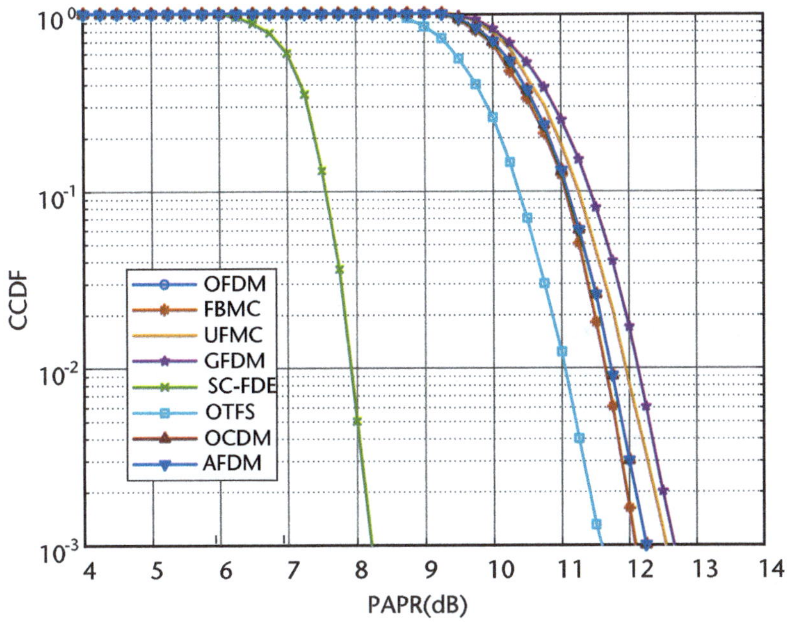

Figure 7.24 PAPR comparison of potential modulation schemes.

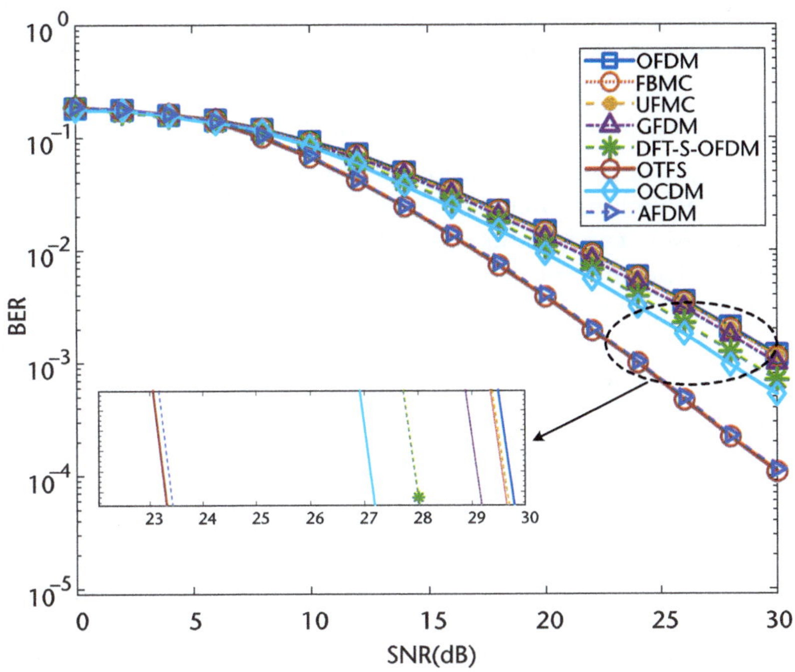

Figure 7.25 Performance comparison of potential carrier waveforms with LEO satellite communication channel.

compares the performance of OFDM, DFT-S-OFDM, FBMC, UFMC, GFDM, OTFS, OCDM, and AFDM schemes relying on a message-passing detection algorithm in the LEO satellite communication scenario. In the simulation, the carrier center frequency is set as 30 GHz, the LEO satellite velocity is 7.9 km/s, the Doppler frequency shift is 790 KHz, the subcarrier spacing is 15 KHz, the baseband modulation is 16-QAM, the channel fading distribution is assumed as a Rice distribution, the elements of channel noise obey Gaussian distribution noise, and the number of propagation paths is 3. It is observed from Figure 7.25 that the performance of UFMC, FBMC, and GFDM schemes is comparable to that of OFDM counterparts. Owing to its capability for partial time diversity, DFT-S-OFDM and OCDM schemes exhibit slightly better performance. Moreover, benefiting from the full diversity in the doubly dispersive channels, OTFS and AFDM perform best, constituting appealing candidates for LEO satellite communications.

7.5 Design Guidelines

7.5.1 Irregular Constellation Configuration Design

Higher-order modulation is capable of improving spectral efficiency by increasing the number of bits carried by each symbol. Constellation configuration plays a vital role in higher-order modulation in terms of PAPR and BER performance. In view of the sensitivity of the nonlinear distortion of the existing regular constellations, it is necessary to optimize the amplitude as well as phase of irregular constellations. First, the influence of channel fading and Gaussian noise on the construction of constellations has to be explored. Second, it is important to derive the analytical expression of PAPR associated with the amplitude and phase of the designed constellation. Subsequently, the amplitude and phase can be optimized by maximizing channel mutual information under the constraint of low PAPR value.

7.5.2 Integrated Coding and Modulation

The integrated code-modulation design is capable of significantly simplifying the transceiver with the potential of breaking through the performance bottleneck of existing disjoint architectures, which is eminently suitable for LEO satellite communication. The inherent flexibility of irregular constellation modulation constitutes an appealing alternative for the integration design. However, the research on irregular constellation modulation is still in its infancy, and the integrated design with coding is an interesting open research problem. Explicitly, it is worth exploring the intricate interplay between coding and modulation. Understanding this intrinsic mechanism helps us fully exploit the effect of the channel polarization on coding and modulation. This is illustrated in Figure 7.26, showing that both irregular constellation design and information mapping method play crucial roles in integrated coding and modulation.

7.5.3 Versatile Carrier Waveform Design

The current satellite systems of communication, navigation, and remote sensing have been developed independently, each serving a single function. This fragmented

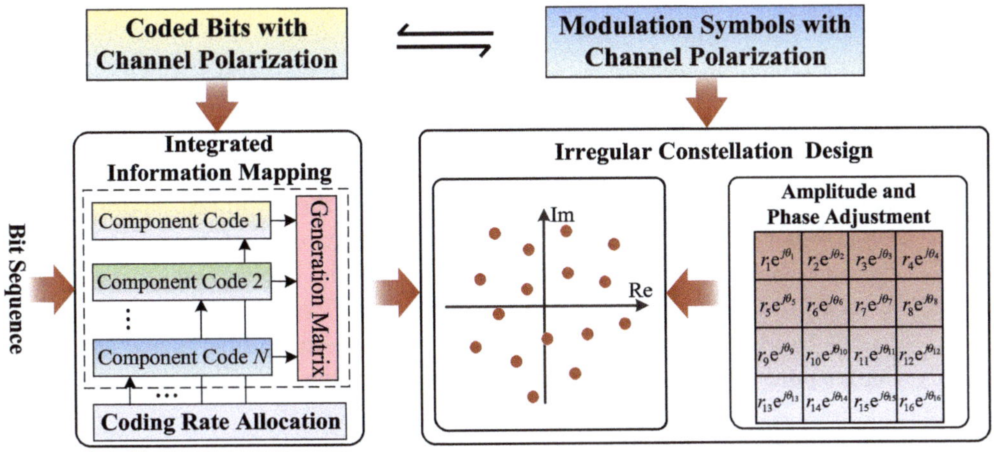

Figure 7.26 Integrated design of coding and modulation.

approach makes it impossible to achieve integrated navigation-communication-remote sensing functions, resulting in underutilization of valuable satellite orbit resources. With the integration of communication and navigation, communication signals are helpful for extending the coverage of navigation signals to achieve high-precision positioning, while navigation signals are able to assist in synchronizing and simplifying signal detection [23]. Meanwhile, the integration of communication and sensing has the potential to target signal tracking, which is beneficial for detecting and tracking objects in space to provide safe and reliable communications [24]. With the ever-growing demands of users in the future, the versatile carrier waveform design merits further exploration.

7.5.4 AI-Aided Adaptive Waveform

As mentioned earlier the LEO satellite communication channel has the characteristics of high mobility, multipath effect, and rain fading, which significantly degrade

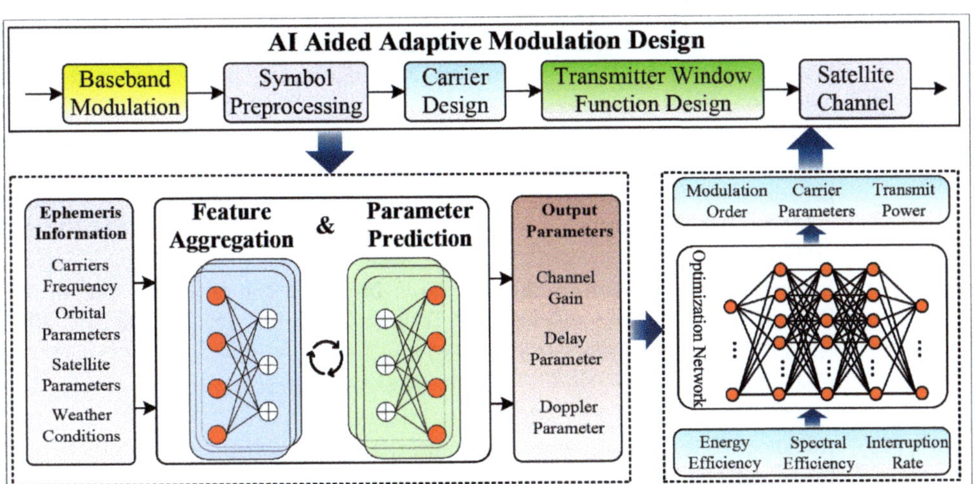

Figure 7.27 AI-aided adaptive waveform design.

reliable communication [25]. Adaptive modulation is a prominent approach to address those issues by adjusting waveform parameters. However, the existing adaptive modulation needs to take into account the multidimensional feature parameters including considerable delay as well as strong Doppler shift, giving rise to limited communication capacity. With the development of artificial intelligence technology, AI-aided adaptive modulation has the potential to improve the capacity of LEO satellite communications, as illustrated in Figure 7.27. Specifically, a high-precision channel prediction network model can be established based on deep reinforcement learning to achieve accurate prediction of channel parameters. Based on the estimated channel parameters, the system coding rate, modulation order, carrier parameter, and transmit power can be dynamically adjusted.

References

[1] Divsalar, D., and M. K. Simon, "Multiple-Symbol Differential Detection of MPSK," *IEEE Transactions on Communications*, Vol. 38, No. 3, March 1990, pp. 300–308.

[2] Pfau, T., S. Hoffmann, and R. Noe, "Hardware-Efficient Coherent Digital Receiver Concept with Feedforward Carrier Recovery for M-QAM Constellations," *Journal of Lightwave Technology*, Vol. 27, No. 8, April 2009, pp. 989–999.

[3] Gokceli, S., I. Peruga, E. Tiirola, K. Pajukoski, T. Riihonen, and M. Valkama, "Novel Tone Reservation Method for DFT-s-OFDM," *IEEE Wireless Communication Letters*, Vol. 10, No. 10, October 2021, pp. 2130–2134.

[4] Ghais, A., G. Berzins, and D. Wright, "INMARSAT and the Future of Mobile Satellite Services," *IEEE Journal on Selected Areas in Communications*, Vol. 5, No. 4, May 1987, pp. 592–600.

[5] Huemer, M., H. Witsching, and J. Hausner, "Unique Word Based Phase Tracking Algorithms for SC-FDE Systems," *IEEE Globecom*, August 2003, pp. 70–74.

[6] Gokceli, S., I. Peruga, E. Tiirola, K. Pajukoski, T. Riihonen, and M. Valkama, "Novel Tone Reservation Method for DFT-s-OFDM," *IEEE Wireless Communications Letters*, Vol. 10, No. 10, October 2021, pp. 2130–2134.

[7] Zaidi, A. A., R. Baldemair, H. Tullberg, et al., "Waveform and Numerology to Support 5G Services and Requirements," *IEEE Communications Magazine*, Vol. 54, No. 11, November 2016, pp. 90–98.

[8] European Telecommunications Standards Institute, Digital Video Broadcasting (DVB): Framing Structure, Channel Coding and Modulation for Digital Terrestrial Communication Television, 1999.

[9] European Telecommunications Standards Institute, Digital Video Broadcasting (DVB)-Second Generation Framing Structure, Channel Coding and Modulation Systems for Broadcasting, Interactive Services, News Gathering and Other Broadband Satellite Applications (DVB-s2), European Standard (Telecommunications Series) EN 302 307, V1. 2.1, 2009.

[10] European Telecommunications Standards Institute, Digital Video Broadcasting (DVB); DVB; Implementation Guidelines[R], Technical Report, ETSI TR 102 377, 2005.

[11] Xiao, Z., J. Yang, T. Mao, et al., "LEO Satellite Access Network (LEO-SAN) Toward 6G: Challenges and Approaches," *IEEE Wireless Communications*, Vol. 31, No. 2, April 2024, pp. 89–96.

[12] Advanced Televisions Systems Committee, Final Report on ATSC 3.0: Next Generation Broadcast Television, ATSC PT2-046r11 Final Report, 2011.

[13] Katoh, H., "Transmission System for ISDB-S," *Proceedings of the IEEE*, Vol. 94, No. 1, January 2006, pp. 289–295.

[14] Park, S. J., "Triangular Quadrature Amplitude Modulation," *IEEE Communications Letters*, Vol. 11, No. 4, April 2007, pp. 292–294.

[15] Xiao, L., X. Zhai, Y. Liu, G. Liu, P. Xiao, and T. Jiang, "A Unified Bit-to-Symbol Mapping for Generalized Constellation Modulation," *China Communications*, Vol. 20, No. 6, June 2023, pp. 229–239.

[16] Chen, D., W. Wang, and T. Jiang, "New Multicarrier Modulation for Satellite-Ground Transmission in Space Information Networks," *IEEE Network*, Vol. 34, No. 1, January 2020, pp. 101–107.

[17] Buzzi, S., C. D. Andrea, D. Li, and S. Feng, "MIMO-UFMC Transceiver Schemes for Millimeter-Wave Wireless Communications," *IEEE Transactions on Communications*, Vol. 67, No. 5, May 2019, pp. 3323–3336.

[18] Michailow, N., M. Matthe, I. S. Gaspar, et al., "Generalized Frequency Division Multiplexing for 5th Generation Cellular Networks," *IEEE Transactions on Communications*, Vol. 62, No. 9, September 2014, pp. 3045–3061.

[19] Xiao, L., S. Li, Y. Qian, D. Chen, and T. Jiang, "An Overview of OTFS for Internet of Things: Concepts, Benefits, and Challenges," *IEEE Internet of Things Journal*, Vol. 9, No. 10, May 2022, pp. 7596–7618.

[20] Shi, J., Z. Li, J. Hu, et al., "OTFS Enabled LEO Satellite Communications: A Promising Solution to Severe Doppler Effects," *IEEE Network*, Vol. 38, No. 1, January 2024, pp. 203–209.

[21] Ouyang, X., and J. Zhao, "Orthogonal Chirp Division Multiplexing," *IEEE Transactions on Communications*, Vol. 64, No. 9, September 2016, pp. 3946–3957.

[22] Zhu, J., Q. Luo, G. Chen, P. Xiao, and L. Xiao, "Design and Performance Analysis of Index Modulation Empowered AFDM System," *IEEE Wireless Communication Letters*, Vol. 13, No. 3, March 2024, pp. 686–690.

[23] Wei, Q., X. Chen, C. Jiang, and Z. Huang, "Time-of-Arrival Estimation for Integrated Satellite Navigation and Communication Signals," *IEEE Transactions on Wireless Communications*, Vol. 22, No. 12, December 2023, pp. 9867–9880.

[24] Yang, J., D. Li, X. Jiang, S. Chen, and L. Hanzo, "Enhancing the Resilience of Low Earth Orbit Remote Sensing Satellite Networks," *IEEE Network*, Vol. 34, No. 4, July 2020, pp. 304–311.

[25] Jiang, P., T. Wang, B. Han, et al., "AI-Aided Online Adaptive OFDM Receiver: Design and Experimental Results," *IEEE Transactions on Wireless Communications*, Vol. 20, No. 11, November 2021, pp. 7655–7668.

CHAPTER 8

Multiantenna Technique for Satellite-Terrestrial Integrated Communication

Multiantenna technology has been widely used in terrestrial communications [9–11], but in satellite communications, the limited payload capacity of satellites limits the application of multiantenna technology. The development of multiantenna technology is crucial to the future integration of satellite-terrestrial communications to meet the growing diversified communication needs. This chapter briefly introduces antenna technology and then discusses multiantenna technology in satellite communications from the user, feed, and intersatellite links.

8.1 Antenna Technology Introduction

8.1.1 Satellite Antenna Classification

With the rapid growth in demand for communication services, high throughput satellite systems in GEO have become a hot research topic in space communication technology. In this system, onboard antennas are critical components, and the multibeam approach effectively addresses the problems caused by demand growth. In addition, multibeam antennas (MBAs) offer remarkable flexibility and strong resistance to interference in processes such as beamforming, beam reconstruction, and beam scanning [12]. As a result, they are also widely used in medium-/low-Earth orbit (MEO/LEO) communications satellites. Table 8.1 shows the different multibeam antenna configurations used by communications satellites in different orbits.

Currently, multibeam antennas mainly include satellite phased array multibeam antennas, satellite reflector multibeam antennas, and satellite lens-type multibeam antennas.

8.1.1.1 Spaceborne Reflector Multibeam Antennas

To enable simultaneous multibeam coverage with increased power on high-orbit satellites, it is typically essential to use a larger aperture antenna. When compared to phased array and lens antennas, reflector antennas offer benefits such as being lightweight, having a straightforward design, and utilizing established technology. They represent the most effective solution for achieving high gain, low sidelobe levels, and multibeam configurations. Therefore, multibeam reflector antennas have been widely used in the launched high-flux satellite systems. Spaceborne reflector multibeam antennas can be broadly divided into two categories: single feed per beam (SFB) and multiple feed per beam (MFB).

Table 8.1 Multibeam Antenna Configuration for Different Orbital Communication Satellites

Satellite Orbit	*Name of Satellite Constellation*	*Antenna Scheme*
IGEO	Inmarsat-4/5,MUOS,Thuraya-2/-3,DBSD-G1, SkyTerra-1/-2,Alphasat-I-XL,TeereStar/-1/-2, MEXSAT-1/-2/-3	Single aperture large spread reflector antenna
GEO	DireCTV-14/-15,EUTELSAT-65,WestA,ABS-2/-3A, Eutelsat-3B,AsiaSat-6/-8,MEXSAT-3b, Express-AM5/-AM7,Amos-3/-4,Intelsat-19/- 22, SATMEX-7,Astra-2E/-5B,YahSat-1A/-1B	Multi-aperture reflector antenna
GEO	WINDS,WGS,AEHF,Space-way3	Phased array antenna
MEO	03b,ICO	Reflector antenna
LEO	Iridium-NEXT,Globalstar-1/-2,Orbcomm2, Oneweb,Telesat,Starlink	Phased array antenna

The reflector antenna is the simplest structure, the lightest weight, and has been around long enough so that the technology is more mature. The multibeam reflector antenna is even simpler and ensures the system's performance in a scenario where there is no need for a large number of spot beams generated by the spaceborne antenna.

In SFB implementation mode each feed reflector radiates from a specific antenna aperture, has high radiation efficiency, and can realize transceiver sharing. However, because each feed requires a separate reflector, the total number of reflectors is large, the multicolor multiplexing requires a larger space, and the beam directivity is relatively poor.

SFB is one of the most direct and straightforward beamforming methods; when a feed hits the reflector, a beam is formed in each feed array. The beams are spaced apart because the feed aperture is more significant in the same antenna. The seamless coverage of high gain, low sidelobe, and multiple beams is achieved through the synergy of multiple antennas, meeting the requirements of high flux satellites. This type of antenna has good performance, but the number of reflecting surfaces will occupy the satellite's limited space, and the antenna's installation accuracy and the pointing accuracy in orbit are more demanding.

The implementation scheme of MFB is to generate multiple spot beams using the wave-shaped network to excite the phase and amplitude of uniformly arranged arrays. The MFB scheme can flexibly adjust the shape and number of beam generation, and only two reflectors are needed to achieve transmission and reception. The beam directionality is better than that of the SFB scheme. However, if the MFB scheme wants to form coverage with the SFB scheme and spot beam, it needs far more feeding units than the SFB scheme, and the feeding network is very complex.

The majority of GEO mobile communication satellites utilize L/S band reflector antennas. The L/S band frequency is low, the wavelength is relatively long, the size of the feed array is significant, and MFB scheme beamforming is employed to meet the requirements of power distribution, phase tracking, direction diagram, and other technologies. As mobile satellite communication has developed to the high-frequency band, the Ku/Ka frequency band has been developed. The high frequency and short wavelength of the Ku/Ka band permit a significant reduction in the volume of the feed array, which is well suited to SFB and MFB beamforming schemes.

8.1.1.2 Spaceborne Phased Array Multibeam Antenna

The spaceborne phased array antennas can modify the amplitude and phase of each array element, allowing for dynamic beam adjustment, which includes controlling the beam size and shape, thereby facilitating beam scanning. Due to their lower profile, phased array antennas are better suited for satellite launches compared to reflector antennas. Phased array antennas can be classified into two categories: passive phased array antennas and active phased array antennas. In passive phased array antennas, the high-frequency energy is generated by a central transmitter and distributed to the various radiation units. The receiver uniformly amplifies the reflected signal. In an active phased array antenna, each array element is equipped with an independent transmit/receive (T/R) module, which enables the antenna to perform independent signal-sending and receiving functions.

The phased array antenna is capable of modifying the beam shape, scanning the beam, and distributing the power between the beams by adjusting the phase and amplitude. Incorporating adaptive zeroing technology into anti-interference measures can significantly enhance the space survivability of communication satellites. Furthermore, compared to reflector antennas, phased array antennas exhibit a low profile, which is advantageous for satellite launches. Phased array antennas can be categorized into two distinct types: passive phased array and active phased array.

The passive phased array antenna consists of a central transmitter and a receiver. The high-frequency energy generated by the transmitter is automatically distributed by the computer to each radiation unit of the antenna and uniformly amplified by the receiver, as shown in Figure 8.1. The early low-frequency passive phased array is not suitable for satellite antennae because of the large size, high loss, and high cost of the beamforming network. However, for the Ka-band that is currently being actively developed, the passive scheme is usually used for large-scale array antennas because the passive array element structure is relatively simple and has an easy-to-scale layout. For example, the Ka-band passive phased array antenna on the Spaceway3 satellite in the United States has more than 1,200 array elements, can generate 24 beams simultaneously, and can generate 784 variable beams.

Each radiating unit of an active phased array antenna has a separate T/R module, as shown in Figure 8.2. Each unit is capable of generating and receiving

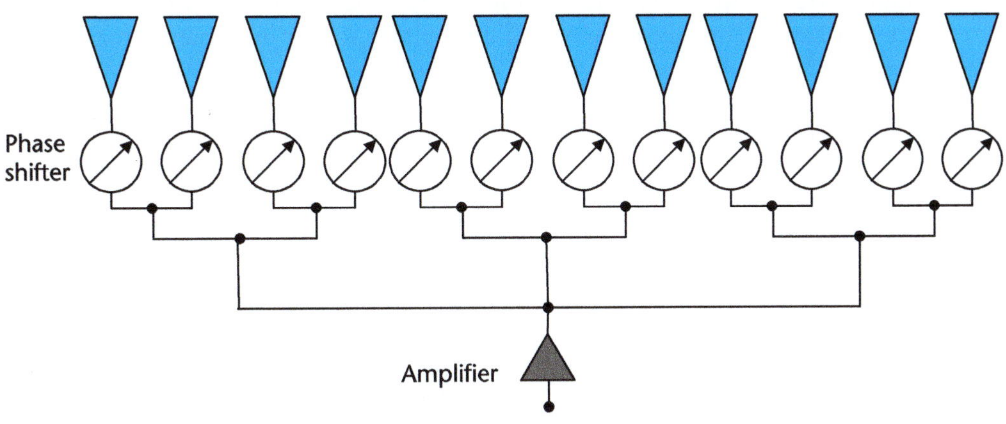

Figure 8.1 The passive phased array.

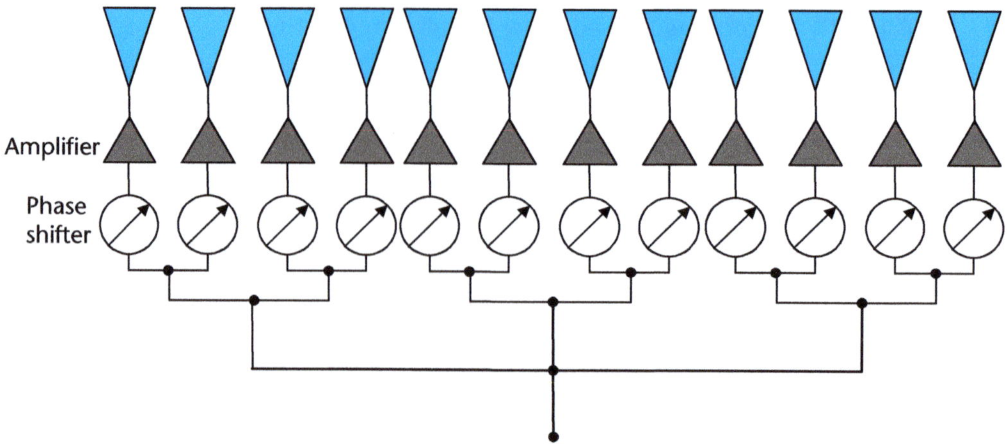

Figure 8.2 The active phased array.

electromagnetic waves independently. If a small number of T/R modules fail, the performance of the phased array antennas is not significantly affected. In addition, the active phased array antennas offer significant advantages over passive ones in aspects such as bandwidth, signal processing, and redundancy design. Consequently, spaceborne multibeam phased array antennas operating in the L/S/X frequency bands typically utilize the active array mode, as illustrated in Table 8.2. However, in the Ka-band, it is not easy to design and calibrate the active array's digital phase shifter and attenuator. In addition, the miniaturization design of active devices is not mature; the power loss is severe, and there are few active cases on board; even if the active array is used, the number of array elements and the beam size are relatively limited. For example, the two-phased array antennas on the Japanese WINDS satellite, each consisting of 128 units, can only generate two beams.

The early L/S/X-band satellite communication used a lower frequency band, analog phased burst interval is proportional to the wavelength to bring the problem of too large volume, active phased array bandwidth is more considerable, signal processing capacity is more robust, and the designed redundancy is lower, and live low-band satellite communication system. However, at present, the satellite communication frequency band is gradually developing into the high-frequency band of the Ka-band. The phased array's digital phase shifter and attenuator are difficult to

Table 8.2 The Application of Phased Array Multibeam Antennas in Satellite Communication

Satellite Constellation	*Year of Launch*	*Frequency*	*Passive/Active*	*Number of Beams*
Telesta	2023	Ka	active	16
Oneweb	2019	Ku,Ka	active	16
Starlink	2018	Ku,Ka	active	16
Iridium-NEXT	2015	L	active	48
WGS-5/-6	2013	X	active	4
Globalstar-2	2010	S/L	active	16
WINDS	2008	Ka	active	2
Spaceway3	2007	Ka	passive	784

calibrate. The power loss is significant, while the passive phased array has a simple structure of the turning point, which makes its application more convenient.

8.1.1.3 Spaceborne Lens Multibeam Antenna

The lens antenna is a technique that applies the principles of geometric optics to the radio frequency range. Compared with the reflector antenna, the multibeam lens antenna has greater design freedom, good rotational symmetry, retains excellent optical characteristics, and there is no need to worry about aperture occlusion. However, such antennas have serious disadvantages, such as heavy weight and a significant loss in low-frequency bands, so they are limited in satellite applications. As the research frequency band is gradually promoted to millimeter waves and submillimeter waves, the shortening of the wavelength offers hope for developing miniaturized lens antennas, so European countries will also pay more attention to this field.

In recent years, the European Space Agency (ESA) has been developing discrete lens antennas. A Ka-band active discrete lens antenna has been constructed. The antenna array employs an active lens with a solid-state microwave amplifier and an aperiodic arrangement. However, the test results indicate that the antenna's radiation efficiency is only approximately 15%, with the majority of the energy converted into heat, thereby posing a significant challenge in heat dissipation. Consequently, further enhancements are currently being developed. Additionally, given that the lenticular waveguide feed array is structurally more straightforward than a multimode feed network, researchers at ASTRIUM GmbH have also researched the lens feed in the Ka-band. However, both models exhibited a wide flat-top in terms of the main lobe, which resulted in a significant shortfall in radiation gain. Consequently, further research and improvement are necessary.

The lens multibeam antenna is more flexible in its design than the reflector antenna. The lenticular multibeam antenna is based on the principle of geometrical optics and exhibits rotational symmetry, thereby avoiding the aperture occlusion problem that afflicts other antenna types. Nevertheless, it shares the same drawbacks as the reflector antenna, namely a considerable size, weight, and loss of efficiency in the low-frequency band.

Under Fermat's principle, a convex lens can produce a plane wave when its refractive index, denoted by the symbol n, exceeds the value of 1. Conversely, if the refractive index of the lens material is situated between 0 and 1, or 1 and 0, such as in the case of metals, a concave lens is required. Furthermore, when the frequency of the electromagnetic radiation changes, the refractive index will also change rapidly, thus affecting the performance of the lens antenna. When the refractive index is n1, there is no dispersion phenomenon, and the lens has a large working bandwidth. Conversely, dispersion occurs when the refractive index is $0 < n < 1$, and the bandwidth is reduced.

As the frequency band used by the satellite is developed to operate at higher frequencies, the wavelength is gradually reduced, allowing for the lens antenna to be designed more compactly. The ESA has developed a lens antenna for the A-band that employs the active lens of the microwave amplifier as the array element. However, since most of the energy is converted to heat, the radiation efficiency is only approximately 15%.

8.1.2 Beamforming Technique

8.1.2.1 Full-Digital Beamforming

Full-digital beamforming is a beamforming technology based on the full-digital architecture, and the structure is shown in Figure 8.3. In this architecture, each antenna is connected to a radio frequency chain, which can realize the simultaneous adjustment of amplitude and phase. The system model can be expressed as

$$\mathbf{y} = \mathbf{H}\mathbf{F}_{BB}\mathbf{x} + \mathbf{n} \tag{8.1}$$

where $\mathbf{F}_{BB}$ is the full-digital precoder, $\mathbf{x}$ is the transmitted signal, $\mathbf{y}$ represents the received signal, and $\mathbf{n}$ represents the Gaussian white noise.

For the full-digital beamforming, each antenna oscillator recovers the amplitude and phase from the signal received by each oscillator in the antenna array by converting the RF signal into two binary baseband signal streams of cosine and sine. This technology aims to convert analog signals into digital signals accurately. Each antenna has its own transceiver and data converter, enabling the array to generate multiple beams simultaneously. This makes receiver matching a complex calibration process. This technique applies the amplitude/phase change to the digital signal at the transmitting end, and then digital-to-analog conversion is performed. An analog-to-digital converter and a digital downconverter process the received signals from each antenna in sequence. Nevertheless, in massive MIMO systems, the architecture necessitates a considerable number of RF chains due to the large number of antennas, resulting in high hardware complexity, cost and power consumption, rendering it impractical.

8.1.2.2 Analog Beamforming

Analog beamforming is a technique used for directional signal sending and receiving. It alters the signal amplitude and phase, which helps to adjust power requirements and rotate the beam in the desired direction. As 5G communication systems become established in the global market, the 6- to 100-GHz band, or millimeter wave, will become an integral part of mobile broadband. Concepts such as beamforming and the comparison of analog and digital are frequently mentioned

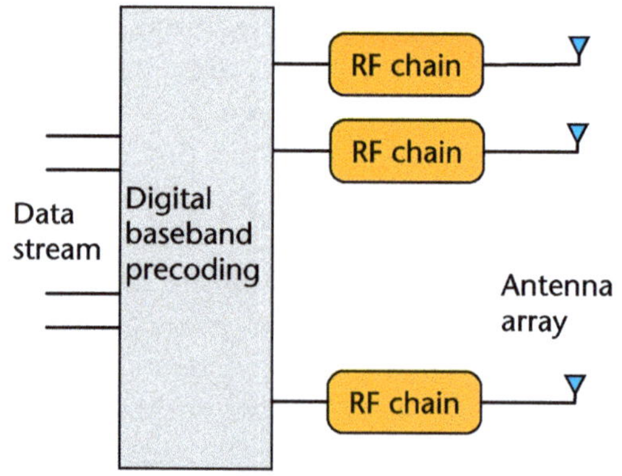

Figure 8.3 The full-digital architecture.

in discussions of next-generation technologies. In high-frequency millimeter wave transmission, the transmission distance is constrained by the substantial path loss during signal propagation.

For the analog beamforming, a single data stream is amplified by a phase shifter and transmitted to the receiving end. Analog beamforming technology represents the most cost-effective beamforming antenna array technology. However, it is essential to note that this technology has the disadvantage of being unable to manage and generate more than one signal beam.

The structure of the analog beamforming is shown in Figure 8.4, where a single RF chain uses a phase shifter to connect multiple antennas and generate a directional beam by changing the phase of each antenna. A common RF chain for all antenna arrays greatly reduces system complexity and power consumption. The system model for simulating beamforming can be expressed as

$$\mathbf{y} = \mathbf{H}\mathbf{F}_{RF}\mathbf{x} + \mathbf{n} \tag{8.2}$$

In analog beamforming, the single signal is amplified by each analog phase shifter and directed to the desired receiver before being fed to each antenna oscillator in the antenna array. The amplitude/phase change is applied to the analog signal at the transmitting end, which adds the signals from different antennas for analog-to-digital conversion. At present, the most cost-effective beamforming antenna array manufacturing technology is analog beamforming technology. However, this technology has the disadvantage that only one signal beam can be managed and generated.

8.1.2.3 Hybrid Beamforming

The traditional hybrid beamforming architecture is based on the phase shifter architecture, divided into two main categories: the fully connected (FC) architecture and the partially connected architecture. In this architecture, the phase shifter hardware is limited, which necessitates the quantification of the phase for analog beamforming.

As illustrated in Figure 8.5, the fully connected architecture comprises each antenna linked to the RF chain through an independent phase shifter, thereby

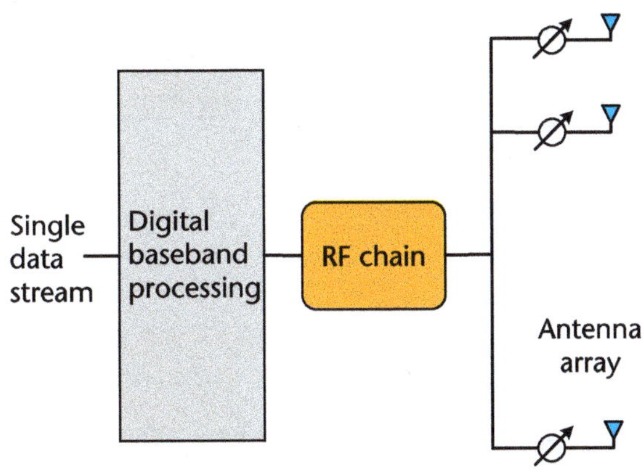

Figure 8.4 The analog beamforming architecture.

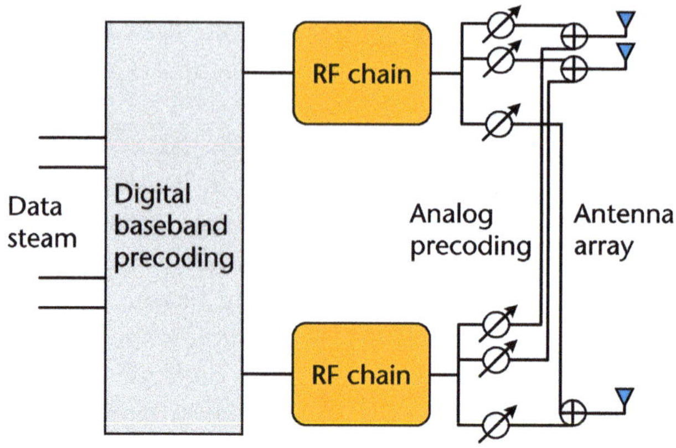

Figure 8.5 The fully connected architecture.

enabling the maximum gain for different transmitted data streams. Nevertheless, a fully connected architecture's intricacy and financial outlay are directly proportional to the number of antennas.

As shown in Figure 8.6, the antenna array of the partially connected architecture is divided into several independent subarrays, with each RF connected to one subarray. At this juncture, the number of phase shifters in the architecture equals the number of antennas. This reduction in the number of components reduces power consumption and cost but also results in a commensurate reduction in the performance gain of the architecture.

For hybrid beamforming, the transmitted signal is first processed by digital beamforming $\mathbf{F}_{BB}$ and then by analog beamforming $\mathbf{F}_{RF}$. The system model of the transmitter can be expressed as

$$\mathbf{y} = \mathbf{H}\mathbf{F}_{RF}\mathbf{F}_{BB}\mathbf{x} + \mathbf{n} \tag{8.3}$$

Similarly, the received signal is received through analog combiner $\mathbf{W}_{RF}$, and then through digital combiner $\mathbf{W}_{BB}$, and the system model can be expressed as

$$\hat{\mathbf{x}} = \mathbf{W}_{BB}^{H}\mathbf{W}_{RF}^{H}\mathbf{H}\mathbf{F}_{RF}\mathbf{F}_{BB}\mathbf{x} + \mathbf{W}_{BB}^{H}\mathbf{W}_{RF}^{H}\mathbf{n} \tag{8.4}$$

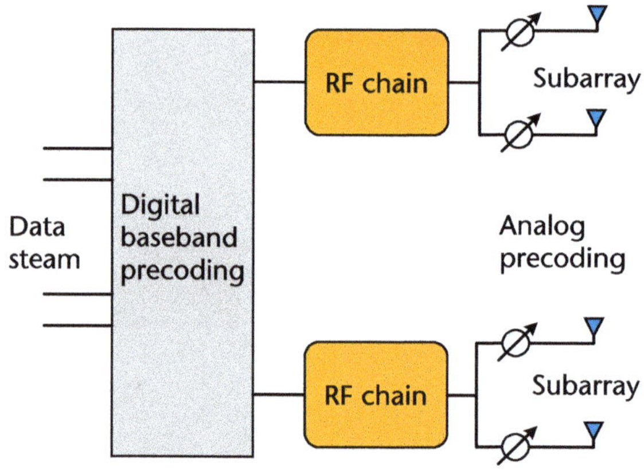

Figure 8.6 The partially connected architecture.

Furthermore, the hybrid precoding family based on phase shifters comprises a range of extended structures that can be employed to attain diverse design objectives, including enhancing energy efficiency, augmenting array gain, and the resolution of nonconvex constraints. These extended structures include hybrid connected structures, overlapping subarrays, and dual phase shifters. In the hybrid connected structure, each subarray can connect multiple RF chains. The fully connected structure and subarray structure represent exceptional cases of the hybrid connected structure. The overlapping subarray structure is based on the traditional one, but the antennas of different subarrays are permitted to overlap. This results in improved precoding gain, although this is accompanied by increased complexity. Finally, the dual phase shifter structure refers to the fact that each RF chain is connected to the antenna through two sets of phase shifters, with the arrangement of phase shifters being either cascaded or parallel. Although each phase shifter must satisfy the nonconvex constraint, the use of two sets of phase shifters allows each complex coefficient to be represented by the sum equivalent of the two-phase shifters, effectively circumventing this nonconvex constraint and making the design of hybrid precoding more convenient.

Another hybrid precoding architecture is based on switching networks, as illustrated in Figure 8.7. While the energy consumption of phase shifters is almost negligible, there is still a specific energy and cost overhead due to the number of phase shifters, which is typically equal to the number of antennas. Furthermore, the hardware circuitry required to implement phase shifters in high-frequency bands is relatively complex. A hybrid precoding architecture based on a switching network is proposed to address these issues, in which an utterly passive switch replaces the phase shifter. Furthermore, the switching network architecture applies to multiple mapping modes between RF chains and antennas. For instance, each RF chain may be connected to all antennas; alternatively, each RF chain may be connected to only some antennas. In this structure, the analog precoded phase is quantified as 0 and 1, the switched connected phase is 1, and the switched disconnected phase is 0. In this case, integer programming can be used to solve the optimization problem of the hybrid beamforming matrix.

In addition, there is an analog precoding structure, which is based on a lens antenna. This structure employs a low-cost, low-power lens antenna and

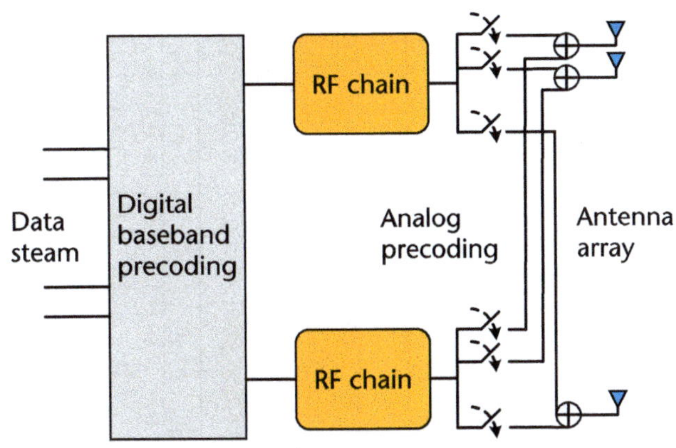

Figure 8.7 The switch network-based architecture.

selection network to achieve analog precoding. The specific structure is depicted in Figure 8.8. The lens array contains electromagnetic lenses with directional energy focusing capabilities and a matching array of antennas, each located at the lens focal point. The lens focuses signals from different directions onto different antennas, thereby converting the spatial channel into a beam-domain channel. Mathematically, a lens array can be regarded as a spatial DFT matrix, where the columns of the DFT matrix correspond to orthogonal beam vectors in predefined directions covering the entire angular space. Due to the sparse nature of millimeter wave channels, hybrid precoding based on lens antennas can achieve approximately optimal performance at a low cost and power consumption. However, due to the limitations of the lens hardware, this structure cannot guarantee similar performance to all-digital precoding in millimeter-wave channel scenarios with more scattering paths because the accuracy of the beam is fixed. Furthermore, since the hardware implementation of the lens is fixed prior to actual communication, there will be gaps between different antennas, which may result in power leakage issues.

8.2 Satellite-Terrestrial User Link Antenna Technology

Several types of satellite communication antennas exist, including multibeam antennas, phased array multibeam antennas, and lens multibeam antennas. In the field of satellite communication antenna technology, the primary focus is on single-satellite beamforming and multisatellite collaborative beamforming. This chapter commences with a classification of satellite antennas and subsequently introduces satellite communication antenna technology, divided into single-satellite beamforming and multisatellite collaborative beamforming.

8.2.1 Single Satellite Beamforming

The beamforming network (BFN) generates the requisite beam through beamforming technology. Beamforming technology can be categorized into two distinct categories: analog and digital. Analog beamforming employs a power splitter and phase shifter to regulate amplitude and phase, thereby enabling higher gains and reducing sidelobes on a single aperture antenna. The principal advantage of digital

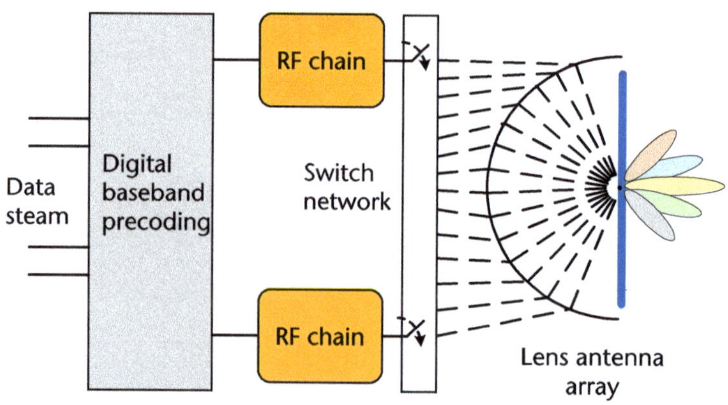

Figure 8.8 The lens antenna architecture.

beamforming in comparison with analog beamforming is that its weight and power consumption are solely determined by the bandwidth and the number of radiating elements, irrespective of the number of beams generated [1]. Furthermore, it can achieve amplitude and phase error compensation with greater flexibility since the third generation of satellite mobile communication systems began to utilize digital beamforming technology.

The signal processing process of beamforming can be completed on the satellite, designated space-based beamforming (SBBF) technology. SBBF technology can enhance the antenna's EIRP and G/T, facilitate rapid beam adjustment, and support single-hop service on the satellite. It is employed extensively in the third generation of satellite mobile communication systems. Single-satellite beamforming is shown in Figure 8.9. However, the system's signal processing capabilities are constrained by the limitations of the satellite payload, satellite power, and susceptibility to interference from space radiation.

The signal processing process of space-based beamforming technology is completed on the satellite, with the beamforming network representing the core of the satellite payload. Currently, Asia Cellular Satellite (ACeS) and Inmarsat-4 employ SBBF technology. ACeS utilizes analog beamforming, whereas Inmarsat-4 employs multiple beamforming.

In the context of the analog scheme for realizing SBBF, the beamforming network can be generated using a Butler matrix and a power splitter phase shifter. This allows for the subsequent realization of phase or amplitude processing of beamforming. The multiport amplifier ensures the power distribution of satellite beam generation, thereby reducing hardware and energy consumption costs. Nevertheless, the analog beamforming structure is fixed, and the limitation is significant when the number of beams is vast.

The schematic diagram of space-based digital beamforming technology is presented in the accompanying figure. In the SBBF implemented by digital beamforming technology, the system's multibeam pointing is generated once the onboard digital unit has completed the signal sampling, channelization, orthogonalization, and beamforming processing [2]. Employing weighted sampling, a directional beam is formed at each feed source.

Beamforming networks can be categorized into two distinct types: fixed and programmable. It is impossible to adjust a fixed beamforming network's coverage area or performance. In contrast, the programmable beamforming network can update the prior knowledge's weighted coefficient against the element and adjust

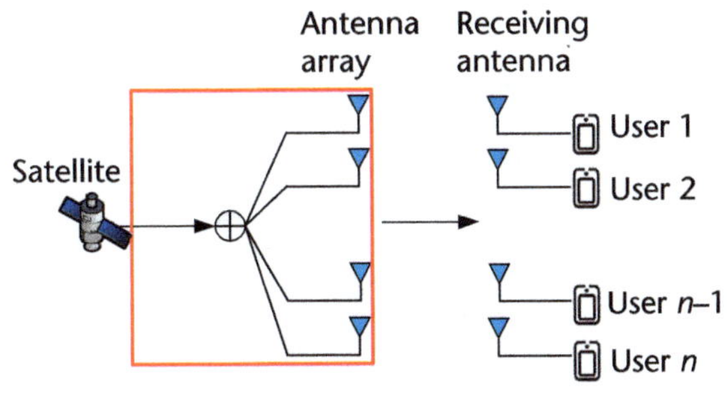

Figure 8.9 Illustration of single-satellite beamforming.

the beam direction to adapt to any changes in the orbit. The Inmarsat-4 system, for instance, employs programmable beamforming to update its weights periodically, thereby ensuring the formation of a fixed multipoint beam on the ground.

The SBBF technology places the beamforming process on the satellite, which has the potential to enhance the EIRP and G/T of the antenna. The bandwidth of the feed link between the SBBF station and the satellite is smaller than that of GBBF technology, which can significantly reduce the number of terrestrial stations required.

8.2.2 Multisatellite Beamforming

Multisatellite cooperative beamforming is a technology that employs the coordinated action of multiple satellites to achieve efficient communication [3]. The technology combines multisatellite antenna arrays and beamforming technology to enhance satellite communication systems' coverage, signal quality, and system capacity. Multisatellite cooperative beamforming is shown in Figure 8.10. China's Tianlian satellite system employs multisatellite cooperative beamforming technology to achieve efficient ground coverage and communication capabilities. This section presents an overview of the principle, critical technology, and prospective applications of multisatellite collaborative beamforming.

8.2.2.1 System Model

Multisatellite cooperative communication system adopts space division multiplexing; the system contains N_s satellites and serves N_u users in the same time slot and frequency. Each satellite is equipped with N_t antennas, and the total number of antennas is $N_{Tx} = N_t N_s > N_u$. The intersatellite collaborative beamforming is utilized to serve users. Satellites can serve UEs directly or indirectly by sending data to relay nodes. It is assumed that the Doppler shift due to the high relative velocity of the satellite has been compensated at the satellite compared to the nonterrestrial terminal.

The signals of N_s UEs can be expressed as

$$\mathbf{s} = [s_1, \ldots, s_{N_u}]^T \tag{8.5}$$

The signals are preprocessed by the digital beamforming matrix $\mathbf{G}_j \in \mathbb{C}^{N_t \times N_u}$ at the satellite terminal, and the transmitted signal is obtained by

$$\mathbf{x} = \mathbf{G}_j \mathbf{s} \in \mathbb{C}^{N_t} \tag{8.6}$$

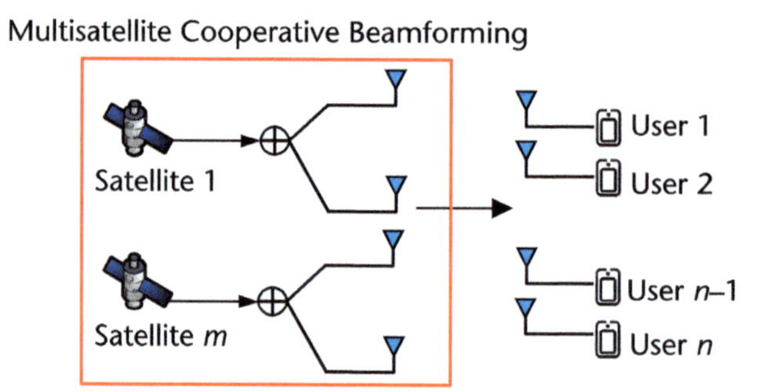

Figure 8.10 Multisatellite cooperative beamforming.

Therefore, the collaborative beamforming matrix can be expressed as

$$\mathbf{G} = [\mathbf{G}_1^T, ..., \mathbf{G}_{N_s}^T] = [\mathbf{g}_1, ..., \mathbf{g}_{N_u}] \in \mathbb{C}^{N_{Tx} \times N_u} \tag{8.7}$$

where $\mathbf{g}_u = [\mathbf{g}_{1,u}^T, ..., \mathbf{g}_{N_s,u}^T]^T \in \mathbb{C}^{N_{Tx}}$ represents the beamforming vector of all satellites to the uth user, so the received signal at the uth user can be expressed as

$$y_u = \mathbf{h}_u^H \mathbf{g}_u s_u + \sum_{v \neq u} \mathbf{h}_u^H \mathbf{g}_v s_v + n_u = \mathbf{h}_u^H \mathbf{G}\mathbf{s} + n_u \tag{8.8}$$

where $\mathbf{h}_u = [\mathbf{h}_{1,u}^T, ..., \mathbf{h}_{N_s,u}^T]^T \in \mathbb{C}^{N_{Tx}}$ is the instantaneous frequency flat channel vector from all transmitting antennas to the uth terminal. $n_u \sim \mathcal{CN}(0, \sigma_n^2)$ represents independent and identically distributed (i.i.d.) complex Gaussian distributed white noise.

Since the signal is affected by path loss during transmission, the uth terminal only rescales the received signal by $\beta_u > 1$. Therefore, the estimated signal before the hard decision of the uth terminal can be expressed as

$$\hat{s}_u = \beta_u(\mathbf{h}_u^H \mathbf{G}\mathbf{s} + n_u) \tag{8.9}$$

In addition, the channel matrix from the JTH satellite to all users can be expressed as $\mathbf{H}_j$. Obtaining accurate CSI on the satellite is challenging due to the long delay and relative speed between the satellite and the NTN terminal. Therefore, it is assumed that only the estimated channel matrix $\hat{\mathbf{H}}_j$ is available in the jth satellite, and $\hat{\mathbf{H}}_j$ can be expressed as

$$\hat{\mathbf{H}}_j = \mathbf{H}_j + \Delta_j, \forall j \in \{1, ..., N_S\} \tag{8.10}$$

where $\Delta_j \sim \mathcal{CN}(0, N_U \sigma_{\tilde{h}} \mathbf{I})$ is the i.i.d. additive channel estimation error.

8.2.2.2 Problem Formulation

By minimizing the mean square sum of all terminals between the estimated symbol $\{\hat{s}_u\}_{u=1}^{N_U}$ and the expected symbol $\{s_u\}_{u=1}^{N_U}$ under the single power constraint, we can transform the precoding design problem into the constrained optimization problem as

$$\begin{aligned} &\min_{\{\mathbf{G}_j\}_{j=1}^{N_S}, \{\beta_u\}_{u=1}^{N_U}} \sum_{u=1}^{N_U} E\{\| s_u - \hat{s}_u \|_2^2\} \\ &s.t.\ tr\{\mathbf{G}_j \mathbf{G}_j^H\} \leqslant P_j \forall j \in \{1, ..., N_S\} \end{aligned} \tag{8.11}$$

where $tr\{\cdot\}$ is the tracking operator and P_j is the maximum transmitting power of the jth satellite. The optimal precoded matrix $\{\mathbf{G}_j\}_{j=1}^{N_S}$ depends on the factor $\{\beta_u\}_{u=1}^{N_U}$. Therefore, the precoded matrix is alternately updated and the factor is rescaled [4–7].

Let $\mathbf{B} = diag(\beta_1, ..., \beta_{N_U})$ be the rescaling matrix, so that the corresponding channel matrix is

$$\mathbf{H} = [\mathbf{H}_1, ..., \mathbf{H}_{N_S}] = [\mathbf{h}_1, ..., \mathbf{h}_{N_U}]^H \in \mathbb{C}^{N_U \times N_{Tx}} \tag{8.12}$$

Similarly, the estimated global channel matrix can be obtained as

$$\hat{\mathbf{H}} = \left[\hat{\mathbf{H}}_1, ..., \hat{\mathbf{H}}_{N_S}\right] = \left[\hat{\mathbf{h}}_1, ..., \hat{\mathbf{h}}_{N_U}\right]^H \in \mathbb{C}^{N_U \times N_{Tx}} \tag{8.13}$$

The estimated multiuser data vector $\hat{\mathbf{s}} = [\hat{s}_1, ..., \hat{s}_{N_U}] \in \mathbb{C}^{N_U}$ can be expressed as

$$\hat{\mathbf{s}} = \mathbf{B}(\mathbf{HGs} + \mathbf{n}) \tag{8.14}$$

where $\mathbf{n} = [n_1, ..., n_{N_U}]^T$. Then, the optimization problem can be further simplified as

$$\begin{aligned} &\min_{\mathbf{G},\mathbf{B}} E\{\|\mathbf{s} - \hat{\mathbf{s}}\|_2^2 \\ &\text{s.t tr}\{\,\mathbf{G}_j\mathbf{G}_j^H\} \leqslant P_j \forall j \in \{1, ..., N_S\} \end{aligned} \tag{8.15}$$

For the sake of solving the optimization problem from the viewpoint of each satellite, **HG** can be represented by

$$\mathbf{HG} = \sum_{j=1}^{N_S} \mathbf{H}_j\mathbf{G}_j \tag{8.16}$$

8.2.2.3 Multisatellite Collaborative Beamforming Principle

Let $\;= diag(\lambda_1, ..., \lambda_{N_S}) \otimes \mathbf{I}$ and $\mathbf{P} = 1/N_t diag(P_1, ..., P_{N_S}) \otimes \mathbf{I}$, where $\lambda_j \geqslant 0$ is the Lagrange multiplier associated with the power constraint of the jth satellite and $\otimes$ represents the Kronecker product. The Lagrange of (8.15) can be expressed as

$$\begin{aligned} \mathcal{L} &= (\mathbf{G}, \mathbf{B}, \;) = E\left\{\|\mathbf{s} - \hat{\mathbf{s}}\|_2^2\right\} + tr\left\{\;(\mathbf{GG}^H - \mathbf{P})\right\} \\ &= \sum_{j=1}^{N_S} (tr\left\{E\left\{\sum_{i=1}^{N_S} \mathbf{G}_i^H\mathbf{H}_i^H\mathbf{B}^H\mathbf{B}\mathbf{H}_j\mathbf{G}_j\right\}\right\} - tr\left\{E\left\{\mathbf{G}_j^H\mathbf{H}_j^H\mathbf{B}^H + \mathbf{B}\mathbf{H}_j\mathbf{G}_j\right\}\right\} \\ &\quad + \lambda_j(tr\left\{\mathbf{G}_j\mathbf{G}_j^H\right\} - P_j)) + E\left\{\mathbf{n}^H\mathbf{B}^H\mathbf{B}\mathbf{n}\right\} + N_U \end{aligned} \tag{8.17}$$

By defining $r = \sum_{u=1}^{N_u} \beta_u^2$, we can obtain

$$tr\left\{E\left\{\sum_{i=1}^{N_S} \mathbf{G}_i^H\mathbf{H}_i^H\mathbf{B}^H\mathbf{B}\mathbf{H}_j\mathbf{G}_j\right\}\right\} = tr\left\{\sum_{i=1}^{N_S} \mathbf{G}_i^H\hat{\mathbf{H}}_i^H\mathbf{B}\mathbf{B}\hat{\mathbf{H}}_j\mathbf{G}_j\right\} + r\sigma_{\tilde{h}}^2 \text{tr}\left\{\mathbf{G}_j\mathbf{G}_j^H\right\} \tag{8.18}$$

According to KKT conditions, for the optimal precoding matrix $\left\{\mathbf{G}_j^*\right\}_{j=1}^{N_S}$, we rescale the factor $\{\beta_u^*\}_{u=1}^{N_U}$ and the Lagrangian multiplier $\left\{\lambda_j^*\right\}_{j=1}^{N_S}$, and the Lagrangian first derivative must be zero, we have

$$\begin{aligned}\nabla_{\mathbf{G}_j}\mathcal{L}(\mathbf{G}^*, \mathbf{B}^*, \lambda^*) = \mathbf{0}\ \forall j \in \{1, ..., N_S\}\\ \nabla_{\beta u}\mathcal{L}(\mathbf{G}^*, \mathbf{B}^*, \lambda^*) = 0\ \forall u \in \{1, ..., N_U\}\end{aligned} \tag{8.19}$$

Therefore, we can obtain the optimal solution of (8.15) $\left\{\mathbf{G}_j^*\right\}_{j=1}^{N_S}$ and $\{\beta_u^*\}_{u=1}^{N_U}$

$$\mathbf{G}_j^* = \mathbf{T}_j^*\hat{\mathbf{H}}_j^H\mathbf{B}^*(\mathbf{I} - \sum_{i\neq j}\mathbf{B}^*\hat{\mathbf{H}}_i\mathbf{G}_i^*)\ \forall j \tag{8.20}$$

$$\beta_u^* = \frac{\mathrm{Re}\left\{\hat{\mathbf{h}}_u^H\mathbf{g}_u^*\right\}}{\left\|\hat{\mathbf{h}}_u^H\mathbf{G}^*\right\|_2^2 + \sigma_{\tilde{h}}^2\|\mathbf{G}^*\|_F^2 + \sigma_n^2}\ \forall u \tag{8.21}$$

where $\mathbf{T}_j^* = (\hat{\mathbf{H}}_j^H\mathbf{B}^*\mathbf{B}^*\hat{\mathbf{H}}_j + (r^*\sigma_{\tilde{h}}^2 + \lambda_j^*)\mathbf{I})^{-1}$. For $j \in \{1, ..., N_S\}$, according to [8], the KKT conditions are

$$\left\|\mathbf{G}_j^*\right\|_F^2 - P_j \leqslant 0\ \forall j \tag{8.22}$$

$$\lambda_j^* \geqslant 0\ \forall j \tag{8.23}$$

$$\lambda_j^*\left(\left\|\mathbf{G}_j^*\right\|_F^2 - P_j\right) = 0\ \forall j \tag{8.24}$$

According to (8.20), the optimal precoding matrix $\mathbf{G}_j^*$ for the jth satellite depends on the Lagrange multiplier λ_j^*, the interference matrix $\left\{\hat{\mathbf{H}}_i\mathbf{G}_i^*\right\}_{i\neq j}$ from other satellites, and the rescaling matrix $\mathbf{B}$, and it can be expressed as

$$\mathbf{G}_j^* = f\left(\left\{\hat{\mathbf{H}}_i\mathbf{G}_i^*\right\}_{i\neq j}, \lambda_j^*, \mathbf{B}^*\right) \tag{8.25}$$

Therefore, we alternately update the precoding matrix, the rescaling factor, and its Lagrange multiplier. We do not update the precoded matrix directly according to (8.20), and the convergence of the DiP algorithm is ensured by introducing the relaxation parameter $0 < \omega < 1$. At the jth satellite, iteration $k = 0, ..., K-1$ by

$$\begin{aligned}\mathbf{G}_j^{(k+1)} &= (1-w)\,\mathbf{G}_j^{(k)} + wf\left(\left\{\hat{\mathbf{H}}_i\mathbf{G}_i^{(k)}\right\}_{i\neq j}, \lambda_j^{(k)}, \mathbf{B}^{(k)}\right)\\ &= \mathbf{G}_j^{(k)} - w\mathbf{T}_j^{(k)}\left(\hat{\mathbf{H}}_j^H\mathbf{B}^{(k)}\sum_{i=1}^{N_S}\hat{\mathbf{H}}_i\mathbf{G}_i^{(k)} + \left(r^{(k)}\sigma_{\tilde{h}}^2 + \lambda_j^{(k)}\right)\mathbf{G}_j^{(k)} - \hat{\mathbf{H}}_j^H\mathbf{B}^{(k)}\right)\end{aligned} \tag{8.26}$$

At the kth iteration, the rescaling factor $\beta_u^{(k)}$ of each NTN terminal is determined by precoding the global matrix

$$\beta_u^{(k)} = \frac{\mathrm{Re}\left\{\hat{\mathbf{h}}_u^H \mathbf{g}_u^{(k)}\right\}}{\left\|\hat{\mathbf{h}}_u^H \mathbf{G}^{(k)}\right\|_2^2 + \sigma_{\tilde{h}}^2 \left\|\mathbf{G}^{(k)}\right\|_F^2 + \sigma_n^2} \quad \forall u \tag{8.27}$$

Then, the Lagrange multiplier $\lambda_j^{(k)}$ must satisfy the KKT condition. As shown in (8.22) to (8.24), either $\lambda_j^{(k)}$ must be zero or the power constraint must satisfy the equation. Therefore, A can be computed in each iteration using a root-finding algorithm like Newton's method or dichotomy

$$\left\|\mathbf{G}_j^{(k+1)}\left(\lambda_j^{(k)}\right)\right\|_F^2 = P_j \tag{8.28}$$

If no positive $\lambda_j^{(k)}$ is found, the multiplier is set to zero.

The precoding matrix $\mathbf{G}_j$ for the jth satellite is initialized by the local minimum mean square error (MMSE) preencoder regardless of the precoding matrix $\{\mathbf{G}_i\}_{i \neq j}$ for other satellites, and it is assumed that all NTN terminals have the same rescaling factor $\beta = \beta_u^{(0)} = \beta_v^{(0)}, \forall u, v \in \{1, \ldots, N_U\}$, we have

$$\mathbf{G}'_j = \left(\hat{\mathbf{H}}_j^H \hat{\mathbf{H}}_j + N_U \left(\sigma_{\tilde{h}}^2 + \frac{\sigma_n^2}{P_j}\right)\mathbf{I}\right)^{-1} \hat{\mathbf{H}}_j^H \tag{8.29}$$

$$\beta^{-1} = \sqrt{\frac{P_j}{tr\left\{\mathbf{G}'_j \mathbf{G}'^H_j\right\}}} \tag{8.30}$$

$$\mathbf{G}_j^{(0)} = \beta^{-1}\mathbf{G}'_j \tag{8.31}$$

In each iteration, each satellite requires the complete matrix $\hat{\mathbf{H}}\mathbf{G}^{(k)}$ in order to update its preceded matrix $\mathbf{G}_j^{(k+1)}$ as well as the Lagrange multiplier $\lambda_j^{(k)}$ and the rescaling matrix $\mathbf{B}^{(k)}$. Thus, in each iteration, each satellite sends its matrix $\hat{\mathbf{H}}_j\mathbf{G}_j^{(k)} \in \mathbb{C}^{N_U \times N_U}$ to all the others via ISL. To reduce communication overhead, only $\mathbf{G}_j^{(k)} \in \mathbb{C}^{N_t \times N_U}$ can be sent. However, if only the precoding matrices are shared, each satellite must carry out extensive matrix multiplication in every iteration, which greatly increases the computational complexity.

The total radiated power of the satellite can be approximated as

$$\left\|\mathbf{G}^{(k)}\right\|_F^2 \approx \sum_{j=1}^{N_S} P_j \tag{8.32}$$

Therefore, the rescaling matrix $\mathbf{B}^{(k)}$ can be computed locally at each satellite without any further exchange of information.

After calculating the final preceded matrix $\mathbf{G}^{(K)}$, each NTN terminal must estimate its rescaling factor $\hat{\beta}_u$, which can be done either by automatic gain control (AGC) or by estimating the effective channel $\mathbf{h}_u^H \mathbf{G}^{(K)}$ with the pilot signal. It is much easier to get an accurate CSI at the terminal than on the satellite. Therefore, it is assumed that each NTN terminal has a known effective channel, and its rescaling factor is calculated by

$$\hat{\beta}_u = \frac{\mathrm{Re}\left\{\mathbf{h}_u^H \mathbf{g}_u^{(K)}\right\}}{\left\|\mathbf{h}_u^H \mathbf{G}^{(K)}\right\|_F^2 + \sigma_n^2} \tag{8.33}$$

The principle of multisatellite cooperative beamforming is to focus the beam on the target area through cooperative work between multiple satellites, thereby achieving high-quality signal transmission. The beamforming and beam tracking process is achieved through the exchange of information and cooperation between satellites, enabling a broader coverage area and more robust signal quality.

8.2.3 Characteristics of User Terminal Antennas

In satellite-to-ground links, the design and performance of user terminal antennas play a critical role in determining the overall quality of the communication system. First, user terminal antennas typically feature high gain, which is essential for compensating the signal attenuation caused by long-distance transmission. Second, the beam steering capability allows the antenna to dynamically adjust its beam direction to track fast-moving LEO satellites, ensuring a stable and continuous connection.

To balance high gain with a compact design, many modern user devices employ phased array antennas. These antennas can electronically adjust the beam direction rapidly without any mechanical movement, offering higher reliability. Additionally, user terminal antennas often support dual-band or multiband operations to accommodate different satellite systems, such as Ka-band and Ku-band. Moreover, polarization capabilities, such as circular and linear polarization, are commonly implemented to enhance the system's resistance to interference and improve spectrum efficiency.

Modern user terminal antennas are also integrated with a low-noise front end, reducing noise levels to enhance the signal-to-noise ratio of the received signal. Furthermore, improved interference suppression is another crucial aspect, as the design must consider potential interference from ground sources and other satellites. In summary, these characteristics enable user terminal antennas to provide stable and efficient communication performance across various applications, including satellite internet access, television reception, and navigation.

8.3 Satellite-Terrestrial Feeder Link Antenna Technology

As the range of communication services continues to expand, the demand for communication capacity is rising, necessitating the deployment of more spot beams and larger antennae arrays in satellite mobile communication systems. This, in turn, drives the development of more complex beamforming networks. However, the advancement of space-based beamforming technology is constrained by the payload and the weight it can carry on the spacecraft. To address this challenge, the beamforming process is proposed to be partially or wholly transferred to the ground, enabling multipoint beam coverage through the terrestrial waveforming network. Consequently, ground-based beamforming (GBBF) technology is primarily employed in the satellite-based feed link, where beamforming is completed at the terrestrial gateway station. The signal processing process of GBBF technology is carried out on the ground, effectively saving onboard resources and improving the system's reliability and flexibility. It is increasingly being used in fourth-generation and later satellite mobile communication systems.

The GBBF technology employs beamforming technology in the terrestrial station, with the antenna, RF module, and a few signal processing functions retained on the satellite. This results in a significant reduction in the processing requirements at the satellite end and a reduction in the weight and volume of the satellite. Furthermore, the cost of signal processing on the ground is considerably less than that of onboard processing. The cost of equipment and technology has been reduced, and the flexibility has been enhanced, including adaptive processing, interference elimination, and other technologies.

However, in GBBF technology, the bandwidth requirement of the feed link is considerable, resulting in a much larger frequency resource requirement than that of the SBBF scheme. Consequently, management needs multiple gateways.

The reverse link, for example, comprises several key components, including frequency and polarization multiplexing from the satellite-to-uplink signal, terrestrial terminal feed signal demultiplexing, and ground-based signal processing technology. In particular, each feed signal is weighted to compensate for amplitude and phase distortion between the GBBF technology and the feed, thereby realizing the directional multipoint beam. The forward link represents the inverse of the preceding process.

For the GBBF, signal frequency multiplexing and polarization multiplexing result in a significant demand for feed link bandwidth. Furthermore, the propagation of wireless signals through the atmosphere will inevitably attenuate the signal. This necessitates the use of compensation and correction signals and the deployment of accurate terrestrial reference clocks, which in turn leads to an increased demand for bandwidth on the feed link. Consequently, the feed link represents a pivotal factor influencing the efficacy of GBBF.

In order to address the issue of limited bandwidth on the feed link, signal compression technology can be employed to eliminate the correlation between signals. One possible solution is to utilize a higher frequency feed link, whereby the spectrum resources of the feed link are multiplexed through multiple gateways. OneWeb utilizes advanced phased array antenna technology for its satellite-terrestrial feeder links, enabling electronic beam steering and multibeam capabilities for efficient and reliable communication. This technology allows for quick satellite tracking and simultaneous connections with multiple satellites, crucial for OneWeb's global coverage.

References

[1] Zhang, Y., A. Liu, P. Li, and S. Jiang, "Deep Learning (DL)-Based Channel Prediction And Hybrid Beamforming for LEO Satellite Massive MIMO System," *IEEE Internet of Things Journal*, Vol. 9, No. 23, 2022, pp. 23705–23715.

[2] Arti, M. K., "Data Detection in Multisatellite Communication Systems," *IEEE Transactions on Aerospace and Electronic Systems*, Vol. 56, No. 2, 2019, pp. 1637–1644.

[3] Röper M., and A. Dekorsy, "Robust Distributed MMSE Precoding in Satellite Constellations for Downlink Transmission," *2019 IEEE 2nd 5G World Forum (5GWF), IEEE*, 2019, pp. 642–647.

[4] Hu, X., K.-X. Guo, B.-Y. Wu, and X.-Q. Sheng, "A Deterministic Terahertz Channel Model for Inter-Satellite Communication Link," *2021 International Applied Computational Electromagnetics Society (ACES-China) Symposium, IEEE*, 2021, pp. 1–2.

[5] Zhang Z., Y. Li, C. Huang, Q. Guo, L. Liu, and C. Yuen, "User Activity Detection and Channel Estimation for Grant-Free Random Access in LEO Satellite-Enabled Internet of Things," *IEEE Internet of Things Journal*, 2020, Vol. 7, No. 9, pp. 8811–8825.

[6] Liu Y., C. Li, J. Li, and L. Feng, "Robust Energy-Efficient Hybrid Beamforming Design for Massive MIMO LEO Satellite Communication Systems," *IEEE Access*, 2022, Vol. 10, pp. 63085–63099.

[7] You L., X. Qiang, K.-X. Li, C. G. Tsinos, W. Wang, and X. Gao, "Massive MIMO Hybrid Precoding for LEO Satellite Communications With Twin-Resolution Phase Shifters and Nonlinear Power Amplifiers," *IEEE Transactions on Communications*, 2022, Vol. 70, No. 8, pp. 5543–5557.

[8] You L., X. Qiang, K.-X. Li, C. G. Tsinos, W. Wang, and X. Gao, "Hybrid Analog/Digital Precoding for Downlink Massive MIMO LEO Satellite Communications," *IEEE Transactions on Wireless Communications*, 2022, Vol. 21, No, 8, pp. 5962–5976.

[9] Ayach, O. E., S. Rajagopal, S. Abu-Surra, Z. Pi, and R. W. Heath, "Spatially Sparse Precoding in Millimeter Wave MIMO Systems," *IEEE Transactions on Wireless Communications*, Vol. 13, No. 3, March 2014, pp. 1499–1513.

[10] Yu, W., and T. Lan, "Transmitter Optimization for the Multi-Antenna Downlink With Per-Antenna Power Constraints," *IEEE Transactions on Signal Processing*, Vol. 55, No. 6, June 2007, pp. 2646–2660.

[11] Yoo, T., N. Jindal, and A. Goldsmith, "Multi-Antenna Downlink Channels with Limited Feedback and User Selection," *IEEE Journal on Selected Areas in Communications*, Vol. 25, No. 7, September 2007, pp. 1478–1491.

[12] Rossi, T., M. De Sanctis, M. Ruggieri, C. Riva, L. Luini, and G. Codispoti, "Satellite Communication and Propagation Experiments Through the AlphasatQ/V Band Aldo ParaboniTechnology Demonstration Payload," *IEEE Aerospace and Electronic Systems Magazine*, Vol. 31, No. 3, March 2016, pp. 18–27.

[13] Palacin, B., et al., "Multibeam Antennas for Very High Throughput Satellites in Europe: Technologies and Trends," *2017 11th European Conference on Antennas and Propagation (EUCAP), Paris, France*, 2017, pp. 2413–2417.

CHAPTER 9

Multiple Access for Satellite-Terrestrial Integrated Communication

Multiple access (MA) technology is a core component of wireless communication access networks. It allocates limited resources to multiple users to provide efficient, orderly, and reliable multiuser access services for wireless communication systems. This chapter will begin by introducing existing MA technologies, including orthogonal multiple access (OMA) and nonorthogonal multiple access (NOMA). Then, it will cover the MA technologies used in current communication standards. Finally, it will explore MA technologies applicable to the integrated satellite-terrestrial communication systems.

9.1 Classic OMA Schemes

Traditional terrestrial and satellite communication systems utilize OMA technology [1, 2] to provide multiuser access services. OMA effectively reduces interuser interference by placing the signals of different users on mutually orthogonal subchannels. This section introduces various MA technologies in OMA, including FDMA, TDMA, CDMA, orthogonal frequency division multiple access (OFDMA), and single carrier-frequency division multiple access (SC-FDMA).

9.1.1 FDMA

FDMA divides the total bandwidth into several nonoverlapping and orthogonal subchannels. This allows multiple users to communicate simultaneously on different frequencies. To prevent interference between users, FDMA uses a guard band between adjacent subchannels for channel isolation. By leveraging the orthogonality between subchannels, FDMA employs a filter at the receiver end to eliminate out-of-band interference and demodulate the original data [3].

In Figure 9.1, the principle of FDMA is illustrated. In this method, different users communicate using subchannels, with a guard band set between each subchannel to prevent interference from adjacent channels. The size of the guard band needs to be carefully evaluated, taking into account both spectral efficiency and implementation difficulty. If the guard band is too wide, it will reduce the system's spectral efficiency, while if it is too narrow, it becomes difficult to implement in hardware.

In satellite communication systems, the nonlinear effect of the high power amplifier (HPA) is the main factor that limits the performance of FDMA. The nonlinear effect occurs when the input signal power is below a certain level (saturation point), causing the HPA to work approximately in a linear region. However, when

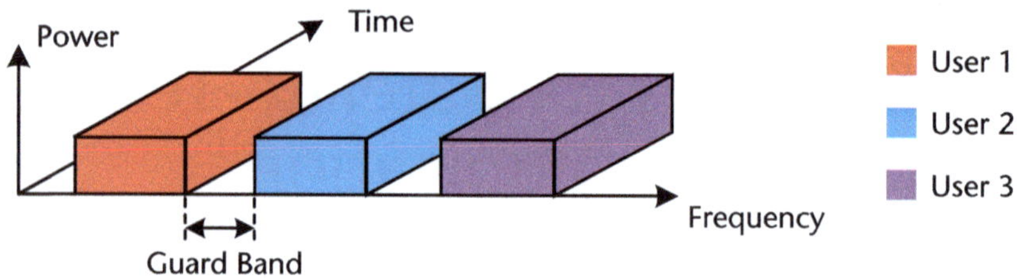

Figure 9.1 FDMA.

the input signal power exceeds this level, the HPA enters the saturation or supersaturation region, which generates a large amount of intermodulation (IM) interference. In satellite systems, most transponders use HPAs, and during operation, most of the HPA is close to the saturation region to improve power efficiency. When the onboard transponder processes multiple truncated signals simultaneously, its HPA amplifies multiple carriers at the same time, causing the HPA to enter the supersaturated region and create cross-modulation interference. This interference results in the creation of new frequency components, which worsens the receiver's signal-to-noise ratio (SNR) and ultimately degrades the communication quality.

9.1.2 TDMA

TDMA divides time into multiple nonoverlapping time slots, allowing all users to communicate on different time slots using the same carrier frequency. The principle is shown in Figure 9.2. To avoid interference between users, TDMA requires a strict time synchronization control mechanism to ensure that the signals sent by each user do not overlap in time [4].

In a TDMA system, users need to use the burst synchronous reference signal to synchronize time and send the service signal in the assigned time slot. Each user's receiver can detect all burst signals on the channel and communicate in the specified time slot based on the detection results. Since only one signal can be transmitted in each specified time slot, TDMA system needs to design reasonable frame structure and accurate time synchronization algorithm. This is to ensure that the multiuser signals at the receiver do not overlap while being distinguishable. However, these additional designs increase the implementation complexity of TDMA.

Compared to the FDMA system, the satellite TDMA system possesses several distinct characteristics:

1. TDMA activates only one carrier, thereby avoiding the IM interference caused by nonlinear effects in FDMA. This allows the HPA to operate in a saturated state for extended periods, reducing the requirements for HPA.

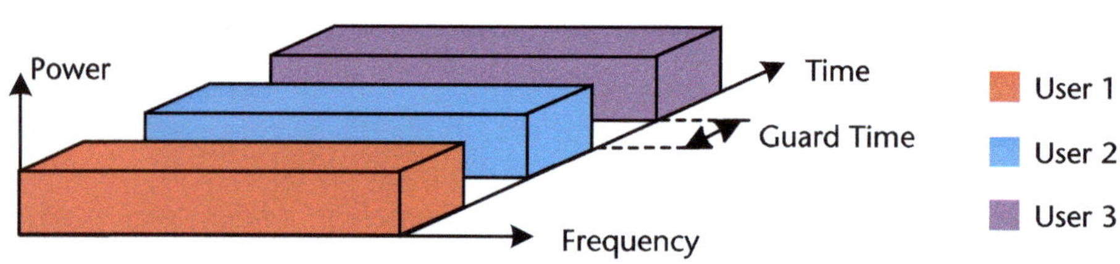

Figure 9.2 TDMA.

2. The restriction of TDMA on the equivalent omnidirectional radiated power change of earth station is more relaxed and more accessible to realize than FDMA.
3. TDMA can adjust the corresponding time slot length according to the throughput size of the satellite station, providing greater flexibility than FDMA.
4. TDMA separates the time of each earth station and repeater, simplifying circuit structure as it does not require multiple frequency conversions like FDMA.
5. In comparison to FDMA, TDMA has stricter requirements for timing synchronization. Precise time synchronization is essential in ensuring that users transmit data within their assigned time slots without any overlap.

9.1.3 CDMA

In CDMA, each user transmits signals in the same time-frequency resource, but is distinguished from one another by independent signature codes [5]. At the receiver, the earth station first identifies its unique code from all the signals sent by each station in order to extract the signals sent to the station. Subsequently, the transmitted information is obtained through demodulation and decoding. CDMA is suitable for systems with many users and low link transmission rates, such as satellite mobile communication systems or thinly routed VSAT satellite systems.

In the direct sequence spread spectrum (DSSS) CDMA system, a signature code is typically used, which utilizes an orthogonal code set with strong autocorrelation and weak cross correlation. There are plenty of codes available, such as the m-sequence in the pseudonoise code (PN code) or the Gold code derived from it. Figure 9.3 illustrates the block diagram of the system. Initially, the source code of rate R_b undergoes an XOR operation for the first modulation of the signature code, where the chip rate of the signature code is R_c. As $R_c \gg R_b$, this first modulation,

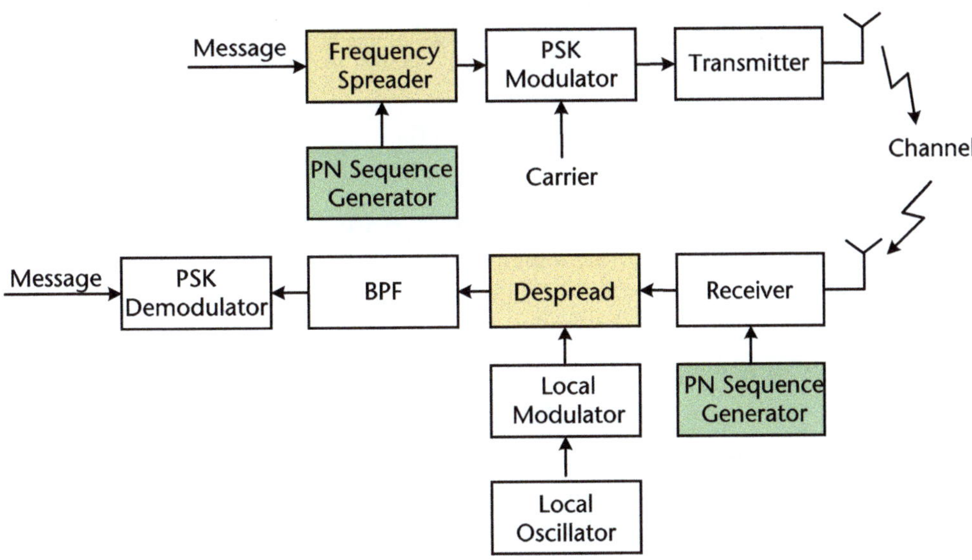

Figure 9.3 CDMA.

known as the spectrum spread, broadens the signal spectrum after modulation. The signal after the spectrum spread is then modulated by phase shift keying (PSK) on the intermediate frequency carrier and uplink transmitted.

The local signature code and the received signal are first used to perform correlation operations, which is called despreading. After despreading, the signal components related to the local signature code are reduced to narrowband signals through autocorrelation, while the signal components of other stations remain as wideband signals. The narrowband bandpass filter then extracts the useful signal and reduces interference from other stations. Finally, demodulating the bandpass-filtered signal ensures high SNR operation. The demodulator's performance depends on the spread spectrum processing gain, which is the ratio of the bandwidth of the spread spectrum signal to the bandwidth of the information signal. Assuming chip rate R_c and information rate R_b represent RF bandwidth and information bandwidth, respectively, the processing gain G_p is calculated as

$$G_p = \frac{RF\ Bandwidth}{Information\ Bandwidth} = R_c / R_b \tag{9.1}$$

In a CDMA system, the quality of end-to-end signal transmission can still be characterized by SNR. However, the noise includes channel Gaussian noise and the system's own MA interference. Assuming I is the sum of the two and I_0 is its spectral density, the ratio of signal power to the total noise interference power is given by

$$C/I = \frac{R_b E_b}{I_0 B} \tag{9.2}$$

where E_b is the energy per bit and B is the spread spectrum bandwidth.

Consider a CDMA system with n channels working simultaneously. This means that n earth stations are allowed to transmit signals with different signature codes on the same carrier. When the signal power and noise interference power of n earth stations reaching a receiver are equal, the SNR can be expressed as

$$C/I = \frac{1}{n-1} \tag{9.3}$$

If $n \gg 1$, then

$$n = \frac{B/R_b}{E_b/I_0} = \frac{G_p}{E_b/I_0} \tag{9.4}$$

When the SNR E_b/I_0 is given, to increase the system capacity n, we must find ways to improve the processing gain G_p.

CDMA systems have the following characteristics:

1. *Flexible access*: Each link in the CDMA system is relatively independent, and the network can accept new users without channel allocation control. The transmission quality of CDMA systems only decreases when the system capacity increases.
2. *Strong antiinterference and anti-interception ability*. Due to the high processing gain generated by spread spectrum, the anti-interference capability of CDMA is significantly enhanced. Meanwhile, spectrum spread reduces

the power spectral density of the CDMA signal, enhancing its resistance to interception. Moreover, the pseudorandom nature of the signature code improves the confidentiality of CDMA.

3. *Strong adaptability to multipath channels.*
4. *Low spectrum utilization makes CDMA suitable only for low-rate data transmission.* Given the fixed transponder bandwidth and spread spectrum processing gain, CDMA is unable to support high-rate data transmission due to the extended signal spectrum.

9.1.4 OFDMA

OFDMA is a multiuser version of OFDM. In OFDM, all subcarriers of the OFDM symbol are assigned to the user station or base station for the entire duration of the connection. On the other hand, in OFDMA, all subcarriers are divided into several groups, known as subchannels, and these subchannels are assigned to each user based on their specific resource requirements [6].

Specifically, OFDM technology utilizes different time slots to distinguish users. In OFDM, each user occupies all subcarriers in a time slot and sends a complete packet. When OFDM is combined with multiuser transmission technology to form OFDMA, it enables multiuser sharing of channel resources and enhances spectrum utilization. OFDMA defines a resource unit (RU) as the minimum subchannel and distinguishes users based on RU. Initially, OFDMA divides the channel resources into fixed-size RU time-frequency resource blocks and then carries the data of corresponding users on each RU, as depicted in Figure 9.4. In general, in OFDMA, multiple users may transmit information simultaneously at any given moment.

An OFDMA system can support more users through subchannelization compared to OFDM. Both OFDM and OFDMA use IFFT and FFT at the transmitter and receiver, respectively. In OFDM, the entire input of the IFFT is completely occupied by the user station or base station. However, in OFDMA, only part of the input value is occupied by the user station, and the rest of the input position is inserted with zero or null values.

In summary, the advantages of OFDMA are as follows:

1. *High spectrum efficiency*: OFDMA allocates subcarrier subsets to different users, making more efficient use of frequency band resources compared to FDMA and OFDM.
2. Better BER performance on fading channels.

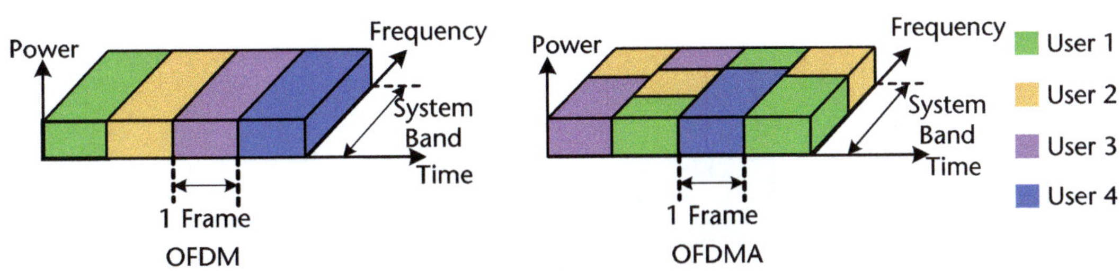

Figure 9.4 OFDM and OFDMA.

3. *The receiver structure is simple.* OFDMA eliminates intracell interference and avoids the need for multiuser detection required in CDMA. As a result, the receiver only needs to perform FFT processing on the received signal to distinguish the information of different users.

However, the disadvantages of OFDMA are as follows:

1. *PAPR is high.* OFDMA is a MA technology based on OFDM. When several orthogonal subcarriers in an OFDM signal have the same phase, the superposition of subcarriers will lead to high PAPR, which will increase the in-band noise caused by the nonlinear effect.
2. *Users have high requirements for time/frequency/channel balance.* OFDMA uses pilot signal and other signal processing techniques to achieve strict equalization, but this increases the overhead of auxiliary signals and the difficulty of implementation.
3. *High complexity of data processing algorithms.* Compared with OFDM, OFDMA needs to consider the arrangement and allocation of subcarriers. Each time a subchannel is opened and ready to send information, OFDMA consumes additional power.
4. *Weak resistance to frequency offset.* This is also due to OFDM technology's susceptibility to frequency offset.

9.1.5 SC-FDMA

SC-FDMA is a solution based on OFDMA, which addresses the issues of high PAPR and sensitivity to frequency offset in OFDMA [7]. Due to its single-carrier characteristics, SC-FDMA has a PAPR that is approximately 1–3 dB lower than that of OFDMA. Lower PAPR can enhance the efficiency of the transmitter signal amplifier and prolong the battery life. Additionally, SC-FDMA employs localized scheduling, which allocates continuous subcarriers to one user, thereby enhancing robustness to frequency offset at the expense of reducing diversity order.

In Figure 9.5, the system architecture of SC-FDMA is depicted. First, the data symbol $\{b_i\}$ is modulated into a complex vector $\{x_i\}$, and then a frequency domain signal X_n is produced through the N-point FFT. Each parallel output of the FFT is then mapped onto a single subcarrier to generate $\hat{X}_k$. The M-point IFFT then

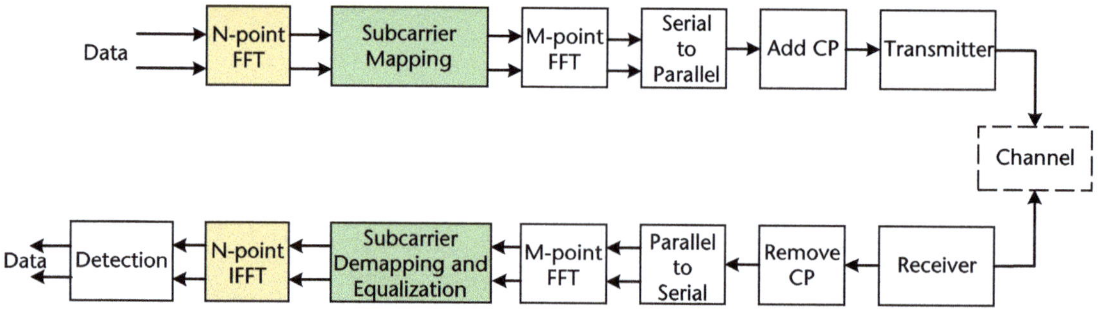

Figure 9.5 SC-FDMA.

converts $\hat{X}_m$ into a time-domain complex signal $\hat{x}_m$, where the conversion from N-point to M-point presents a resource allocation problem, $N < M$. $Q = M/N$ is an integer representing the maximum number of concurrent users in the system without any interference. Finally, the cyclic prefix (CP) is added to the front end of $\hat{x}_m$, and then it is serialized before single-frequency carrier modulation.

Upon reception, the received signal is initially converted to digital signal. Next, the CP is removed, and then the time domain signal is transformed into the frequency domain signal using an M-point FFT. Following channel estimation and equalization, the output signal undergoes an N-point IFFT operation before being sent to the detector to get $\{x_i\}$.

The user's N subcarriers are mapped to M subcarriers in a distributed or localized manner. Figure 9.6 shows an example of two users adopting distributed and localized interleaving schemes, respectively, where $N = 6, M = 12$.

Distributed mapping (interleaved FDMA (IFDMA)) introduces a bandwidth spreading factor by interleaving the subcarriers assigned to users. The time-domain sampled signal $\hat{x}_m$ of IFDMA is equivalent to $x_{\bar{m}}/Q$ and $\bar{m} = m \bmod N$, with mapping starting from the first subcarrier. If the mapping starts from the rth subcarrier $(0 < r < Q)$, then

$$\hat{x}_m = \frac{1}{Q} e^{j2\pi z(r)} x_{\bar{m}} \tag{9.5}$$

where $z(r)$ represents an additional phase rotation.

Localized mapping (localized FDMA (LFDMA)) will assign users adjacent subcarriers. For LFDMA, the time domain sampling signal $\hat{x}_m$ still conforms to $x_{\bar{m}}/Q$. If $r \neq 0$, then

$$\hat{x}_m = \frac{1}{QN} \left(1 - E^{j2\pi y(r)}\right) \sum_{i=0}^{N-1} \frac{x_i}{1 - e^{j2\pi w(r,i)}} \tag{9.6}$$

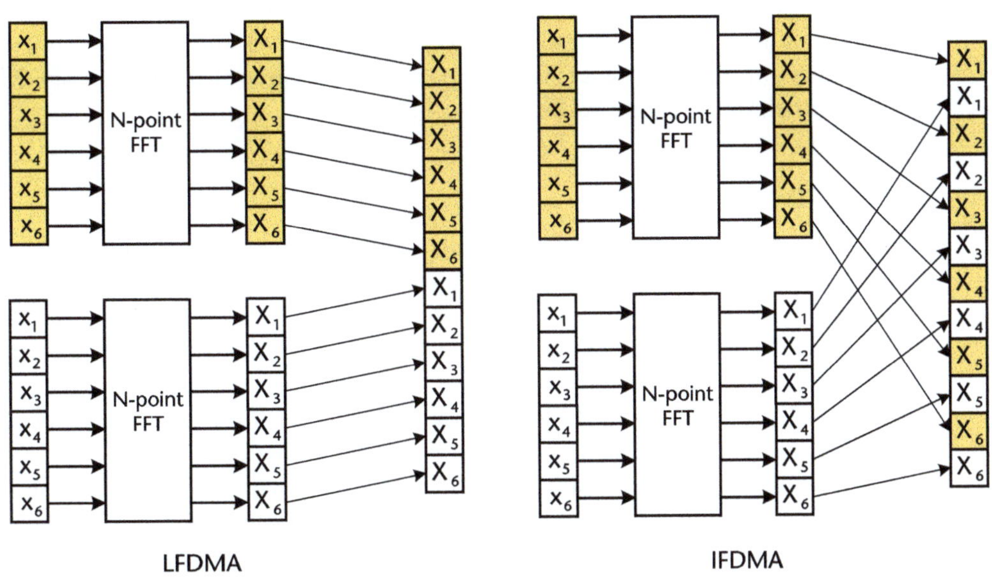

Figure 9.6 LFDMA and IFDMA for two users.

where $y(r)$, $w(r, i)$ are the additional phase and complex weighting coefficients, respectively. In addition, in any case, the time domain sampling signal has a factor $1/Q$ that significantly reduces the peak power of SC-FDMA.

IFDMA primarily utilizes frequency diversity and averages channel differences through interweaving. LFDMA primarily utilizes user diversity and selects subcarrier blocks for each user based on channel characteristics. Due to the uneven distribution of user input symbols, LFDMA has larger peak fluctuations in the time domain than IFDMA. Additionally, LFDMA maintains the orthogonality of the subcarriers with low complexity. Based on these characteristics, IFDMA is suitable for high-mobility environments, while LFDMA is suitable for channel-related scheduling in low-mobility environments.

9.2 Classic NOMA Schemes

The subchannel division in traditional OMA is typically based on the frequency domain, time domain, or code domain. In these approaches, the wireless resources allocated to different users do not overlap, which effectively reduced interference between users. However, due to limited user capacity and resource utilization, OMA cannot meet the needs of low-delay, high-quality, and high-user fairness in future communication systems. Therefore, 5G and its subsequent systems will explore new NOMA technologies [8] based on OFDM as a key technology.

NOMA allows different users to send data at the same frequency and at the same time, thus improving the utilization of resources [4]. However, the overlap of signals creates multiuser interference. As a result, the receiver needs to differentiate the multiuser signal using successive interference cancellation (SIC) or maximum likelihood detection to successfully detect the signal.

The current NOMA technology primarily focuses on two domains: the power domain and the code domain. In the power domain, the power domain NOMA (PD-NOMA) utilizes power variances to differentiate users, while in the code domain, it uses nonorthogonal short sequences. Code domain NOMA encompasses sparse code multiple access (SCMA), pattern division multiple access (PDMA), and multiuser shared access (MUSA).

9.2.1 PD-NOMA

PD-NOMA adjusts the power of each user's signal based on their path loss differences [9]. It allocates lower power to users with good channel conditions and higher power to users with poor channel conditions. This technique improves spectral efficiency by combining the signals of different users on the same time-frequency resources for transmission. In Figure 9.7, the resource allocation principle of PD-NOMA is illustrated, showing how different transmit powers can be allocated to three users on the same time slot and frequency band.

PD-NOMA uses the SIC algorithm to detect signals from multiple users at the receiver. First, SIC considers the low-power user signal as interference and demodulates for high-power users. Subsequently, SIC eliminates the high-power user signal from the received signal, thereby demodulating the low-power user

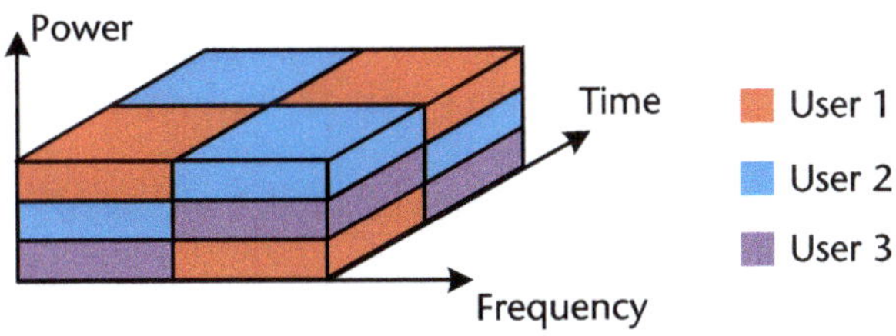

Figure 9.7 PD-NOMA.

signal. In comparison to OMA, PD-NOMA enhances resource utilization but leads to multiuser interference and increases the complexity of the receiver.

9.2.2 MUSA

MUSA uses sequences with low cross-correlation characteristics to distinguish users [10]. The sequence design is shown in Figure 9.8. If MUSA has two users, it initially assigns each user a unique sequence. Subsequently, it proceeds to multiply each user's symbol with its corresponding sequence in order to obtain the extended sequence. Finally, the extended sequences of both users are combined.

Upon reception, the detector initially utilizes the linear detection algorithm to acquire the initial user estimation message. Subsequently, it employs the SIC algorithm to process each user's message. The SIC algorithm demodulates one user's message at a time and treats the message of other users as interference. If the SIC successfully obtains the initial estimate of a single user, the user's message is considered to have been detected. Otherwise, the SIC further performs interference reconstruction elimination for detection.

MUSA measures system user capacity by defining user overload rates as shown below:

$$overload = K/L \tag{9.7}$$

where K is the number of access users and L is the length of the unique sequence. If the user extends the sequence length by L times, the MUSA overload occurs when $K > L$. This results in an overload rate generally exceeding 100%. While a longer extension sequence can accommodate more access users, it also leads

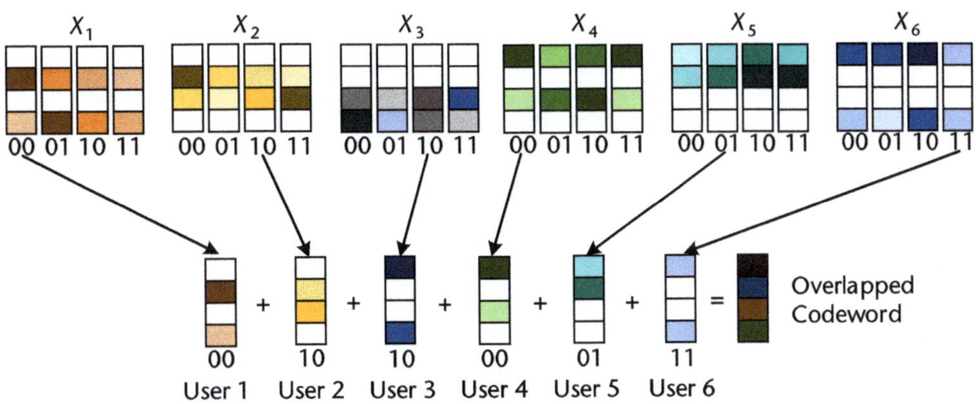

Figure 9.8 MUSA.

to increased system processing delay and computational complexity. Therefore, designing MUSA systems involves a trade-off between sequence length and system processing delay.

In summary, the characteristics of MUSA are as follows:

1. *No scheduling*: MUSA does not require resource allocation, allowing users to send data any time and reducing terminal energy consumption.
2. *The user overload rate is higher at a low data rate*: In MUSA, multiple users use complex domain sequences to transmit information over the same time-frequency resource. Increasing the complexity of the domain sequence reduces the transfer rate of MUSA, but it also increases the number of accessing users.
3. *Short sequence*: The sequence length of MUSA is shorter than CDMA's, but its cross correlation is very low, making it suitable for scenarios with large user access.

9.2.3 SCMA

SCMA uses unique nonorthogonal sparse codebooks to differentiate users [11], and detects multiple users' signals based on the sparsity of the codebooks. In the (4, 6) SCMA system, as shown in Figure 9.9, the message of six users is first mapped to codewords in a nonorthogonal codebook and then superimposed on four subcarriers. Because the codewords between users are nonorthogonal, SCMA achieves a 150% user overload rate, which improves the system's spectral efficiency.

In the SCMA system, the message is mapped directly to the K-dimensional transmission sequence based on the SCMA codebook. Therefore, the design of the codebook is crucial in determining the performance of the SCMA system. SCMA refers to the K-dimensional transmission sequence as the K-dimensional sparse codeword, and the set of them as the user codebook. SCMA encoders are defined as follows:

$$f : B^{\log_2 M} \rightarrow X \tag{9.8}$$

where B is the bit stream each user enters, X is the codebook size, and A is the K-dimensional sparse codeword generated after SCMA coding.

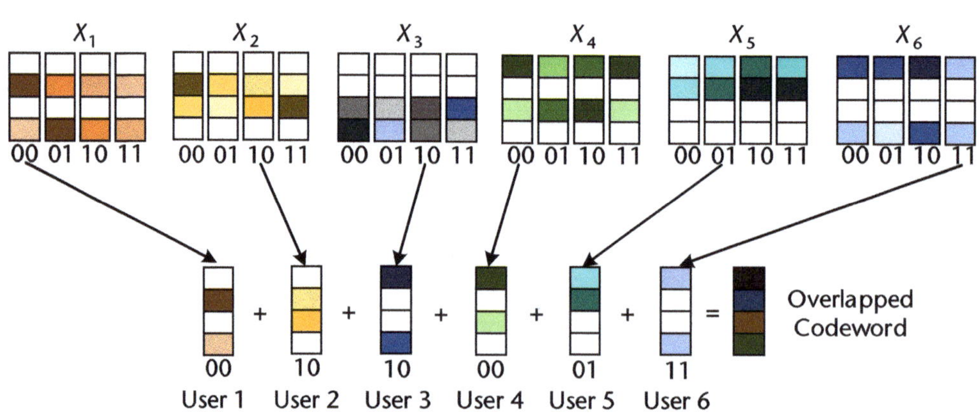

Figure 9.9 (4, 6) SCMA.

The SCMA encoder maps a bitstream of length $\log_2 M$ to a sparse codeword, whose dimension K generally equals the number of subcarriers. Assuming N is the number of nonzero elements per codeword, $N < K$, then each user only transmit signals over these N fixed resource blocks. The allocation of resource blocks between users can be represented by a binary signature matrix $\mathbf{F}$, where $\mathbf{F}[k, j] = 1$ means user j actively transmits signal on the kth RE. For example, for an $(4, 6)$ SCMA system, the signature matrix can be expressed as

$$\mathbf{F}_{4\times 6} = \begin{bmatrix} 1 & 1 & 1 & 0 & 0 & 0 \\ 1 & 0 & 0 & 1 & 1 & 0 \\ 0 & 1 & 0 & 1 & 0 & 1 \\ 0 & 0 & 1 & 0 & 1 & 1 \end{bmatrix} \tag{9.9}$$

According to the signature matrix $\mathbf{F}_{4\times 6}$, the binary mapping matrix of six users is shown as follows:

$$\begin{aligned} \mathbf{V}_1 &= \begin{bmatrix} 1 & 0 \\ 0 & 1 \\ 0 & 0 \\ 0 & 0 \end{bmatrix}, \mathbf{V}_2 = \begin{bmatrix} 1 & 0 \\ 0 & 0 \\ 0 & 1 \\ 0 & 0 \end{bmatrix}, \mathbf{V}_3 = \begin{bmatrix} 1 & 0 \\ 0 & 0 \\ 0 & 0 \\ 0 & 1 \end{bmatrix}, \\ \mathbf{V}_4 &= \begin{bmatrix} 0 & 0 \\ 1 & 0 \\ 0 & 1 \\ 0 & 0 \end{bmatrix}, \mathbf{V}_5 = \begin{bmatrix} 0 & 0 \\ 1 & 0 \\ 0 & 0 \\ 0 & 1 \end{bmatrix}, \mathbf{V}_6 = \begin{bmatrix} 0 & 0 \\ 0 & 0 \\ 1 & 0 \\ 0 & 1 \end{bmatrix} \end{aligned} \tag{9.10}$$

Based on the mapping matrix above, the user's codebook is generated by the following operations:

$$X_j = \mathbf{V}_j \Delta_j A_{\mathrm{MC}} \tag{9.11}$$

where Δ_j is the jth user's constellation calculation, $1 \leq j \leq J$. A_{MC} is the multidimension mother constellation. By performing constellation operations such as phase rotation, complex conjugation, layer power shift, and dimension exchange on the parent constellation, SCMA can enhance the disparity between user codebooks and improve the performance of multiuser detection. Mother constellations can be created in different ways, with the most common method being the generation of complex vector codewords through rotation interleaving:

$$_1 = \left\{ Y_m (1 + i) \,|\, Y_m = 2m - 1 - M, m = 1, \ldots, M \right\} \tag{9.12}$$

The design of the SCMA codebook involves mapping matrix design, constellation operation, and mother constellation design. These are all closely linked to the system's performance. The mapping matrix design decides the time and frequency resources used by each user, the constellation operation determines the arrangement of the constellation points for each user, and the mother constellation design represents the mapping relationship between user data and the complex constellation matrix. The SCMA codebook design is flexible and configurable, allowing for the generation of different codebooks to meet the requirements of various applications and scenarios.

9.2.4 PDMA

PDMA involves creating various diversity patterns for users to enable multidimensional resource allocations, including power, space, and code domains [12]. The base station overlays the user signal encoded by PDMA, which allows for increased multiuser multiplexing and diversity gain. The receiver uses a SIC detection algorithm to separate and demodulate user signals.

The PDMA multiuser pattern mapping design is illustrated in Figure 9.10. Considering a PDMA system with six users and four subcarriers, the system overload rate is 150%. In this configuration, the blank items of each user will not be mapped to RE.

In a PDMA pattern, each user fills a certain number of colors, which is then referred to as diversity. In Figure 9.10, User 1 has a diversity of 4, User 2 has a diversity of 2, and the PDMA pattern matrix is as follows:

$$G = \begin{bmatrix} 1 & 0 & 0 & 1 & 1 & 0 \\ 1 & 1 & 1 & 1 & 0 & 0 \\ 1 & 1 & 0 & 0 & 1 & 0 \\ 1 & 0 & 1 & 1 & 0 & 1 \end{bmatrix} \tag{9.13}$$

The complexity of the receiver increases with the number of "1." In the pattern matrix of (9.13), User 1 has a diversity of 4, so its data is mapped to all four REs. User 2 has a diversity of 2, so its data maps to the second and third RE. This process is repeated for six users, with each user's data being mapped to different REs according to the pattern. The data is then superimposed and finally transmitted in the same time-frequency resource.

User diversity is closely linked to transmission reliability. Increased diversity leads to higher transmission reliability, but also increases detection complexity. Therefore, the base station should properly design the PDMA pattern and user overload rate. It is important to minimize the detection complexity for the receiver while ensuring that the system performance meets the requirements.

There are three key technologies of PDMA.

9.2.4.1 Pattern Design

The pattern design of PDMA needs to consider two principles:

- To enhance multiuser access and spectrum utilization, the PDMA system can create multiple diversity patterns to improve multiplexing capability

Figure 9.10 PDMA.

by increasing the number of groups with different diversity. However, the receiver's complexity increases exponentially with the number of nonzero terms in the pattern matrix. Therefore, the PDMA should minimize the number of patterns to meet the system's requirements.

- PDMA shall minimize interference within the same diversity group. By minimizing the overlap of different pattern sequences and reducing the number of users using the same resource, the BER performance of PDMA can be improved.

9.2.4.2 Pattern Distribution

The level of diversity in the pattern can be adjusted based on the specific business requirements of different users. A high diversity indicates high business requirements, while low diversity suggests lower business requirements. Additionally, the pattern can also be adjusted based on the principle of fairness. This means that a pattern with higher diversity is assigned to a user with poor channel quality, while a pattern with lower diversity is assigned to a user with better channel quality.

In situations where there are strict time delay requirements, such as in vehicle networking, the uplink PDMA system can be used with a scheduling-free scheme to minimize the delay caused by request scheduling. Moreover, in scenarios where short packet service is predominant in the uplink, the no-schedule scheme can help reduce the performance decline caused by delay and signaling overhead, but it can also increase the complexity of the receiver.

9.2.4.3 Power Optimization

PDMA is a MA scheme designed to operate in the power, space, and code domains. It assigns different power optimization schemes to different users based on their requirements. Designing a reasonable power optimization scheme at the transmitter can enhance the BER performance.

9.3 MA for Terrestrial Cellular Communication

The terrestrial mobile cellular communication networks divides the ground area into hexagonal cells, where each cell is served by a base station to provide wide area access service for numerous users.

As outlined in Chapter 1, the terrestrial cellular network standard has gone through five generations of evolution. The first generation (1G) was developed by Bell Labs in the 1970s and primarily transmitted analog voice signals. 1G used FDMA to allocate each user's wireless resources based on a fixed frequency, with each analog user channel being 30 kHz/25 kHz. Its main characteristics included low spectrum utilization, poor security, and limited service types. Additionally, there were issues such as high base station equipment costs, a large number of terminals, and poor call quality.

The second generation of cellular communications (2G) marked the transition from analog to digital communications. The ETSI played a key role in this transition. ETSI proposed the GSM, while Qualcomm proposed the IS-95 standard system. GSM uses TDMA and is mainly deployed in the 900-MHz and 1800-MHz frequency bands, while IS-95 uses CDMA and is mainly deployed in the 800-MHz and 1900-MHz bands.

The third generation of mobile communications (3G) uses CDMA as the core of MA technology. It consists of three global standards:

1. WCDMA, initially proposed by Europe and Japan, builds its core network based on evolving GSM/GPRS network technology. It adopts DSSS broadband CDMA for the air interface.
2. CDMA 2000 (CDMA2000), proposed by North America, builds the core network based on the evolved IS-95 CDMA technology.
3. TD-SCDMA, mainly promoted by China, incorporates smart antennas and time-division duplex methods based on traditional CDMA. It offers unique spectrum utilization and flexibility advantages.

The fourth generation mobile communications (4G) technology standard, LTE-A, was developed by the 3GPP organization. LTE-A uses OFDMA as the MA standard in the downlink and SC-FDMA in the uplink [13]. In the downlink, OFDMA allocates subcarriers to different users based on OFDM, allowing for multiuser multiplexing of channel resources. In the uplink, SC-FDMA effectively reduces the PAPR of the multicarrier transmission system through DFT pre-processing, thereby easing the hardware requirements of user terminals and encouraging the widespread use of 4G devices.

The technical specifications for the fifth generation of mobile communications (5G) are mainly developed by the 3GPP organization and are called NR. In 5G NR, OFDMA is used as the main MA scheme for uplink and downlink, and SC-FDMA is used as a supplement [14]. There are two reasons for adopting OFDMA. First, the reduction of hardware cost enables the user terminal to support OFDMA uplink transmission. Second, 5G base stations use multiantenna technologies such as MIMO to improve performance, and OFDMA is highly adapted to it. However, when the user is at the edge of the cell, the performance of uplink OFDMA is limited by power. Hence, the user terminal adopts SC-FDMA to provide quality service assurance.

9.4 MA for Satellite Communication

9.4.1 MF-TDMA

The current standards for satellite-ground MA are mainly developed by ETSI. These standards include the Digital Video Broadcasting - Return Channel by Satellite (DVB-RCS) and the second generation of DVB-RCS, known as DVB-RCS2 [18]. Both of these standards are proposed as extensions of the second generation of Digital Video Broadcasting Satellite (DVB-S2) to support communications backlink in interactive satellite applications.

Both DVB-RCS and DVB-RCS2 utilize multifrequency TDMA (MF-TDMA) as the MA technology. MF-TDMA combines FDMA and TDMA to allocate communication resources for different users across both time and frequency domains. This approach enhances satellite resource utilization efficiency and ensures flexibility in resource allocation, addressing the limitations of TDMA to a certain extent. As a result, MF-TDMA has emerged as the predominant MA technology for high-throughput satellite communication systems.

The principle of MF-TDMA is shown in the Figure 9.11. MF-TDMA uses time division multiplexing for each carrier in the system. The TDMA rate can be the same or different, and even the carrier rate of the same carrier in different time slots can be changed. Compared with the typical single-carrier TDMA system, MF-TDMA reduces the carrier rate and the transmission capability of the client. By combining different carrier rates, MF-TDMA makes it easier to construct a high-throughput satellite communication system that is compatible with flexible networking.

MF-TDMA can be converted between TDMA and FDMA. When the total number of carriers in the MF-TDMA system gradually decreases to 1 and the air interface rate increases, MF-TDMA is converted to high-speed TDMA. On the other hand, when the total carrier number of the MF-TDMA system gradually increases and the air interface rate decreases to the same as the client rate, the corresponding MF-TDMA is converted to FDMA.

As shown in Figure 9.12, MF-TDMA is classified into static and dynamic, depending on the frequency hopping (FH) capability of the user terminal. In static MF-TDMA, the user terminal cannot change the time slot width, carrier rate, and modulation coding mode while continuously sending signals. It only frequency hops on carriers with the same rate and slot size but at different frequency points. Additionally, other configurations of these carriers, such as modulation coding mode, must be the same. In dynamic MF-TDMA, the user terminal can change the time

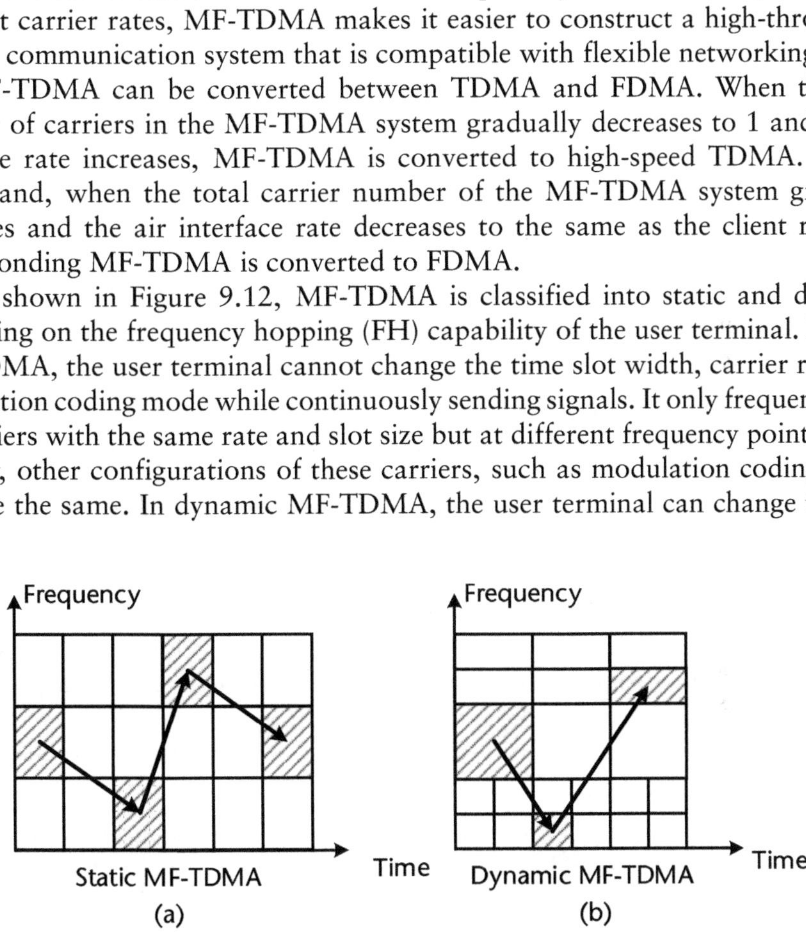

Figure 9.11 MF-TDMA: (a) Static MF-TDMA, and (b) dynamic MF-TDMA.

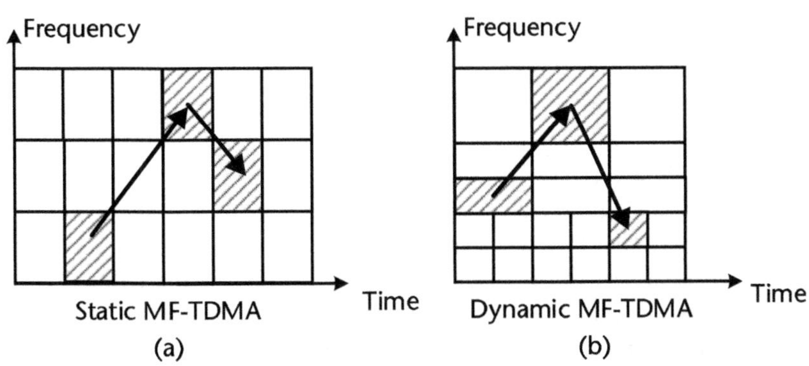

Figure 9.12 Static and dynamic MF-TDMA: (a) Static MF-TDMA, and (b) dynamic MF-TDMA.

slot width, carrier rate, and modulation coding mode at any time while continuously sending signals. This means that dynamic multifrequency TDMA allows FH on different time slot widths and carriers with different rates, making it more suitable for communication services with different broadband requirements.

According to the FH speed of the terminal, dynamic multifrequency TDMA can also be divided into fast hop and slow hop. In fast FH, the terminal continuously executes FH within the time slot and utilizes the protection interval to complete the frequency switch. Typically, the length of the protection interval ranges from a few to a dozen symbols. In slow FH, the terminal cannot continuously switch frequency points, and it takes at least 1 time slot and no more than 1 second to complete FH each time. In order to ensure that FH does not occur within the frame, the slow FH system distributes the time slots within a frame on the same carrier continuously. Compared with slow FH terminals, fast FH terminals are more complex, but also more flexible to adapt to the needs of broadband communication services.

Depending on whether the transmitter and receiver perform FH, MF-TDMA communication systems can be classified into the following three types:

1. *Transmitter FH Only MF-TDMA Network System*: The transmitter FH only (TFHO) mechanism is widely used in the current MF-TDMA satellite communication system. In TFHO, the transmitted carrier hops to different frequency points in each time slot, and each receiving station adopts a different frequency for the received carrier. When setting up the network, all receiving sites are divided into groups, and each group is assigned a fixed receiving carrier known as the on-duty carrier. In the communication between stations, the sending station performs FH according to the duty carrier of the corresponding receiving station by time slot, and the receiving station receives the information sent by other stations on the duty channel. The receiving station then receives the information sent by other stations on the duty channel.
2. *Receiver FH Only MF-TDMA Network System*: Like TFHO, the receiver FH only (RFHO) system also needs to group all receiving stations in the construction of the network, but each group is configured with a fixed-frequency transmission carrier. In a specific time slot, the sending station transmits a signal on a fixed carrier, while the receiver changes time slot according to the corresponding carrier of the sender to receive signal. Compared with TFHO, the maximum carrier rate of a large aperture station in an RFHO-based MF-TDMA system is higher, but the maximum carrier rate of a small aperture station is the same.
3. *Transmitter-Receiver FH (TRFH) MF-TDMA Networking System*: In TRFH-based MF-TDMA, the stations are not grouped, and the signals sent and received by each station can be frequency-hopped across different carriers. MF-TDMA allocates carriers and time slots based on the processing capacity of the transceiver station. This means time slots on different carriers are allocated based on their asymmetric transmission capacity. When setting up a multistation type hybrid network, the appropriate carrier rate

will be selected based on the sending and receiving capability of large and small caliber stations.

The TRFH system is most suitable for multistation hybrid networking, while the TFHO system is the least suitable. In reality, the TRFH and RFHO systems can only build a circuit-switched network based on the time slot, whereas the TFHO system can build a packet-switched network. The TFHO system has the lowest system complexity and is the easiest to implement. Considering the advantages and practical needs of the three MF-TDMA systems, the TFHO system is the most widely used, but its ability to support multistation hybrid networking still needs improvement.

In summary, MF-TDMA offers several advantages, including one-to-many communication capabilities, the flexibility to build various networking structures, high-speed data communication capabilities, dynamic allocation of channel resources, and strong support for internet multimedia services. However, MF-TDMA also has some shortcomings, such as the need for synchronization of the entire network due to the multicarrier configuration, the potential for intermodulation component generation between the multicarriers, and the complexity of the resource allocation algorithm.

9.4.2 Hybrid TDMA/CDMA

In satellite group networking, a hybrid TDMA/CDMA system is commonly used. In the current scenario of intersatellite communication, establishing a direct path between the source and destination satellites proves to be challenging. This difficulty arises from the intermittent connectivity between satellites, and the susceptibility of small satellite networks to unexpected failures [19]. To enhance the network's robustness, the TDMA/CDMA hybrid MA protocol divides the entire satellite network into clusters and implements a primary-secondary mode, as illustrated in Figure 9.13. Each cluster comprises one primary satellite and several secondary satellites. The clusters are represented by orange dotted lines in the figure, with the primary satellite denoted by a red square and the secondary satellites by gray squares.

Intercluster communication involves the secondary satellite sending data to the primary satellite, which then forwards the data to the destination satellite within the same cluster. For intercluster communication, if a dependent satellite needs to

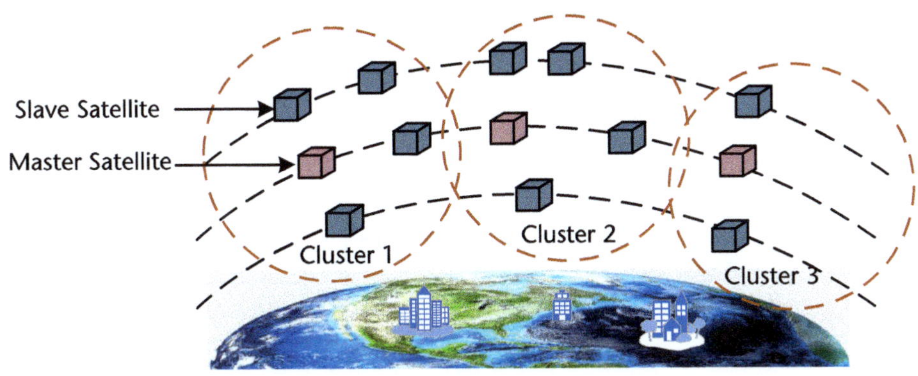

Figure 9.13 Satellite cluster system model.

communicate with satellites in another cluster, it must first communicate with its primary satellite and then communicate with the primary satellite of the cluster where the destination is located. However, multistep data forwarding consumes a lot of energy and significantly increases latency. Therefore, it is important to reorganize the network and consider using the proximity center algorithm to select the primary satellite that meets the minimum power requirements.

In the hybrid TDMA/CDMA system, there are two implementation schemes: TDMA-centered and CDMA-centered. In the TDMA-centered scheme, each cluster is assigned a unique signature code to distinguish different clusters. Additionally, each satellite has specific uplink and downlink time slots for transmitting data to the master satellite. During the same time slot, multiple satellites from different groups transmit signals using different signature codes to avoid interference. Figure 9.14 illustrates the TDMA-centered frame structure.

In the CDMA-centered scheme, satellites in the same cluster are each assigned a unique signature code. As shown in Figure 9.15, dependent satellites can simultaneously transmit data to the primary satellite without interference in the first time slot using their respective orthogonal sequence codes. For the primary satellite, there are unique time slots for transmitting data to neighboring satellites and downlink time slots for receiving data from neighboring satellites.

9.5 Potential MA for Integrated Communication

9.5.1 Rate Splitting Multiple Access

Multibeam satellite systems are usually fitted with multiple feeders and cater to various user groups across multiple cochannel beams. This type of satellite

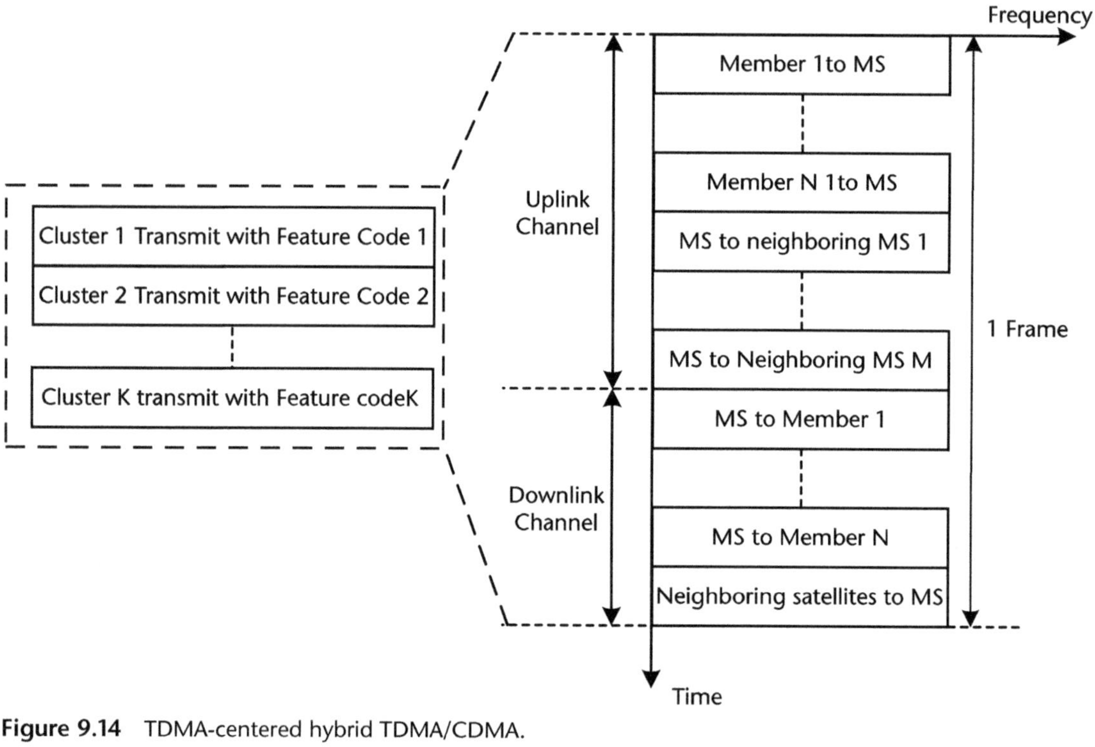

Figure 9.14 TDMA-centered hybrid TDMA/CDMA.

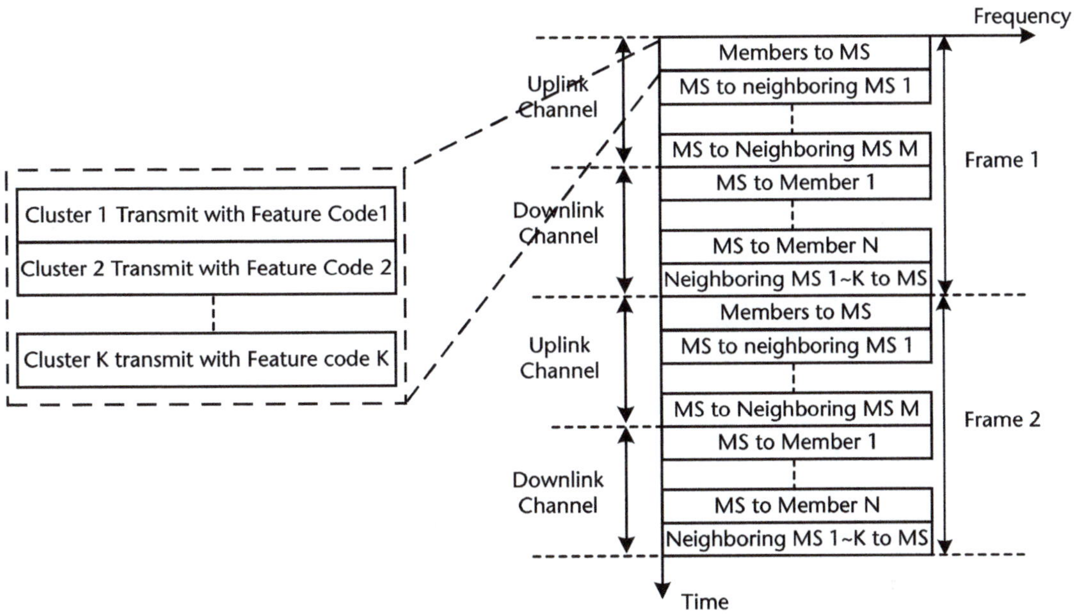

Figure 9.15 CDMA-centered hybrid TDMA/CDMA.

communication operates in a manner similar to cellular multigroup multicast mode, leading to interbeam interference. As a result, rate splitting multiple access (RSMA), which is well-suited for multigroup multicast transmission in cellular networks, holds great promise for the advancement of multibeam satellite communication [20].

RSMA uses low-rate channels to combine coding with scrambling and uses the excellent correlation characteristics of scrambling to distinguish users. As shown in Figure 9.16, the RSMA scheme enables multiple users to occupy all time-frequency resources simultaneously.

Specifically, RSMA divides the message to be transmitted into public and private components at the transmitter. Subsequently, RSMA combines the public component into a unified signal and transmits it using the same time-frequency resources as the private component. At the receiver, each user decodes part of the interference and their message.

As illustrated in Figure 9.17, considering the RSMA process in the single-cell two-user downlink scenario. The base station sends message w_1, w_2 to users 1 and 2. First, the message w_k is split into a private part $w_{p,k}$ and a public part $w_{c,k}$,

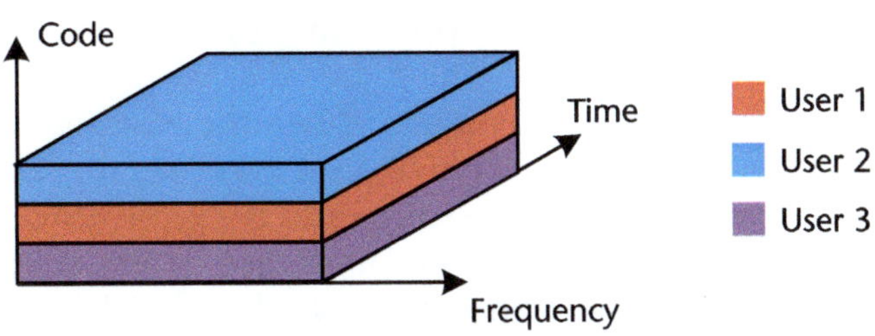

Figure 9.16 RSMA.

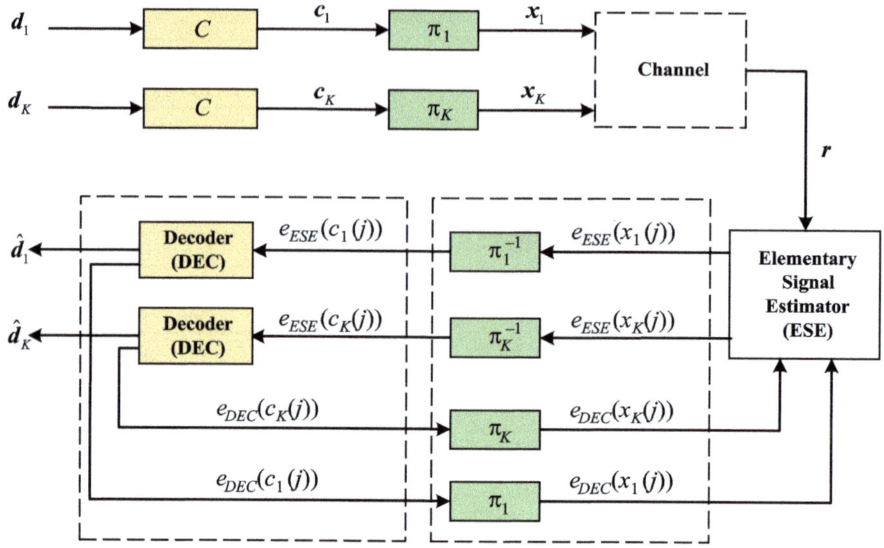

Figure 9.17 RSMA model with two users.

$(k = 1, 2)$. It's important to note that the original message doesn't have to be split, and the ratio of segmentation can be a design parameter. Next, we combine all the separate public parts $w_{c,k}$ into a whole w_c, and then obtain the public signal flow s_c after coding. The private message $w_{p,k}$ is separately encoded to obtain the private signal stream s_k.

At the receiver, user k initially considers all private signals as interference. User k then decodes public signal s_c to obtain public message $\tilde{w}_c$. Next, user k reconstructs signal $\tilde{w}_c$ with coding, precoding, and channel processing. After that, user k subtracts the reconstructed public signal from the received signal. Then, user k treats other users' private signal s_j $(j \neq k)$ as interference and decodes private signal s_k to obtain the user's private message $w_{p,k}$. Finally, user k extracts $\tilde{w}_{p,k}$ from $\tilde{w}_c$ in the public message. After merging, the complete message $\tilde{w}_1$, $\tilde{w}_2$ sent by the base station to user k can be obtained.

To extract its own public message from the overall public message after decoding, user k needs to rely on control signal transmission between the base station and the user. When the public message contains only a single user's public part, RSMA only needs to attach control signaling between the transmitter and the user. However, if the public message contains the public part of multiple users, each user must first decode the public message and then retrieve their own part from the decoded message. This process requires an additional control signal between the transmitter and all users to indicate how to split and recover the original message for each user.

The RSMA connects the two extremes by partially decoding the interference and treating part of it as noise. This means that the interference is either completely decoded or entirely treated as noise. When the base station doesn't separate the public message, the message w_k sent to User k will be fully encoded to s_k. In this case, User k won't decode any interference, and the receiver will treat the interference brought by s_j $(j \neq k)$ as noise, similar to the concept of spatial division multiple access (SDMA). If the base station fully encodes w_2 and w_1 to s_c and s_1, respectively, User 1 has to decode the message it needs after fully decoding User 2's message upon receiving. However, User 2 only needs to decode its own message, which is akin to the concept of PD-NOMA.

As a flexible technology to manage interference, RSMA encompasses three design principles: including regarding interference as noise (e.g., SDMA), treating interference entirely as useful signal (e.g., NOMA), and transmitting the signal of a single user to avoid interference (e.g., OMA). By adjusting the resources allocated to the common stream, RSMA can adapt the level of interference to accommodate different network loads (underload/overload) and user deployment needs (different channel direction/strength). It can also switch modes in different situations. For example, by allocating the power to public and private streams properly, RSMA can be converted to SDMA when the network is underloaded. In this mode, the user channels are orthogonal, and the channel state information at transmitter (CSIT) is perfect. Also, the RSMA transforms to NOMA when the channels for different users are overlapped. In other channel conditions, where channels for different users are neither orthogonal nor completely overlapped, common streams can be utilized by RSMA, which makes RSMA outperform all other MA schemes.

RSMA is well suited for managing multiuser interference caused by imperfect CSITs. Also, RSMA is robust to different sources of damage, such as quantized feedback, lead contamination, channel estimation errors, and CSIT uncertainties resulting from user movement. The overall rate performance of RSMA quickly saturates as the SNR increases, and decreases rapidly as the user speed increases. This is in contrast to other MA schemes (OMA, SDMA, NOMA), which are primarily designed for perfect CSITs and are susceptible to imperfect CSITs.

9.5.2 Interleave Division Multiple Access

Interleave division multiple access (IDMA) is a special case of CDMA. Instead of considering a spread sequence specific to the user, IDMA uses specific interleaves for user segregation.

Figure 9.18 shows the IDMA scheme with K simultaneous users where users are distinguished by their interleavers. The data sequence d_k of user-k is encoded based on a low-rate code c, generating a coded sequence $c_k \equiv [c_k(1), \ldots, c_k(j), \ldots, c_k(J)]^T$, where J is the frame length. Then c_k is permutated by an interleaver π_k, producing $x_k \equiv [x_k(1), \ldots, x_k(j), \ldots, x_k(J)]^T$.

The fundamental principle of IDMA is that the interleavers, denoted as π_k, must be distinct for each user. It is assumed that these interleavers are generated independently and randomly. This differentiation allows the interleavers to spread the coded sequences, resulting in adjacent chips that are approximately uncorrelated. Such dispersion facilitates the simple chip-by-chip detection scheme discussed in the following sections.

Adopting an iterative receiver as illustrated in Figure 9.18, which consists of an elementary signal estimator (ESE) and K decoders (DECs). The MA and coding constraints are considered separately in the ESE and DECs. The outputs of the ESE and DECs are LLRs about $\{x_k(j)\}$ defined below:

$$e(x_k(j)) \equiv \log\left(\frac{p(y\,|x_k(j) = +1)}{p(y\,|x_k(j) = -1)}\right), \forall k, j \tag{9.14}$$

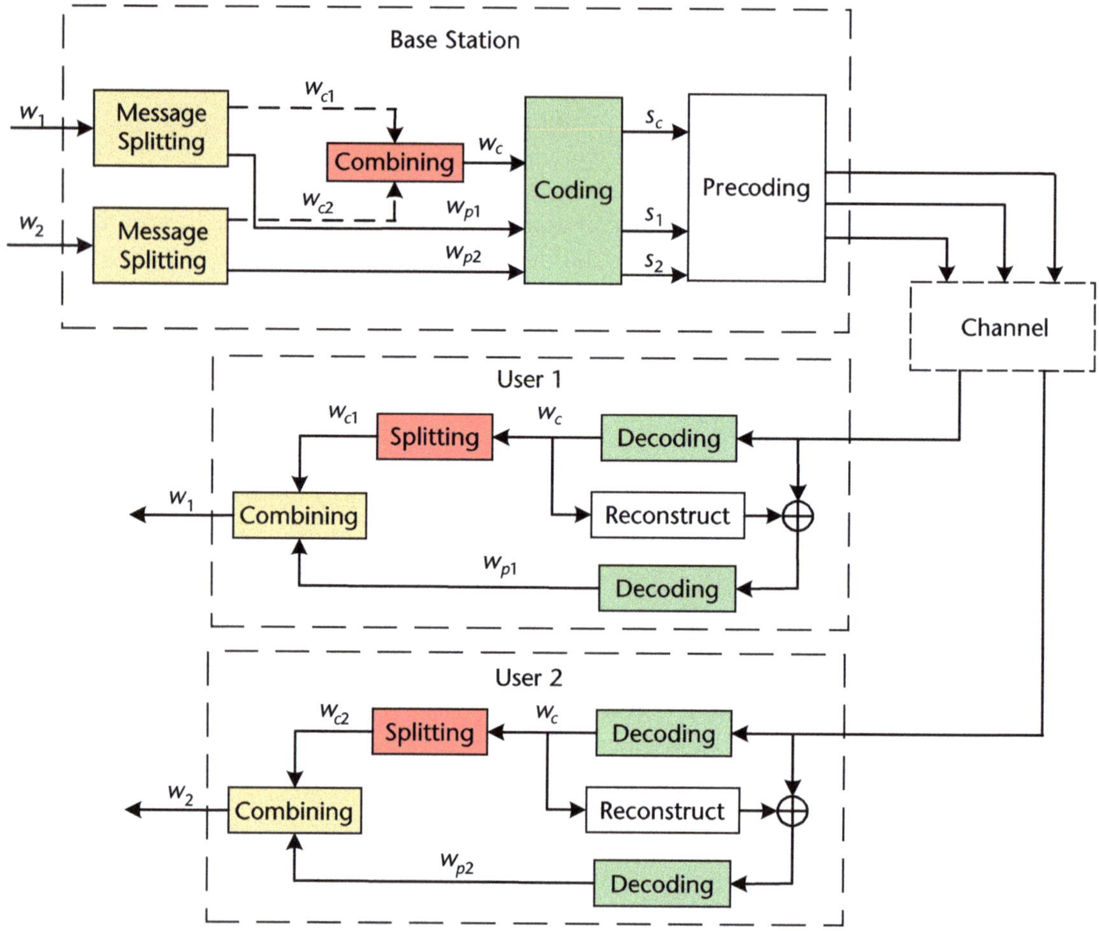

Figure 9.18 IDMA with K simultaneous users.

These LLRs are different between ESE and DEC. As a result, $e_{ESE}(x_k(j))$ and $e_{DEC}(x_k(j))$ represents they are separately generated by the ESE and DECs. For the ESE, y in equation (9.14) denotes the received signal. For the DECs, y in (9.14) is formed by the deinterleaved version of the outputs of the ESE. A global turbo decoding is applied to process the LLRs.

9.5.3 Lattice Partition Multiple Access

Lattice partition multiple access (LPMA) is a novel downlink nonorthogonal multiuser superposition transmission scheme for future cellular networks, where the base station transmits multilevel lattice codes for multiple users. Each user's code level corresponds to a distinct prime and is weighted by a product of all distinct primes of the other users excluding its own. Due to the structural property of lattice codes, each user can cancel out the interference from the other code levels by using the modulo lattice operation in a successive/parallel manner. LPMA can offer improved user fairness in symmetrical broadcast channels compared with NOMA.

Assuming the mapping function of superposition coding is designed based on construction π_A, given by

$$\mathbf{x} = \beta \left[w\left(\mathbf{v}_1, \ldots, \mathbf{v}_L\right) + \mathbf{u} \right] \tag{9.15}$$

where the symbol $\mathbf{x}$ represents the resulting lattice codeword for L UE, $\mathbf{u}$ denotes a fixed dither that minimizes the average transmission power to ensure the entire constellation has zero mean, β is the scaling factor used to meet the power constraint, and w signifies the mapping function, given by

$$w\left(\mathbf{v}_1, \ldots, \mathbf{v}_L\right) \triangleq \left[\sum_{l=1}^{L} \mathbf{v}_l \prod_{l'=1, l' \neq l}^{L} \theta_{l'} \right] \bmod \prod_{l=1}^{L} \theta_l R \tag{9.16}$$

which maps $\mathbf{v}_l$ into a lattice codeword in an element-wise manner, where $\mathbf{v}_l$ is generated from C_l over the finite field $\mathbf{F}_{ql}$. Similar to the power allocation scheme of NOMA, LPMA ensures fairness through proper power distribution. More specifically, users with poor channel status get more power, while those with good channel status get a relatively small share. For example, assuming that UE 1 is further away from the BS than UE 2, let $\theta_1 = 2$ and $\theta_2 = 7$, then $\mathbf{x} = \beta \left[(7\mathbf{v}_1 + 2\mathbf{v}_2) \bmod 14 + \mathbf{u} \right]$. In this case, UE 1 with poor channel status is compensated by larger prime $\theta_2 = 7$.

References

[1] Shahen Shah, A. F. M., A. N. Qasim, M. A. Karabulut, H. Ilhan, and M. B. Islam, "Survey and Performance Evaluation of Multiple Access Schemes for Next-Generation Wireless Communication Systems," *IEEE Access*, Vol. 9, 2021, pp. 113428–113442.

[2] Gilhousen, K. S., I. M. Jacobs, R. Padovani, and L. A. Weaver, "Increased Capacity Using CDMA for Mobile Satellite Communication," *IEEE Journal on Selected Areas in Communications*, Vol. 8, No. 4, 1990, pp. 503–514.

[3] Saleh, A. A. M., "Intermodulation Analysis of FDMA Satellite Systems Employing Compensated and Uncompensated TWT's," *IEEE Transactions on Communications*, Vol. 30, No. 5, 1982, pp. 1233–1242.

[4] Bongiovanni, G., D. Coppersmith, and C. Wong, "An Optimum Time Slot Assignment Algorithm for an SS/TDMA System with Variable Number of Transponders," *IEEE Transactions on Communications*, Vol. 29, No. 5, 1981, pp. 721–726.

[5] Gilhousen, K. S., I. M. Jacobs, R. Padovani, A. J. Viterbi, L. A. Weaver, and C. E. Wheatley, "On the Capacity of a Cellular CDMA System," *IEEE Transactions on Vehicular Technology*, Vol. 40, No. 2, pp. 303–312, 1991.

[6] Lopez-Perez, D., A. Valcarce, G. de la Roche, and J. Zhang, "OFDMA Femtocells: A Roadmap on Interference Avoidance," *IEEE Communications Magazine*, Vol. 47, No. 9, 2009, pp. 41–48.

[7] Myung, H. G., J. Lim, and D. J. Goodman, "Single Carrier FDMA for Uplink Wireless Transmission," *IEEE Vehicular Technology Magazine*, Vol. 1, No. 3, 2006, pp. 30–38.

[8] Dai, L., B. Wang, Y. Yuan, S. Han, I. Chih-lin, and Z. Wang, "Non-Orthogonal Multiple Access for 5G: Solutions, Challenges, Opportunities, and Future Research Trends," *IEEE Communications Magazine*, Vol. 53, No. 9, 2015, pp. 74–81.

[9] Maraqa, O., A. S. Rajasekaran, S. Al-Ahmadi, H. Yanikomeroglu, and S. M. Sait, "A Survey Of Rate-Optimal Power Domain NOMA with Enabling Technologies of Future Wireless

Networks," *IEEE Communications Surveys & Tutorials*, Vol. 22, No. 4, 2020, pp. 2192–2235.

[10] Yuan, Z., G. Yu, W. Li, Y. Yuan, X. Wang, and J. Xu, "Multi-User Shared Access for Internet of Things," *2016 IEEE 83rd Vehicular Technology Conference (VTC Spring)*, Nanjing, China, 2016, pp. 1–5.

[11] Mu, H., Z. Ma, M. Alhaji, P. Fan, and D. Chen, "A Fixed Low Complexity Message Pass Algorithm Detector for Up-Link SCMA System," *IEEE Wireless Communications Letters*, Vol. 4, No. 6, 2015, pp. 585–588.

[12] Chen, S., B. Ren, Q. Gao, S. Kang, S. Sun, and K. Niu, "Pattern Division Multiple Access Novel Nonorthogonal Multiple Access for fifth-Generation Radio Networks," *IEEE Transactions on Vehicular Technology*, Vol. 66, No. 4, 2017, pp. 3185–3196.

[13] 3GPP, TS 36.201: V17.0.0; Evolved Universal Terrestrial Radio Access (E-UTRA); LTE Physical Layer; General Description (Release 17), 2022.

[14] GPP, TS 38.201: V17.0.0; NR; Physical layer; General description (Release 17), 2022.

[15] IEEE Std 802.11n-2009, IEEE Standard for Information Technology–Local and Metropolitan Area Networks–Specific Requirements–Part 11: Wireless LAN Medium Access Control (MAC) and Physical Layer (PHY) Specifications Amendment 5: Enhancements for Higher Throughput, 2009.

[16] IEEE Std 802.11ac-2013, IEEE Standard for Information Technology–Telecommunications and Information Exchange Between Systems Local and Metropolitan Area Networks–Specific Requirements–Part 11: Wireless LAN Medium Access Control (MAC) and Physical Layer (PHY) Specifications–Amendment 4: Enhancements for Very High Throughput for Operation in Bands Below 6 GHz, 2013.

[17] IEEE Std 802.11ax-2021, IEEE Standard for Information Technology–Telecommunications and Information Exchange Between Systems Local and Metropolitan Area Networks–Specific Requirements Part 11: Wireless LAN Medium Access Control (MAC) and Physical Layer (PHY) Specifications Amendment 1: Enhancements for High-Efficiency WLAN[S], 2021.

[18] ETSI EN 301 545-2 V1.3.1, Digital Video Broadcasting (DVB); Second Generation DVB Interactive Satellite System (DVB-RCS2); Part 2: Lower Layers For Satellite Standard, 2020.

[19] Radhakrishnan, R., W. W. Edmonson, F. Afghah, R. Martinez Rodriguez-Osorio, F. Pinto, and S. C. Burleigh, "Survey of Inter-Satellite Communication for Small Satellite Systems: Physical Layer to Network Layer View," *IEEE Communications Surveys & Tutorials*, 2016, Vol. 18, No. 4, pp. 2442–2473.

[20] Mao Y., O. Dizdar, B. Clerckx, R. Schober, P. Popovski, and H. V. Poor, "Rate-Splitting Multiple Access: Fundamentals, Survey, and Future Research Trends," *IEEE Communications Surveys & Tutorials*, 2022.

CHAPTER 10

Resource Management for Satellite-Terrestrial Integrated Communication

10.1 Overview of Multidimensional Resources

10.1.1 Spectrum Resources

The spectrum resource of satellite communication refers to the frequency range of electromagnetic waves used in satellite communication, which is a limited, invaluable, and nonrenewable natural resource. The specific use of frequency band and frequency division may vary from country to country and region. Telecommunications regulatory agencies in different countries will plan and allocate spectrum according to domestic needs and international agreements. Therefore, the specific use and regulations of the frequency band should refer to local authorities or relevant spectrum management regulations. As shown in Figure 10.1, satellite communication spectrum resources are divided into the following frequency bands [1].

10.1.1.1 L-Band: 1–2 GHz

The L-band is mainly used in mobile communications, broadcasting, remote sensing, and military communications. The L-band is used in satellite mobile communication systems, such as satellite phone systems, which provide users with voice calls and text messaging services worldwide through satellite networks; it is used in some satellite broadcasting service systems to provide satellite TV and broadcasting services over a large area; it is used in remote sensing systems to receive and transmit data from earth observation satellites, monitor and study changes in the earth's surface and atmosphere, such as weather, oceans, and climate; in addition, the L-band can also be used for military communications.

10.1.1.2 S-Band: 2–4 GHz

The S-band is mainly used for meteorology, marine radar, and satellite communications. When electromagnetic waves with a wavelength of 10 cm are used, their band is defined as the short (S) band, which means electromagnetic waves are shorter than the original wavelength. According to the ITU satellite mobile services can use the 1,980–2,010/2,170–2,200 MHz uplink and downlink frequency bands with a bandwidth of 30 MHz and the 2,483.5–2,500 MHz downlink frequency

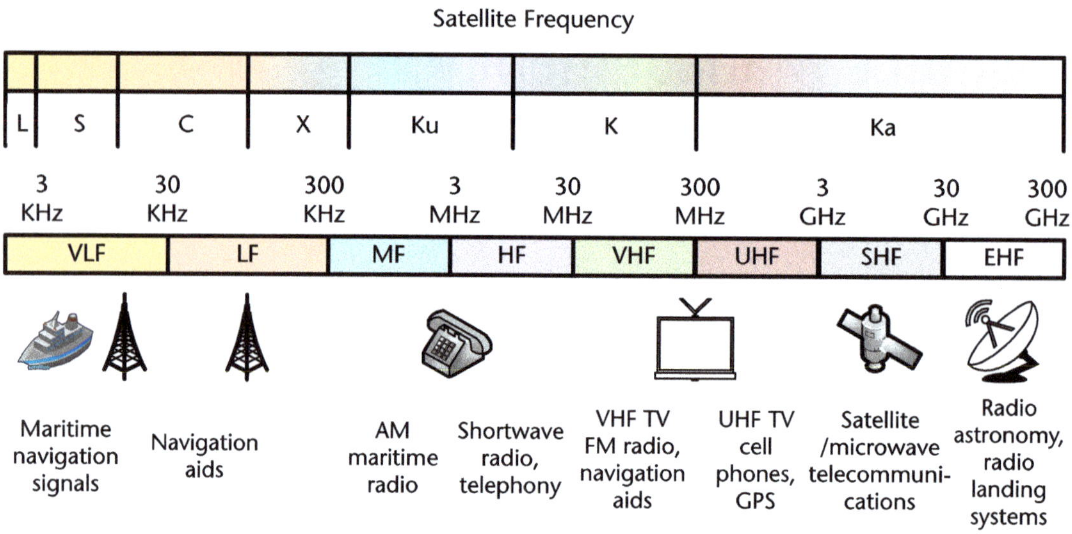

Figure 10.1 Satellite spectrum resource allocation diagram.

band with a bandwidth of 16.5 MHz. Still, their priority is lower than that of terrestrial services. The S-band communication frequency range commonly used by deep-space stations of NASA and the ESA is 2,025–2,120 MHz for uplink and 2,200–2,300 MHz for downlink. Inmarsat [2] and Eutelsat use the 1.98–2.01/2.17–2.20 GHz frequency bands for satellite mobile services. NASA uses the S-band for satellite relay services between the space shuttle, the International Space Station, and the terrestrial terminals. The Federal Communications Commission (FCC) stipulates that the 2.31–2.36 GHz frequency band is for satellite sound broadcasting. The bandwidth resources of the S-band are minimal, and the terrestrial terminal antenna is not very directional.

10.1.1.3 C-Band: 4–8 GHz

The C-band refers to the 4–8 GHz band. It is mainly used for satellite communications, full-time satellite TV networks, or raw satellite signals. The C-band is usually used in tropical rainfall areas because it has stronger resistance to rain attenuation than the Ku-band. The C-band is the first band allocated to commercial telecommunications via satellite, and the terrestrial microwave radio relay chain has already used the same frequency. Almost all C-band communication satellites use the 3.7–4.2 GHz band for downlinks and the 5.925–6.425 GHz band for uplinks.

10.1.1.4 X-Band: 8–12 GHz

The X-band has good characteristics regarding anti-interference and rain attenuation, antenna terminal size, transmission rate, and coverage in remote areas. It is a band with a relatively comprehensive solid performance, and the X-band is mainly used as a reserved band for government use, so it is very suitable as a military communication band. The military widely uses X-band satellite communications for beyond-line-of-sight communications. Satellite communications widely use the 7.9–7.4/7.25–7.75 GHz band, referred to as the 8/7 GHz band.

10.1.1.5 Ku-Band: 12–18 GHz

The Ku-band is widely used in the field of satellite communications. Since the Ku-band was applied by the National Broadcasting Corporation (NBC) in the United States in 1983, it has been a popular choice for direct broadcasting. The Ku-band can also be used as a communication band between satellites, such as the Tracking and Data Relay Satellite (TDRS) and SpaceX Starlink satellites for communications between the International Space Station and the Space Shuttle. The Ku-band is also used for other satellite communication applications, including data transmission, voice communications, broadband internet access, and military communications. Its high frequency allows for greater bandwidth, achieving faster data transmission rates and facilitating various multimedia services. Satellite communications are divided into fixed satellite services (FSS) and broadcast satellite services (BSS). In the Asia-Pacific region, fixed satellite services generally use the 14.0–14.25/12.25–12.75 GHz band (abbreviated as 14/12 GHz band); fixed satellite services can use extended bands with uplinks of 13.75–14 GHz and downlinks of 10.7–10.95 and 11.45–11.7 GHz. The Ku-band offers several advantages for satellite communications, including high data rates and relatively small antenna sizes. However, compared with lower bands such as the C-band, the Ku-band is more susceptible to signal attenuation caused by adverse weather conditions such as rain fade.

10.1.1.6 Ka-Band: 26.5–40 GHz

The Ka-band is mainly used for uplinks in the 27.5 and 31 GHz bands of communication satellites and high-resolution, close-range targeting radars on military aircraft. In satellite communications, the Ka-band allows communications with a broader bandwidth. The Ka-band was first used in the experimental ACTS Gigabit Satellite Network. Currently, the Ka-band is used by Inmarsat [2] and Kacific for internet access in geostationary high-throughput satellites. The SES O3b system also uses it in medium orbit, the Space Starlink system, and Iridium Next [2] satellites in low orbit. Satellite projects currently using or planning to use the Ka-band include Amazon's LEO Project Kuiper satellite internet constellation, SES's GEO SES-17 satellite multiorbit satellite internet system, and the MEO O3b mPOWER [2] constellation. Compared with the Ku-band, the Ka-band is more susceptible to rain attenuation.

With the development of satellite communication technology and the growth of application demand, prominent satellite constellations such as Oneweb [3], Starlink [4], and Telesat [5] have been deployed in Earth orbit. Satellite communication services have also gradually diversified and scaled up, and the demand for spectrum resources has steadily increased to support performance indicators such as high speed, high capacity, and low latency. However, due to the physical characteristics of electromagnetic waves and the constraints of international regulations, the frequency bands and bandwidths available for satellite communication are limited. In addition, different frequency bands have different propagation characteristics and applicable scenarios and cannot be replaced at will. As a result, satellite communication spectrum resources are facing increasingly severe tension and competition. The scarcity of satellite spectrum resources has become one of the main obstacles to the development of satellite communication [6]. Due to the global coverage and spatial coexistence of satellite communication systems, satellite communication spectrum resources are easily interfered with by other satellite systems, terrestrial systems, or the natural environment, affecting the reliability and security of satellite communication systems [7]. At the same time, due to historical reasons and the complexity of international coordination, satellite communication spectrum resources are unbalanced and underutilized among different countries, regions, and systems. This has also caused resource waste and conflicts of interest, exacerbating the consumption of spectrum resources.

To ease the tension and competition of satellite communication spectrum resources at the spectrum management level, it is necessary to strengthen the planning, management, and protection of satellite communication spectrum resources, avoid duplication and abuse, and prevent illegal occupation and malicious interference, promote international cooperation and coordination, establish fair, reasonable, and transparent satellite communication spectrum resource allocation mechanisms and rules, and strengthen information exchange and technical support. At the technical level, innovate satellite communication technologies and methods, improve the utilization efficiency and optimization performance of satellite communication spectrum resources, such as using cognitive radio, multiple access, beamforming, dynamic allocation, and other technologies to achieve spectrum sensing, sharing, reuse, and other functions, explore new satellite communication frequency bands and space domains, develop higher frequency or lower orbit satellite communication systems, expand the spatial and temporal dimensions of satellite communication spectrum resources, and provide more diversified and personalized satellite communication services.

10.1.2 Power Resources

There are three primary sources of energy for driving equipment on spacecraft in space: solar energy, chemical batteries, and nuclear energy. Solar panels convert solar energy into electrical energy and store it in photovoltaic cells. They are the most common and widely used satellite energy source. For example, the International Space Station (ISS) has eight solar panels that can provide an average of 75 kw of power. Mars rovers can also use solar panels, such as the Mars Exploration Rover (Spirit), Phoenix Lander, and InSight Lander, as well as some probes far from the sun. However, they require a considerable area to collect enough light energy. Chemical batteries are generally used as satellites' secondary or backup power resources, especially when solar panels are unavailable or insufficient. The most common type of battery currently used in satellites is lithium-ion batteries, which have high energy density and long cycle life [8]. However, nuclear power has the advantages of being independent of external conditions, having high reliability, and being high power compared to solar cell and chemical power. However, nuclear power has intense radiation. Therefore, if nuclear power is used as the energy supply of satellites, the on-satellite equipment must take corresponding shielding measures, which increases the mass and design complexity of satellites. At the same time, nuclear power costs are high, and there are safety hazards. Therefore, this power supply is rarely used on satellites, mainly in deep-space probes, such as Voyager 1 and 2 probes and Galileo probes.

As the primary energy source for Earth-orbiting satellites, the collection efficiency of solar energy is not only related to the environment but also limited by the satellite's conditions. The satellite's size determines the satellite's surface area and volume, which affects the power resources and payload capacity of the satellite. Generally speaking, the larger the satellite, the larger the surface area, and the more solar panels can be installed to obtain more power. At the same time, the larger the volume, the more batteries, fuels, instruments, and other payloads can be accommodated to provide more functions and performance. However, larger satellites also mean higher manufacturing and launch costs. In recent years, the low-Earth orbit satellite constellation system has developed rapidly, including OneWeb, SpaceX, and Telesat systems. Their single LEO satellites run fast and have a small coverage area. Therefore, many satellites need to be launched to provide broadband internet access services to the world. To control costs, the size of the launched satellites will be reduced as much as possible, thereby increasing the number of satellites carried by the rocket. However, a smaller size also means weaker solar energy acquisition capabilities and fewer power resource reserves. Meteorological satellites operating in geosynchronous orbits are about 36,000 km from the earth's surface and can provide a few hundred watts to a few kilowatts. Some examples are, China's Fengyun series of meteorological satellites, the United States' GOES series of meteorological satellites, and Europe's Meteosat series of meteorological satellites. Communication satellites operating in geosynchronous orbits or medium orbits are about 20,000 to 36,000 km from the earth's surface and can provide a few kilowatts to tens of thousands of watts of power. Some examples are, China's Dongfanghong series of communication satellites, the United States' Intelsat series of communication satellites, and Europe's Eutelsat series of communication satellites. Remote sensing satellites operating in low orbits about hundreds to thousands of kilometers from the earth's surface can generally provide tens of watts to hundreds of watts of power. Some examples are, China's Gaofen series of remote sensing satellites, the United States' Landsat series of remote sensing satellites, and Europe's Sentinel series of remote sensing satellites. In SpaceX's Starlink low-orbit satellite constellation, the average power requirement of each satellite is 0.8 kw, of which the communication antenna accounts for 0.6 kw. However, this power requirement varies with factors such as the satellite's operating mode, mission execution, and orbital parameters. The power that each satellite can provide is approximately between 0.5 and 1.5 kw. In summary, the power of a satellite is not only related to factors such as the intensity of solar radiation and the type and efficiency of solar panels but also to the type of work of the satellite. For example, although a meteorological satellite operates in a synchronous orbit, its power is less than that of a communication satellite operating in a medium orbit. Therefore, satellite designers need to comprehensively consider multiple factors, such as the satellite's mission objectives, quality constraints, and cost budgets to optimize the satellite's power resource allocation and utilization.

The limited power load on board brings a series of challenges to the design and optimization of satellite communication systems: how to improve the spectral efficiency and energy efficiency of satellite systems under limited power resources, how to reasonably allocate satellite power resources while ensuring user service quality and satisfaction, and how to achieve flexible adjustment of satellite power resources while considering satellite channel characteristics and user demand diversity. A multibeam satellite [9] is a communication system with multiple beams. Traditional satellite communication systems usually have only one or a few beams, while multibeam satellites can provide services to various terrestrial stations or users simultaneously. Multibeam satellites have multiple independent beam transmission and reception systems, and each beam can be independently directed and adjusted. This enables multibeam satellites to provide services to geographically dispersed users simultaneously, improving the communication system's capacity and efficiency. Multibeam satellites can provide higher communication capacity, better coverage, and more flexible communication services by utilizing multiple beams. However, in multibeam satellite systems, the radiation power between different beams may interfere with each other. When the radiation directions between beams are close or overlapping, the receiver may receive signals from multiple beams simultaneously, resulting in signal quality degradation or data transmission errors. Therefore, it is usually necessary to adopt an appropriate allocation method to reasonably allocate limited on-satellite power resources to maximize the performance and efficiency of the system while ensuring the service quality and satisfaction of users.

The traditional power allocation method evenly distributes the total power transmitted by the satellite to all beams or coverage areas or in a particular proportion. Although this fixed power allocation strategy is simple and direct, it cannot fully take into account the differences and demand changes between different users or regions. It cannot dynamically respond to the changing conditions of the system. For example, when the number of users, channel quality, or interference level changes, the traditional method cannot adjust the power allocation strategy in time to adapt to the new situation, which may lead to resource waste, performance degradation, or failure to meet user needs. In response to these shortcomings, related research proposes new power allocation algorithms and strategies to improve the performance and efficiency of satellite communication systems. These new methods include intelligent power allocation, adaptive power control, and optimization algorithms, aiming to provide more flexible, adaptive, and efficient power allocation solutions. The following introduces some related research on multibeam satellite power allocation.

10.1.2.1 Power Allocation Algorithm Based on Business Demand

According to the business demand of different spot beams, the power allocation of spot beams is dynamically adjusted so that spot beams with high demand get more power and spot beams with low demand get less energy, thereby improving the capacity and efficiency of the system. In the literature, Ottersten proposed a power allocation algorithm for multibeam satellite systems based on multiobjective optimization [10]. It is divided into two stages: the first stage minimizes the unmet system capacity according to the business demand of users; the second stage minimizes the total power consumption of the satellite while ensuring the results of the first stage. A genetic algorithm and nondominated sorting genetic algorithm are used to solve this multiobjective optimization problem, and the performance and trade-off of the system are demonstrated through the Pareto frontier.

10.1.2.2 Power Allocation Algorithm Based on Channel State Information

This method dynamically adjusts the power allocation of spot beams according to the channel state information of different spot beams so that spot beams with poor channel conditions get more power and spot beams with good channel conditions get less power, thereby improving the reliability and performance of the system. Durand proposed a multibeam satellite communication power allocation method based on heuristic particle swarm optimization (PSO) in [11], aiming to solve the power allocation problem in multiple narrow beams to provide the minimum signal-to-noise ratio (SINR) required for earth station users to establish reliable communication, optimize power allocation, and improve the energy efficiency and performance of the system.

10.1.2.3 Power Allocation Algorithm Based on Nonorthogonal Multiple Access Technology

Nonorthogonal multiple access technology (NOMA) is used to realize spectrum multiplexing of various users in the same spot beam. By reasonably allocating the power ratio between users, the signal serial interference cancellation (SIC) between users is realized, thereby improving the throughput and flexibility of the system. In [12], the offered capacity to requested traffic ratio (OCTR) is used to measure the matching of requested and offered data rates in multibeam satellite systems. NOMA is used to alleviate intrabeam interference, while precoding is used to reduce inter-beam interference. The power, decoding order, and terminal time slot allocation are jointly optimized to improve the fairness of OCTR.

10.1.2.4 Beam Hopping

Beam hopping (BH) is a multibeam satellite-to-terrestrial communication technology that can use satellite resources to provide services to specific locations or users. It allocates resources in four dimensions: space, time, frequency, and power to adapt to the dynamic changes and uneven distribution of terrestrial services [13]. This technology is based on time-slicing technology and uses fewer beams to achieve traditional multibeam coverage, thereby reducing the number of antennas required by the satellite [14]. By adjusting the duration and period of the beam lighting, the satellite can provide different capacity values for other cells to balance the requirements of varying beam coverage areas. In addition, beam hopping can also place the unlit beam position as an isolation area between cofrequency beams to reduce cofrequency interference. Beam-hopping technology provides a basis for the flexible allocation and efficient use of satellite resources. It is considered a key technology for the new generation of high-throughput satellites. Beam-hopping technology will be introduced in detail in the subsequent sections.

10.1.3 Time Slot Resources

In satellite communication systems, available bandwidth resources are divided into two dimensions: frequency and time, and different time and frequency resources are allocated to other users or services. Using TDMA, other users can send or receive signals in various periods on the same frequency band, thereby realizing spectrum resource sharing among multiple users. In the beam-hopping system shown in Figure 10.2, different beam-hopping illumination modes provide services to the cells in its coverage area in a time-division multiplexing manner.

The advantages of TDMA are that it can avoid cochannel interference, simplify signal processing, and improve spectrum utilization. Its disadvantages are that it requires precise clock synchronization, has time slot gap loss, and is sensitive to multipath fading. In satellite-to-terrestrial communication systems, reasonable management and scheduling of time slot resources and reasonable allocation and utilization of time slot resources according to the needs of different users or services can improve system performance, efficiency, and reliability and reduce interference and cost. According to the way time slots are divided, there are three types of time slot resource management and scheduling methods:

1. *Slot-based scheduling*: Divide the two frequency and time dimensions and allocate one or more fixed time slots to each user or service. This method is simple and easy to implement, but it has low flexibility and may cause waste or shortage of resources.
2. *Nonslot-based scheduling*: That is, the time granularity of resource allocation is allowed to be less than one-time slot, and the starting point can be at any OFDM position. This method can improve resource utilization and adapt to delay-sensitive services, but the implementation complexity is relatively high.
3. *Scheduling based on minislot and slot aggregation*: This combines the scheduling methods based on time slots and nonslots and can be flexibly allocated according to factors such as business volume, priority, and quality of service (QoS). This method can balance resource utilization and implementation complexity and supports multiple access and frequency-hopping technology.

Depending on the type of satellite communication system, there are also different methods and strategies for time slot resource allocation. For example, in FSS, time slot resource allocation mainly considers factors such as user traffic, priority, and QoS; in Mobile Satellite Service (MSS), time slot resource allocation also

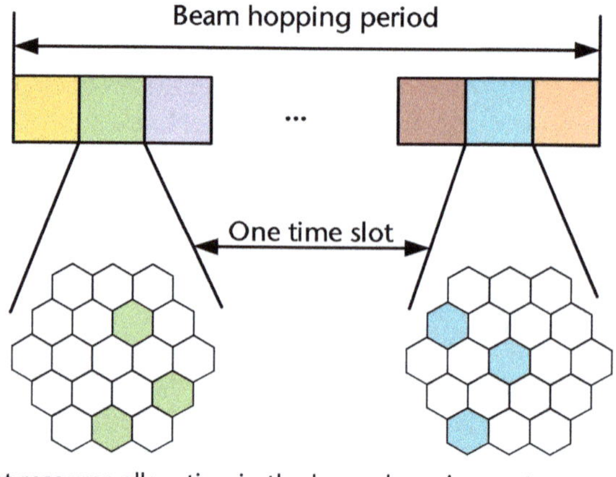

Figure 10.2 Time slot resource allocation in the beam-hopping system.

considers factors such as user mobility, location, and speed; in low Earth orbit satellite network (LEOSN), time slot resource allocation also considers factors such as satellite orbit dynamics, intersatellite links, and terrestrial gateways. Time slot resource allocation is a complex optimization problem involving multiple objectives and constraints. Generally speaking, the goal of time slot resource allocation is to maximize system throughput, minimize transmission delay, optimize resource utilization, and best meet user QoS; the constraints of time slot resource allocation include spectrum resources, power resources, modulation and coding schemes, multiple access technologies, and interference restrictions. To solve the problem of time slot resource allocation, scholars have proposed many algorithms and methods, such as heuristic algorithms, genetic algorithms, particle swarm algorithms, and neural network algorithms. These algorithms and techniques have advantages and disadvantages and must be selected and designed according to specific application scenarios and system requirements.

10.2 Resource Management Technology

10.2.1 Frequency Reuse Technology

A multibeam antenna system with multiple beams is used in applications such as satellite communications, radar systems, and mobile communications. Compared with traditional single-beam satellites, satellites equipped with multibeam antennas can serve multiple users or regions simultaneously, provide higher communication capacity and broader coverage, and have the following key features and advantages. First, multibeam satellites can transmit and receive multiple independent beams at the same time. Each beam can be independently pointed to different users, regions, or directions to achieve parallel communication connections. In this way, the capacity and efficiency of the communication system are greatly improved. Second, different beams can transmit data simultaneously on the same spectrum resources through multibeam technology. This realizes spatial and frequency multiplexing, increases the utilization efficiency of the spectrum, and provides more communication capabilities. In addition, multibeam satellites can realize flexible formation and beam adjustment by controlling the antenna array's unit elements. By changing the direction and width of the antenna beam, it can adapt to different communication needs and coverage and provide directional transmission or broadcast services. Multibeam satellites can also achieve interference suppression and isolation by adjusting the angle and power distribution between different beams. This reduces interference between users or regions and improves communication quality and system performance. In addition, multibeam satellites allow dynamic resource allocation, allocating the energy and bandwidth of the antenna to different beams on demand. Resource allocation can be adjusted in real time according to user needs and the environment, providing flexible communication services. Multibeam satellite systems are widely used in satellite communications. They can provide high-capacity and high-efficiency communication services, supporting a wide range of application scenarios, including satellite broadcasting, internet access, mobile communications, aviation, and maritime communications. The development of multibeam satellite systems provides an essential solution for improving satellite communication capabilities and meeting the growing communication needs. Multibeam satellite systems have three main frequency reuse methods: multicolor frequency reuse, soft frequency reuse, and nonorthogonal multiple access technology.

10.2.1.1 Multicolor Frequency Reuse Technology

Multibeam satellite frequency reuse technology aims to improve satellite communication systems' spectrum utilization efficiency and capacity. This technology is based on frequency division multiplexing (FDM), which enhances spectrum utilization and communication capacity by dividing the available spectrum into multiple nonoverlapping frequency bands and allocating each frequency band to different beams for independent communication. The principle is that at the transmitting end, a multibeam or array antenna is used to divide the same frequency band into multiple subbands, which are allocated to beams covering specific areas or targets to form different colors. Taking three-color multiplexing as an example, the available frequency band is divided into three mutually orthogonal subbands, and then the subbands are allocated to different beams. All beams constitute three nonintersecting sets, and the beams in the same set are given the same color. At the same time, to reduce cochannel interference, adjacent beams need to use subbands of different colors. In addition, to ensure that each beam's transmission power in its designated frequency band meets the communication requirements, appropriate power allocation is required. Power allocation can be optimized based on the interference between beams and the balance of system performance to maximize the communication quality. Common frequency band allocation strategies include uniform allocation based on user priority and data transmission volume, and dynamic allocation at the receiving end based on real-time communication load and user demand changes. At the receiving end, the corresponding multibeam antenna or array antenna is used to separate and demodulate the signal according to the color of the beam to obtain a helpful signal. Between different beams, spatial isolation, polarization isolation, coding isolation, and other technologies are used to suppress or eliminate cochannel interference. Multicolor frequency multiplexing technology can simultaneously transmit multiple signals at the same frequency, improve the spectrum efficiency and system capacity of the satellite coverage area, and realize the interconnection between different beams. The fewer the number of colors in multicolor multiplexing, the higher the efficiency of frequency multiplexing. Therefore, three-color multiplexing is the most efficient frequency multiplexing method, but the cochannel interference faced by three-color multiplexing is also the most serious. Standard multiplexing methods include three-color, four-color, and six-color multiplexing, and their schematic diagrams are given in Figure 10.3.

Although spatial isolation between beams of the same color can effectively reduce cochannel interference, the receiver will still receive signals from different beams and satellites, causing interference to the receiver and decreasing communication quality, affecting communication efficiency and reliability. Therefore, further solutions are needed to suppress co-channel interference. Commonly used technologies are as follows:

1. *Precoding technology*: Precoding technology is a technology that uses CSI to preprocess the transmitted signal. By using the satellite uplink to obtain

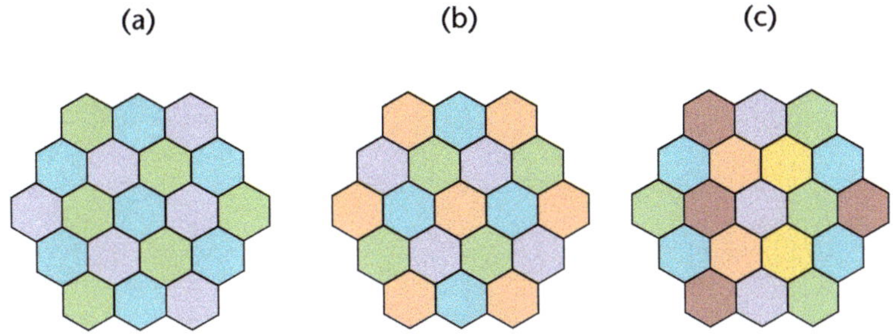

Figure 10.3 Schematic diagram of (a) three-color, (b) four-color, and (c) six-color frequency multiplexing.

channel state information and preprocess the downlink signal, cochannel interference can be effectively reduced, and the signal energy can be concentrated near the target user so that the receiving end can obtain a better signal-to-noise ratio (SNR) and improve the system channel capacity [15]. The main problems precoding technology faces are the acquisition of CSI and the design of the precoding matrix. With the application of large-scale antennas, the dimensions of the channel matrix and the precoding matrix will also increase, thereby increasing the complexity of the algorithm and the difficulty of system hardware implementation.

2. *Beamforming technology*: Beamforming can divide the space into multiple nonoverlapping logical channels, suppress interference signals in nontarget directions, and enhance signals in the target direction. The exact frequency can be used for communication within and between spot beams with spatial isolation characteristics. Beamforming technology can be divided into explicit beamforming and implicit beamforming. The former requires the terminal to provide feedback channel information. At the same time, the latter uses the interchange of the time division duplex system to calculate the channel information in the sending direction.
3. *Cell-splitting technology*: The original larger cell is split into three or four smaller cells, and the channel groups are rearranged according to a specific rule to reduce the cofrequency reuse distance and increase the number of channels available in the same cell, thereby suppressing cofrequency interference and improving system capacity.

10.2.1.2 Soft Frequency Reuse Technology

Soft frequency reuse (SFR) is an improvement and development of traditional frequency reuse technology. Figure 10.4 is a schematic diagram of the principle of SFR. A frequency is no longer determined as being used or not used in a cell, but the transmission power threshold determines the use of the frequency in the cell. The principle of soft frequency reuse is to divide the available frequency band into N parts, and for each cell, one part is used as the primary carrier, and the other is used as the subcarrier. The power threshold of the primary carrier is higher than that of the subcarrier, the leading carriers of adjacent cells do not overlap, the primary carrier can be used in the entire cell and the subcarrier is only used inside

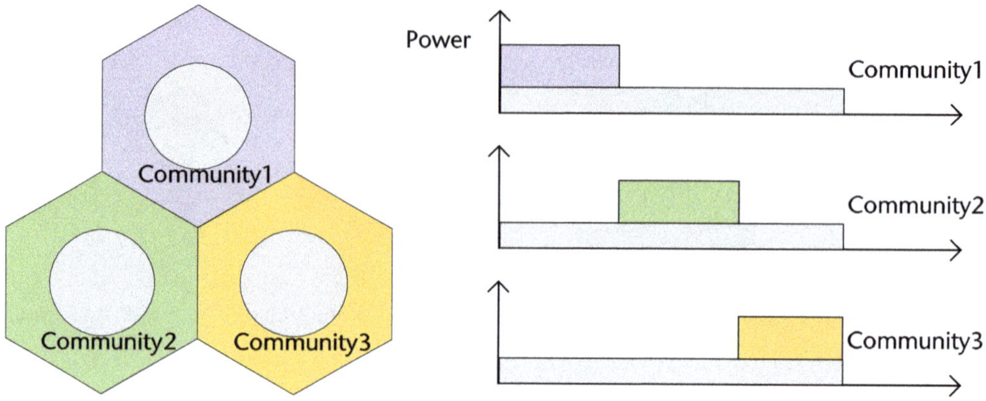

Figure 10.4 SFR principle diagram.

the cell, and by adjusting the ratio of the power threshold of the subcarrier to the primary carrier, the distribution of the load inside the cell and at the edge of the cell can be adapted.

In [16], a frequency reuse method of vectorizing the cell using directional antennas is proposed based on the SFR scheme. The basic principle is to use directional antennas to divide the entire cell into several sector areas, divide the main frequency band into an equal number of frequency bands, and assign them to each sector area. The main frequency band can be used in the entire cell, and the secondary frequency band is shared by multiple cells. A lower transmission power is used in the cell center. This maintains the advantages of SFR and enables each cell to access all frequency band resources, effectively improving spectrum utilization.

10.2.2 Beam Hopping Technology

In traditional multibeam satellite systems, onboard resources (such as power or bandwidth) are evenly distributed, and all beams are permanently illuminated, making it impossible to match the provided capacity with the heterogeneous service allocation between beams. In this case, beams that cannot meet service needs will have poor service quality problems, while beams with excess capacity will cause resource waste. To solve the problem of the fixed beam allocation method not being flexibly adapted to the dynamic changes in terrestrial service needs, resulting in low resource utilization and limited capacity, beam-hopping technology was proposed [18]. It can dynamically adjust the illumination time and position of the beam according to different traffic requirements, thereby improving the utilization and flexibility of satellite resources. Beam-hopping technology can bring the following three benefits: First, in the BH system, beam scheduling is optimized based on the requested traffic, which can effectively reduce unmet and unused capacity. Second, when all beams are not illuminated simultaneously, the number of RF links required is more minor, thereby reducing power consumption and payload quality. Third, by illuminating beams that are far apart, the adjacent beams of these illuminated beams are inactive, which can significantly reduce cochannel interference. Reference [19] shows that compared with satellite communication systems without BH antenna payloads, satellite systems equipped with BH antennas can significantly reduce energy consumption and increase system capacity.

Figure 10.5 is a diagram of a BH satellite system. The BH satellite system consists of four parts: a network control center (NCC), a gateway (GW), a satellite equipped with a flexible payload, and a user terminal (UT). The NCC is responsible for managing and controlling the satellite system, including generating and distributing beam hopping patterns and routing and scheduling of data streams. The GW is responsible for bidirectional communication with the satellite, forwarding the data stream of the terrestrial network to the satellite and receiving the data stream forwarded by the satellite. The satellite is responsible for realizing beam hopping and switching according to the instructions of the NCC and forwarding the uplink carrier to the corresponding downlink carrier and beam. The UT is responsible for bidirectional communication with the satellite, sending and receiving data streams.

In a BH satellite system, K beams serve cells in a time-division manner. In each time slot, the satellite's multibeam antenna will adjust the beam and select K cells for illumination. The illumination direction of the activated beam is different in each time slot, which is called the BH pattern. The BH illumination pattern of all time slots in a beam-hopping window constitutes the beam-hopping time plan (BHTP).

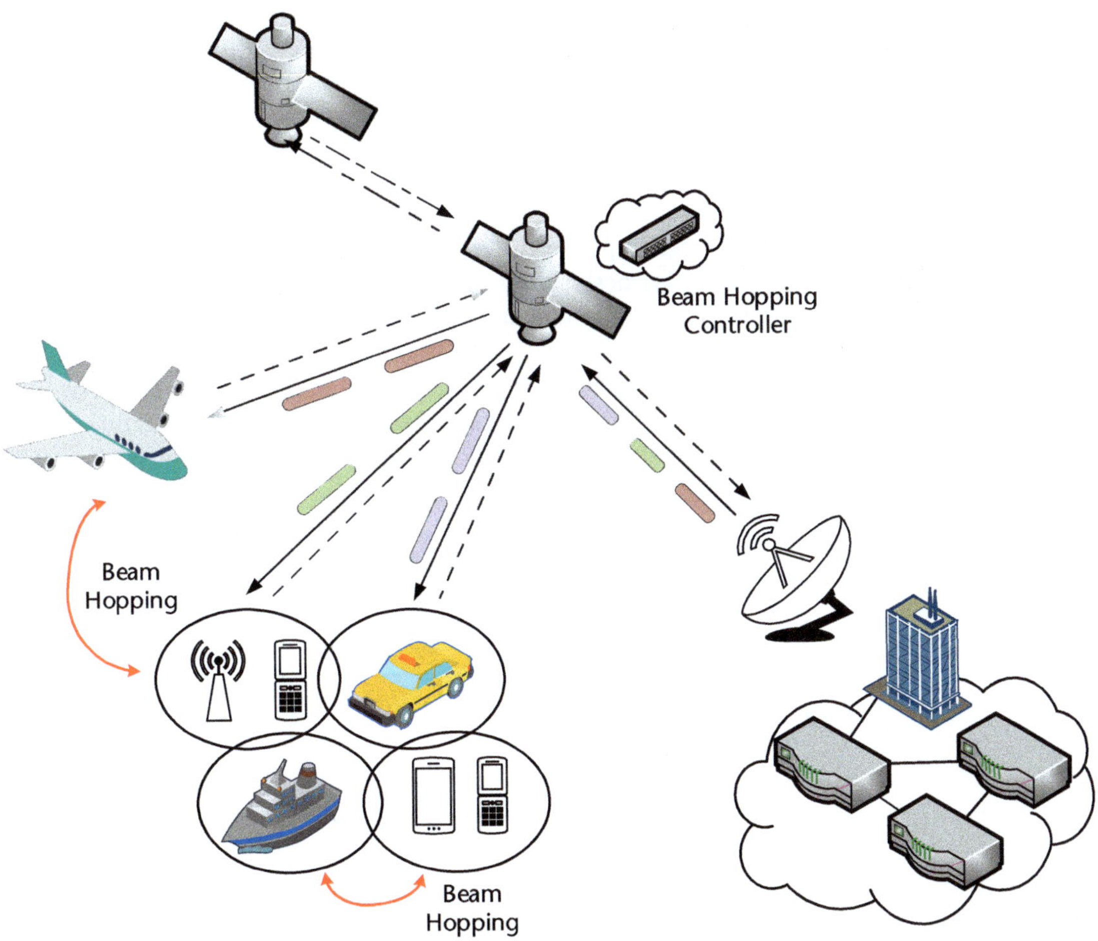

Figure 10.5 BH satellite system composition.

The satellite will select the corresponding beam for illumination in each time slot according to the BHTP. The BHTP diagram of a beam-hopping window is shown in Figure 10.6.

To improve the performance of BHTP satellite systems, most research focuses on how to design effective methods to determine BHTP [20]. The design of BHTP needs to consider the following factors: beam capacity requirements and channel status to ensure service quality and resource utilization, interference between beams and satellite power constraints to improve system performance and reliability, and frequency and complexity of beam hopping to reduce system overhead and complexity. The design methods of BHTP can be divided into two types: static and dynamic. The static method generates a fixed BHTP based on preset business requirements and channel status. In contrast, the dynamic method generates a variable BHTP based on real-time business requirements and channel status. The dynamic method can better adapt to business requirements and channel status changes, but it also requires more information transmission and processing. The design method of BHTP can adopt different optimization algorithms, such as genetic algorithm [19] and deep reinforcement learning algorithm [21]. A primary BHTP design method is as follows: (1) In each beam-hopping cycle, the priority of each beam is calculated according to the capacity requirements and channel status of each beam (i.e., the transmission power required for each beam per unit time); (2) according to the priority, the beams are selected from high to low and assigned to the corresponding time slots and frequency bands until all time and frequency resources are used up; and (3) if there are multiple beams with the same priority, they are randomly or sequentially selected according to specific rules.

The above method can improve the system's total throughput and resource utilization while ensuring fairness. However, there are also some disadvantages: the capacity requirements and channel status information of all beams need to be obtained in real time, which will increase the complexity and overhead of the system; some low-priority beams may not be served for a long time, thus affecting user experience and service quality; and interference between beams and satellite power constraints are not considered, which will reduce the performance and reliability of the system.

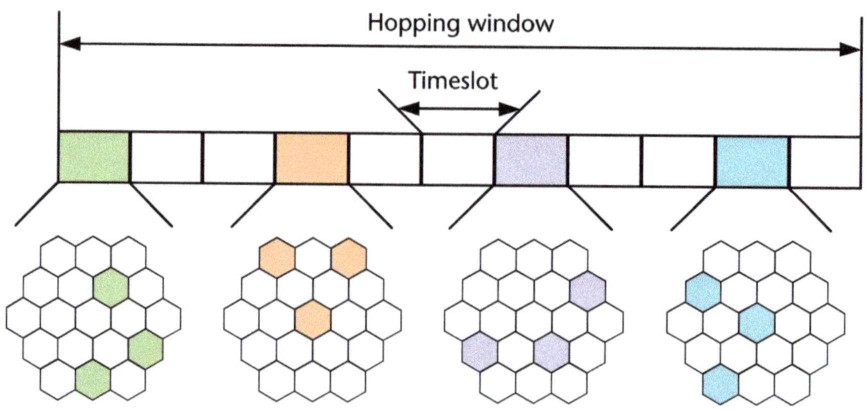

Figure 10.6 BHTP schematic diagram.

10.3 Intersatellite Resource Management

For satellite communication networks, satellites in different orbits have different hardware designs, so satellite performance characteristics and adapted services differ. Effective resource allocation design should be combined with specific satellite characteristics. LEO satellites have a low orbit altitude and the advantages of low transmission delay and slight information transmission loss. This effectively ensures the real-time and communication quality of transmission services. However, due to the limited orbit altitude, the coverage of a single satellite is limited. To effectively achieve seamless coverage, a large number of LEO satellites are required for collaborative transmission. However, to achieve stable connection in high-mobility multisatellite scenarios, the system operation cost will be significantly increased. GEO satellites have the characteristics of extensive coverage and good stability because of their high orbit altitude and relative stationary position concerning the Earth's surface. In theory, three geostationary orbit satellites are deployed to cover the entire world except the two poles. At the same time, geostationary orbit satellites' storage capacity and computing power are generally better than those of low-orbit and medium-orbit satellites. The disadvantage is that the transmission delay and transmission loss caused by the high orbit altitude are significant, which is unsuitable for transmitting services with high real-time requirements such as video conferencing and remote assistance. The performance of MEO satellites is between GEO and LEO satellites. Compared with LEO satellites, MEO satellites have a slower relative motion speed, so the link switching caused is less frequent, the stability is better, the satellite coverage is more extensive, and MEO satellites do not have as long a transmission distance as GEO satellites. Therefore, medium Earth orbit satellite networks mainly provide global mobile communications and navigation services. The specific performance comparison of satellites in different orbits is shown in Figure 10.7.

Resource allocation in satellite-to-terrestrial link scenarios has been widely studied, but intersatellite links have not received much attention. The practical resource allocation of intersatellite links is the key to supporting high-speed and low-latency transmission of the system, and it is also an essential basis for routing design. Limited intersatellite link resources, such as time slots and power, are used to forward data from different satellites, achieving efficient resource utilization.

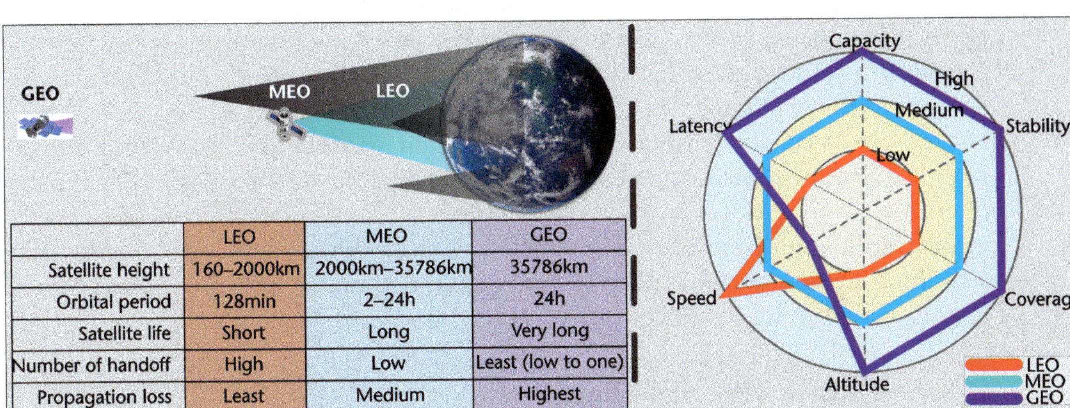

	LEO	MEO	GEO
Satellite height	160–2000km	2000km–35786km	35786km
Orbital period	128min	2–24h	24h
Satellite life	Short	Long	Very long
Number of handoff	High	Low	Least (low to one)
Propagation loss	Least	Medium	Highest

Figure 10.7 Performance comparison of satellites in different orbits.

However, to date, most studies are based on the interconnection of satellite and terrestrial communications, and only a small number of studies on microwave resource allocation in intersatellite communication scenarios have been conducted.

In fact, in microwave resource allocation, the challenges encountered in satellite-to-terrestrial and intersatellite scenarios are similar: limited resources such as power and time slots and performance requirements such as low latency. However, satellites at different orbital altitudes have different characteristics in the intersatellite scenario. With the continuous development of satellite communications and the exponential growth of satellite constellations, satellites in a single orbit will gradually fail to meet the growing performance requirements due to their inherent design defects. Therefore, to meet the application requirements in different business scenarios, it is necessary to consider the intersatellite cross-layer design between satellites at different orbital altitudes, weigh their advantages and disadvantages through efficient design, and improve the satellite network architecture. Based on this, this section mainly discusses the critical issues of intersatellite resource management. It is worth noting that compared with single optimization, multiresource joint optimization has become a standard optimization design in wireless communications. Here, we mainly introduce the main problems each resource faces in intersatellite resource allocation.

10.3.1 Limited On-Satellite Power

Satellite-limited power allocation is one of the keys to determining transmission performance. The construction of Starlink often accompanies the power allocation of intersatellite networks, and its design is more complex. The satellite access scheduling design will directly affect the power allocation of other service terminals. At the same time, affected by the severe attenuation of satellite channels, the effective allocation of limited power among multiple satellite service terminals to achieve stable and optimal transmission capacity is one of the difficulties in intersatellite scenarios. Furthermore, the transmission of other satellites in the intersatellite link will be regarded as interference, and the higher the power, the more serious the interference. Although the interference from distant satellites can be ignored due to severe free-space attenuation, and the interference from adjacent satellites can be alleviated by frequency reuse and other technologies, the intersatellite spectrum resources are also limited, which is not a long-term solution. How to effectively balance the performance between SE and EE is one of the problems that urgently needs to be solved. In addition, many existing power allocation methods are based on centralized methods [22], which are suitable for communication links with slow link state changes. In contrast, distributed power allocation methods are more ideal for high-mobility satellite transmission environments by optimizing the transmission power of each mobile terminal. In addition, considering the substantial growth of traffic in the future, multilayer satellite collaboration will replace the single-layer satellite allocation design, and it is even more necessary to design an efficient distributed link power allocation strategy.

10.3.2 Dynamic Time Slot Allocation

The high mobility of satellites lead to a shortage of time slot resources. In addition, due to the long-distance transmission characteristics of intersatellite links, within a given time slot t, if a satellite access switch occurs, the propagation delay tp and link establishment delay ts must be considered. The remaining delay (t-tp-ts) is the practical information transmission delay [23]. Therefore, based on the different practical propagation delays caused by the unequal intersatellite link distances, improving the overall information transmission volume is a complex problem that needs to be considered in intersatellite time slot allocation. At the same time, the transmission distance of the intersatellite link changes rapidly over time. Still, it is worth noting that the movement of satellite networks is always regular so that the intersatellite distance can be predicted in advance. However, to maximize the effective transmission delay, the number of access switching events needs to be reduced. Still, the increase in the intersatellite distance under a given time slot requires an increase in time slot allocation to complete information transmission. Therefore, solving the contradiction between the effective transmission delay and the number of time slots (achieving an efficient trade-off between the two) is also the key to intersatellite time slot allocation. In addition, time slot allocation is prone to transmission synchronization and conflict problems, essentially caused by the unequal transmission distances in the intersatellite links. Based on this, the literature [24] proposes that when a conflict occurs, the time slices of the affected traffic requests are reallocated with lower bandwidth if there are sufficient bandwidth resources in the relevant ISL. If the execution fails, the subsequent conflicting traffic request is abandoned. This strategy cannot adapt to the global transmission of the entire satellite constellation and cannot meet the extensive transmission needs of satellite communications.

10.3.3 Channel Availability in Short Time Slots

The satellite clusters included in the satellite constellation require many channel resources for information transmission. However, intersatellite communication typically operates in the gigahertz frequency band. The high-frequency band leads to profound free-space transmission loss. The extremely high satellite altitude and long transmission distance also lead to severe channel attenuation. As a result, the limited channel resources are even more strained. In addition, combined with the analysis in the previous paragraph, the high-speed mobility of satellites makes channel allocation only effective within a specific time slot. Although the satellite position can be predicted based on the regularity of satellite movement, the large number and high dynamics of satellites require a more efficient and low-complexity scheme design. As a popular technology in current research, AI has attracted much attention due to its advantages, such as predicting and mastering environmental change trends to achieve active scheduling and its ability to efficiently and intelligently process massive data. Reference [25] proposed an AI-based learning task-oriented channel allocation design to improve the effectiveness and efficiency of multiagent communication. The AI reward mechanism simplifies and predicts the optimal allocation design, significantly reducing the computational complexity, but satellite scenarios have not yet been considered.

10.3.4 Cross-Orbit Multilayer Cooperative Transmission

Generally speaking, the design of the LEO satellite combined with a higher orbit (MEO/GEO) satellite can make up for the defects of the LEO satellite, such as small transmission capacity, weak computing power, and poor stability. At the same time, by building intersatellite links with LEO, the transmission delay of high-orbit satellites can be reduced. Although the existing routing scheme does not consider the establishment of transmission links for satellites in different orbits, the severe Doppler frequency shift caused by the high mobility rate between satellites at different altitudes but in the same direction is an indispensable key factor that needs to be considered in the design of cross-layer resource allocation. In addition, multilayer satellite networks are more complex than single-layer networks. Single resources must consider the interaction and switching between multiple layers of satellites under multilayer satellites, and the access options for transmission services are more diverse. Resource management issues are more complicated. Formulating adaptive resource allocation schemes based on specific application scenarios is necessary, analyzing capacity management issues under complex multilayer satellite networks, and optimizing the long-term utility of three-layer heterogeneous satellite systems [26]. At the same time, it is necessary to divide system resources that meet the needs based on different business types and business resource requirements to achieve the rational use of multilayer satellite network resources. Therefore, a joint optimization design based on multiple limited resources is more necessary.

10.3.5 Intersatellite Transmission Based on OPA

In addition to resource management in the microwave frequency band, the optical phased array (OPA) is a phase-adjustable optical antenna array. Its working principle is similar to that of the microwave-phased array. It controls the phase of light radiated by each optical antenna to perform beam control. It has a larger communication capacity and more concentrated energy than microwave communication. Under the exact rate requirement, its size, weight, and power consumption are also better, making it a strong candidate for ISL and cross-layer links [27]. In addition, the existing intersatellite links mainly use the laser as the transmission medium. Due to its strong directionality, research focuses on tracking and capture and does not involve resource allocation. The application of OPA in intersatellite links will effectively expand the transmission capacity of the intersatellite network. The multioptical beam control method will significantly enrich the design of intersatellite routing and resource allocation. Routing will no longer be limited to a four-link design. Satellites can establish connections with more adjacent satellites, expanding the diversity of routing solutions as shown in Figure 10.8. However, it should be noted that more transmission path options are also accompanied by beam scanning and alignment problems. Correspondingly, the multibeam attribute also challenges the design of intersatellite resource allocation. One of the promising research directions is how to allocate the number of beams between connected satellites and the design of intersatellite limited resource allocation between different optical beams.

10.4 Interference Management

10.4.1 Natural Interference

10.4.1.1 Rain Fade Interference

In satellite communications, electrical signals must pass through the earth's atmosphere to achieve information transmission between terrestrial stations and

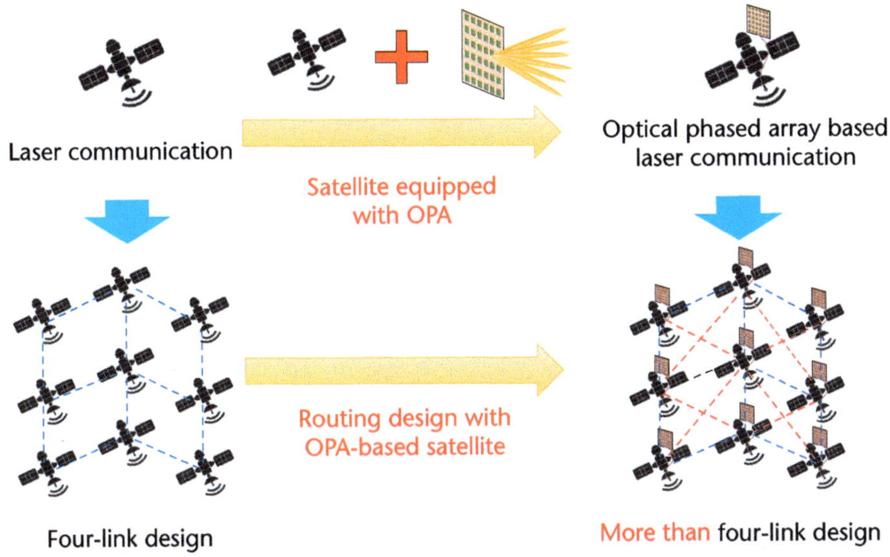

Figure 10.8 Intersatellite transmission design based on OPA.

satellites. The atmospheric medium is complex in composition. When electrical signals are transmitted in this medium, they will inevitably suffer from much energy reflection and absorption of ions and free electrons. In addition, electrical signals will also be affected by a large amount of atmospheric particles such as water vapor and carbon dioxide, as well as molecules such as clouds and rain in the troposphere, resulting in signal energy loss.

Among the above losses, the content of raindrops in the air is higher when it rains, which will absorb or refract a large amount of satellite communication transmission signals, resulting in severe signal attenuation. The attenuation effect of rainfall on signals is mainly reflected in the absorption and scattering of signals by raindrops. When the signal passes through raindrops, the electrons inside and outside the raindrops are affected by the electric field force to resonate, thereby absorbing part of the signal energy, and the energy absorbed by the damping effect evaporates in the form of heat. After evaporation, the signal repeats the above process, resulting in cumulative signal attenuation.

To calculate the specific value of rain attenuation, it is necessary to consider the rainfall rate and the annual probability percentage of actual rainfall. The rainfall rate exceeds γ_R (in mm/h) for 0.01% of the time in a year, which can be recorded as $R_{0.01} = \gamma$.

According to the classical Mir scattering theory, the relationship between unit attenuation γ_R and rainfall rate R is

$$\gamma_R = k(R_{0.01})^a \tag{10.1}$$

where $R_{0.01}$ represents the rainfall rate obtained by taking any 0.01% in a particular year, k and a are constant values related to the frequency, and the specific calculation method of k and a has been introduced in the ITU-R standard.

ITU-R proposed a rain attenuation prediction model based on the equivalent path length concept, which evens out nonuniform rainfall and introduces a shortening factor that can play an equivalent role. The shortened effective path length L_E (in km) multiplied by the unit path attenuation is the rainfall attenuation; that is,

$$A_{0.01} = \gamma_R L_E \tag{10.2}$$

The formula for converting the rainfall attenuation with a time percentage of 0.01% to the rainfall attenuation with a time percentage of $p\%$ is

$$A_p = A_{0.01}\left(\frac{p}{0.01}\right)^{-(0.655+0.033\ln A_{0.01}-\gamma(1-p)\sin\theta)} \tag{10.3}$$

where A_p is the attenuation expected to exceed the annual average probability of p in decibels; p is the time percentage, θ is the antenna elevation angle, which is determined by the longitude and latitude of the terrestrial station and the location of the satellite, and γ is expressed as

$$\gamma = \begin{cases} 0 & p \geq 1\%, |\varphi| \geq 36^\circ \\ -0.005(|\varphi - 36|) & p \leq 1\%, |\varphi| < 36^\circ, \theta \geq 25^\circ \\ -0.005(|\varphi - 36|) + 1.8 - 4.25\sin\theta & \text{others} \end{cases} \tag{10.4}$$

where φ is the latitude of the earth station.

The effective rain attenuation path L_E is equal to the product of the rainfall geometric path L_S and the path shortening factor r_P:

$$L_E = L_S \times r_P \tag{10.5}$$

As shown in Figure 10.9, the calculation formula for L_S, r_P is

$$L_S = \frac{h_R - h_S}{\sin\theta} \tag{10.6}$$

$$r_P = \frac{1}{1 + L_G/L_O} \tag{10.7}$$

$$L_O = 35\exp(-0.015r_p) \tag{10.8}$$

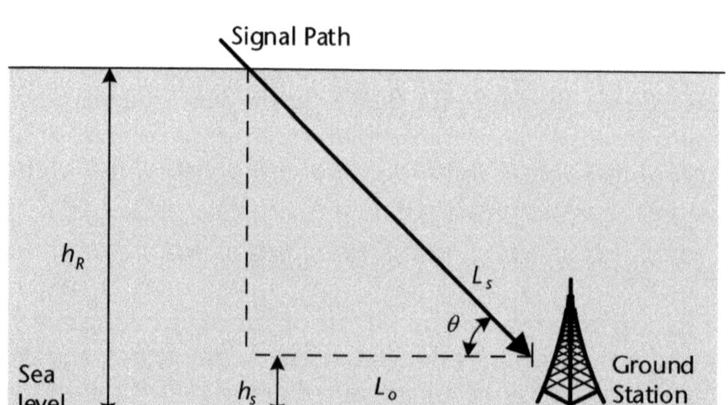

Figure 10.9 Schematic diagram of satellite signal passing through rain layer.

The specific manifestations of rain attenuation include level attenuation, increased system noise, and reduced signal cross-polarization rate.

Signal Level Attenuation

Rainfall will seriously interfere with signal transmission. Although the impact on low-frequency bands is small when the signal operating frequency is above 15 GHz, the signal in the Ka-band may suffer from the cliff effect, completely interrupting the signal transmission. The reasons for signal attenuation in this band include many situations: macroscopically, it consists of the height, shape, and water content of the rain cloud layer; microscopically, it consists of the size, shape, and wavelength of the scattered radio waves of the raindrops themselves, which leads to signal level attenuation and reduced signal transmission power.

Increase System Noise Temperature

$$\Delta T = (1 - 10^{\frac{-A}{10}}) \cdot T \tag{10.9}$$

where ΔT represents the noise temperature when it rains, A represents its attenuation, T represents the effective temperature of the rain medium, which can be considered unchanged (when the rain falls for a long time), and ΔT changes in the same direction as the rain attenuation. Through the transmission of the downlink-terrestrial station, the increase of ΔT will cause the noise component of the total system to increase and the downlink signal-to-noise ratio to decrease.

Reduce the Signal Cross-Polarization Discrimination Rate

To solve the problem of limited satellite communication frequency band, frequency reuse technology can be used in the channel. In theory, two different signals can be orthogonally polarized to achieve no interference between signals and a high isolation level. However, in the actual signal transmission process, due to factors such as rainfall, the theoretically strictly orthogonal signals are not entirely orthogonal, causing cross-polarization interference and reducing the signal cross-polarization rate; that is, signal depolarization. Specifically, the circular raindrops that have just condensed from water vapor are affected by air resistance during free fall, and their lower surface changes to an elliptical shape, resulting in a certain angle between the polarization plane and the central axis of the raindrop when the signal passes through the raindrop, thereby producing a depolarization effect.

Solar Eclipse Interference

In satellite communications, the terrestrial station is assumed to rotate synchronously with the earth. When the sun, the earth, and the satellite move to a straight trajectory, and the earth is on the same side of the sun and the terrestrial station, the antenna parabola of the earth station is aimed at the satellite and the sun at the same time (as shown in Figure 10.10). The electromagnetic waves generated by the sun are strongly projected directly into the antenna beam range of the

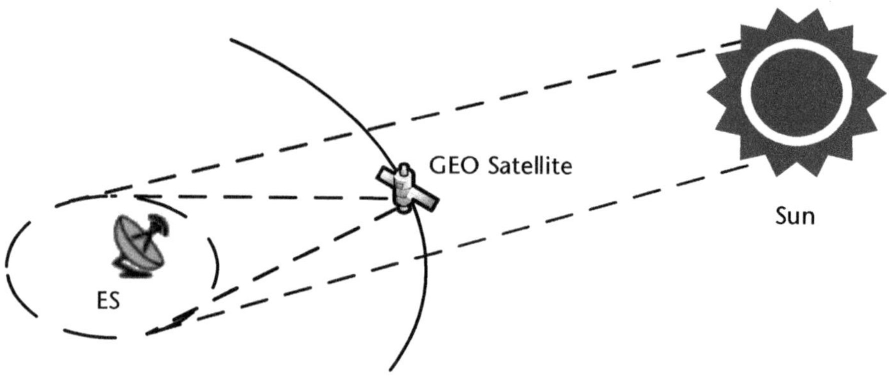

Figure 10.10 Schematic diagram of satellite solar eclipse interference.

earth station, and the spectrum is broad. Compared with the earth station, solar electromagnetic waves can be regarded as a considerable noise source, causing the signal-to-noise ratio of the signal received by the earth station to decrease to varying degrees. This is the solar eclipse interference phenomenon. When the communication quality deteriorates seriously, the performance quality of the communication link drops sharply. In severe cases, the communication is even interrupted, which is called solar eclipse interruption [28].

Next, we study the factors that affect the earth station by solar eclipse interference through the causes of solar eclipse interference. There are three main aspects.

Geographical Location of the Earth Station

The longitude and latitude of the earth station determine the duration of the interference caused by solar eclipse on the station. Combining the positional relationship between the earth, satellite, and sun, through the analysis of monitoring data over the years, it can be seen that under the same conditions of the earth station antenna aperture and working frequency band in the same hemisphere, the lower the latitude of the earth station, the earlier the first solar eclipse interference date; conversely, the higher the latitude of the earth station (the closer to the poles), the later the first solar eclipse interference date. At the same time, it was found that the intensity of the interference caused by the solar eclipse in spring for earth stations located in the northern hemisphere is less than that caused by the solar eclipse in autumn; conversely, the intensity of the interference caused by the solar eclipse in spring for earth stations located in the southern hemisphere is more significant than that caused by the solar eclipse in autumn.

Half-Beam Width of the Earth Station

For an earth station, its solar eclipse duration starts when the sun enters its antenna's half-beam width (i.e., 3-dB beam) and ends when it leaves its half-beam width. Therefore, the duration of the solar eclipse of the earth station is related to its operating frequency band and the size of the antenna aperture. Specifically, the calculation formula for the half-beam width of the antenna is as follows:

$$\theta_{1/2} = 70\frac{\lambda}{D} \tag{10.10}$$

where λ is the wavelength of the electromagnetic wave corresponding to the working frequency band of the earth station, and D is the antenna aperture of the earth station. From the above formula, it can be seen that when the antenna aperture is constant, the higher the working frequency band, the narrower the antenna half-beam width, and the shorter the solar eclipse duration. When the working frequency band is constant, the larger the antenna aperture, the narrower the half-beam width, and the shorter the solar eclipse duration.

Other Interference

Other satellite communication interference losses include, antenna pointing loss, atmospheric loss, and mainly free space transmission loss. Details are as follows:

Free-Space Transmission Loss

The received power flux density is known to be

$$W = P_T G_T / 4\pi d^2 \tag{10.11}$$

where P_T is the transmission power, G_T is the antenna gain at the transmitting end, and d is the transmission distance. If the effective receiving surface of the receiving antenna is A_η, the received power P_R is

$$P_R = W A_\eta = \frac{P_T G_T A_\eta}{4\pi d^2} \tag{10.12}$$

If the gain of the receiving antenna is represented by G_R, then

$$P_R = P_T G_T G_R \left(\frac{\lambda}{4\pi d}\right)^2 = \frac{P_T G_T G_R}{L_f} \tag{10.13}$$

where $L_f = \left(\frac{4\pi d}{\lambda}\right)^2$ is the free-space loss, and λ is the wavelength:

$$[L_f] = 20\lg\left(\frac{4\pi d}{\lambda}\right) = 92.45 + 20\lg(d \cdot f) \tag{10.14}$$

where f is the operating frequency in gigahertz.

Pointing Loss

Due to the low accuracy of satellite attitude pointing control, beam pointing swing caused by atmospheric refraction, and limited tracking accuracy of terrestrial station antenna pointing, the antenna gain of the satellite-to-terrestrial link is not the maximum value, resulting in signal loss, which is defined as

$$L_{\mathrm{T}} = \frac{G(0)}{G(\theta)} \tag{10.15}$$

where $G(0)$ is the gain value in the direction of maximum antenna gain, $G(\theta)$ is the gain value corresponding to the antenna pattern function, and θ is the deviation angle between the direction of maximum antenna gain and the satellite direction. Usually, $G(\theta)$ can be approximately expressed as

$$G(\theta) \approx G(0) \cdot \mathrm{e}^{-2.77\times\left(\frac{\theta}{\theta_{1/2}}\right)^2} \tag{10.16}$$

where $\theta_{1/2}$ is the half-power width of the antenna. Therefore, the pointing loss can generally be calculated by the following formula:

$$L_{\mathrm{T}} \approx \mathrm{e}^{-2.77\times\left(\frac{\theta}{\theta_{1/2}}\right)^2} \tag{10.17}$$

10.4.2 Space Interference

The complex space in which satellite communications are located is an electromagnetic space environment composed of many electromagnetic signals distributed in space, time, frequency, and energy. The satellite communication pattern is complex and the satellite distribution is dense. It has the characteristics of extensiveness, diversity, dynamics, and relativity. Its internal composition is intricate and complicated to perceive and accurately describe directly. This complex electromagnetic environment includes natural electromagnetic radiation, unintentional human radiation, and intentional radiation. Natural electromagnetic radiation is generated by nonhuman factors, including lightning, static electricity deposition, and planetary and cosmic noise. Unintentional human radiation refers to unintentional radiation generated when the satellite itself, surrounding satellites, or earth station electronic equipment is working. Intentional radiation refers to targeted interference with the uplink, downlink, intersatellite link, and transponder of satellite communications through various deception and interference suppression artificial attack methods. These different types of radiation sources lead to a complex electromagnetic environment for satellite communications, which poses a severe test to the regular operation of satellite communications [28].

In addition, with the continuous increase in the number of satellites, frequency bands, as nonrenewable resources, seriously restrict the future development of satellite networks [29]. Especially in LEO-GEO satellite systems, due to the scarcity of frequency band resources, different satellite constellations will work in the same frequency band and share spectrum resources. As the number of satellites and terrestrial terminals increases, it will become more common for multiple satellites in different orbits to cover the same terrestrial area. Therefore, designing the optimal resource allocation scheme under different satellite constellations, especially between satellite constellations in different orbital planes, is an urgent problem that needs to be solved in future satellite communications. There are four joint spectrum coexistence scenarios [30]:

1. Terrestrial systems reuse satellite spectrum resources;
2. Satellite systems use terrestrial spectrum resources;
3. Satellites serve as relays to expand terrestrial communication networks;
4. Two satellite systems share the same frequency band.

Due to the characteristics of the Ka-band, such as the wide available frequency band, low interference, and small antenna size, broadband low-orbit satellite systems and GEO satellites usually use the Ka-band, resulting in cofrequency interference between different satellite systems. Compared with GEO satellites, LEO constellation communication satellites are more numerous and have a more extensive coverage rate. Their positions relative to the earth change over time. When the two use the same frequency band, non-geostationary orbit (NGSO) satellites fly over the GEO satellite receiver, causing cofrequency interference to the GEO satellite and its receiver. The interference changes quickly and is of high intensity. Especially when the NGSO satellite is on the line connecting the GEO satellite and the GEO satellite terrestrial station, the interference of the NGSO satellite to the GEO satellite is the greatest (i.e., inline interference) [31].

If the spectrum-sharing problem between satellites cannot be adequately solved, their mutual interference will seriously affect the transmission stability, thereby affecting the availability of the entire space network. Existing communication satellites are used for communication services, measurement and control links, and feeder links. The operating frequencies cover UHF, L, S, C, X, Ku, Ka, and Q/V bands. The orbit altitude covers low, medium, and high orbits. They have the characteristics of multifrequency services, wide frequency range, and total space spectrum support. They are exposed in space, and the channels are open, making satellite transponders, uplinks, and down-links extremely susceptible to interference. Therefore, this section introduces the common space interference satellites encounter during signal transmission.

10.4.2.1 Frequency Band Interference

The explosive development of NGSO satellite constellations has led to a shortage of satellite frequency and orbit resources [32]. It is foreseeable that the continuous development of the global space information network and the large-scale deployment of NGSO satellite systems will not only cause interference to GEO satellites in orbit but also cause cofrequency severe interference between different NGSO systems. On the one hand, NGSO constellations often have global coverage characteristics, and the communication links of different systems usually cannot meet the requirements of angular isolation, resulting in collisions in the space domain; on the other hand, the leading frequency bands of typical NGSO constellations currently under construction are all located in the Ku/Ka bands, and all have applied for Q/V/E band reserve resources. It is difficult to avoid situations where multiple systems share the same frequency, resulting in potential cofrequency interference risks for various NGSO satellite systems. Therefore, during the constellation design stage and before deployment, it is necessary to conduct interference analysis and interference avoidance technology research based on relevant ITU regulations and frequency divisions.

In addition, the rapid development of large-scale NGSO communication constellations has also brought new challenges to frequency-orbit resource coordination technology [33]. On the one hand, as the number of satellites increases, the mutual interference between these constellation systems containing thousands or even tens of thousands of satellites becomes more severe and complex. The increase in the number of satellites not only makes the total interference signal power received by the receiver stronger but also dramatically increases the probability of inline interference. On the other hand, most of the existing interference avoidance methods are often based on traditional, relatively static interference scenarios. For highly dynamic NGSO satellite constellations, traditional interference avoidance methods have significant limitations in dealing with frequent changes in the relative positions of satellites and severe interference.

Current interference management solutions include interference coordination, elimination, and mitigation to solve the problem of satellite cofrequency interference. Interference coordination involves coordinating and managing system resources in space, time, frequency, and other dimensions. Specific technologies include frequency reuse technology and space protection zone technology. Interference elimination refers to eliminating or reducing interference between users through mathematical theory, precoding user signals according to CSI, and suppressing interference from other user information. Specific technologies include beamforming and intelligent antenna technology, which can reduce useless or harmful directional interference in dense networks but require complex communication equipment and interfering terminal location information. Interference mitigation technology generally adopts interference power control methods to reduce the transmission power of the interfering transmitter to within the tolerable range of the receiver. The power control problem is to maximize the network throughput while meeting the SINR requirements of the satellite and terrestrial link signals.

The above technologies are all based on interference between the same communication systems. Commonly used methods for suppressing interference between satellite systems are primarily based on geographical factors, such as setting up terrestrial station protection restricted areas [34]. Still, this method is not suitable for the scenario of satellite resource shortage in the future. It is necessary to consider how the interfering party and the interfered party can reduce interference at the network layer and physical layer, which requires the application of technologies to suppress cofrequency interference, such as power control technology.

With the establishment of broadband low-orbit satellite constellations, interference between GEO satellite systems and NGSO satellite systems is inevitable. Analysis of interference scenarios and interference avoidance between satellite systems are urgent issues that must be addressed. Here is a simple scenario example:

The basic architecture of a multibeam satellite system is shown in Figure 10.11, where N spot beams of a GEO satellite serve the coverage area. A TDM transmission strategy is adopted, assuming that each beam forms a forward link with a single user terminal at a given time [35]. For simplicity, it is assumed that there is an ideal feeder link between the terrestrial gateway station and the satellite, and the system adopts a conventional frequency reuse factor of 4, so when multiple beams

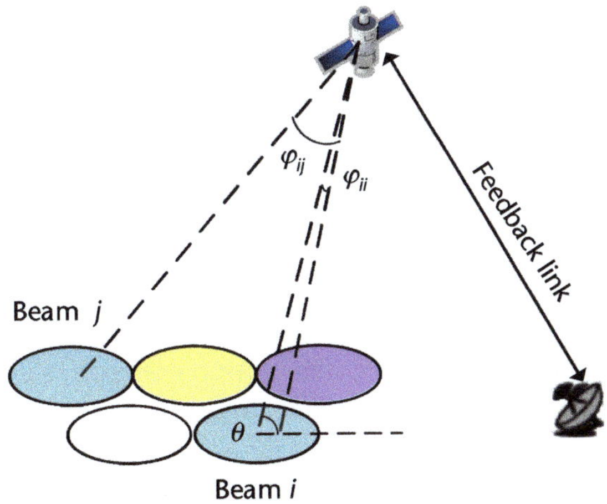

Figure 10.11 Multibeam satellite communication system.

communicate with all users operating in the same frequency band at the same time, there is a certain amount of cochannel interference. θ represents the elevation angle of the desired user terminal, φ_{ii} represents the off-axis angle of the ith desired beam, and φ_{ij} represents the off-axis angle from the ith desired beam to the center of the jth interference beam.

The equivalent total channel gain between the jth beam and the ith user can be expressed as

$$h_{ij} = \hat{h}_j G(\varphi_{ij})^{\frac{1}{2}}, i, j = 1, \cdots, N \tag{10.18}$$

where $G(\varphi_{ii})$ is the antenna gain between the ith beam and the ith user calculated according to φ_{ii}. $G(\varphi_{ij})$ is the observed leakage antenna gain between the jth interfering beam and the ith user calculated according to φ_{ij}. According to the radiation pattern of [35], the antenna beam gain observed by the desired user can be expressed as

$$G(\varphi) = G_{\max}\left(\frac{J_1(u)}{2u} + 36\frac{J_3(u)}{u^3}\right)^2 \tag{10.19}$$

where $u = 2.07123\frac{\sin\varphi}{\sin\varphi_{3dB}}$, φ_{3dB} represents the angle corresponding to half power loss, J_1 and J_3 are the first-order and third-order Bessel functions of the first kind, respectively, and $G_{\max} = \left(\frac{\lambda}{4\pi}\right)^2 \frac{1}{d_0^2}$ represents the maximum antenna gain, where λ is the wavelength and $d_0 \approx 35{,}786$ km is the satellite height. Assuming that P_i is the transmit power of the ith beam, x_i is the transmit symbol from the ith beam, then the received signal of the ith user can be expressed as

$$y_i = h_{ii}\sqrt{P_i}x_k + \sum_{j\in\Phi_i} h_{ij}\sqrt{P_j}x_j + n_i \tag{10.20}$$

where Φ_i represents the cochannel interference beam set for the ith beam, and n_i is the Gaussian noise at the ith beam. Subsequent chapters will further analyze satellite interference in different scenarios.

10.4.2.2 Inline Interference

Due to the high demand for ultralow latency and broadband massive data in real-time systems, the deployment of LEO/MEO satellite systems in several frequency bands has increased significantly. As the number of available NGSO satellites (i.e., LEO/MEO in space) increases, frequency coexistence between NGSO satellite systems and GEO satellite networks is inevitable.

In the case of the coexistence of GEO and NGSO networks, inline interference can become a severe problem. Whenever an NGSO satellite passes through the line-of-sight path between an ES and a GEO satellite, inline interference will occur, as shown in Figure 10.12. Earth stations aligned with GEO and NGSO satellites may receive and cause interference through their main beams. Considering the case of O3b satellites, inline interference will cause potential transmission interference problems for GEO networks operating near the equator [36].

Figure 10.13 shows a GEO-NGSO ES downlink transmission model, taking the GEO ES receiver as an example, where d_{GG} and d_{NG} represent the spatial distances from the GEO satellite and NGSO satellite to the GEO ES, respectively, and θ^{t}_{NG} and θ^{r}_{NG} represent the off-axis angles of the NGSO satellite facing the GEO ES at the transmitting and receiving ends, respectively. Similarly, d_{GN} and d_{NN} represent the spatial distances from the GEO satellite and NGSO satellite to the NGSO ES, respectively, and θ^{t}_{GN} and θ^{r}_{GN} represent the off-axis angles of the GEO satellite facing the NGSO ES at the transmitting and receiving ends, respectively. Assuming that the GEO ES and NGSO ES are very close, the transmission interference caused by satellites in different orbits should be considered simultaneously.

Taking GEO ES as an example, it receives useful signals from GEO satellites and interference signals from NGSO satellites simultaneously. Therefore, the

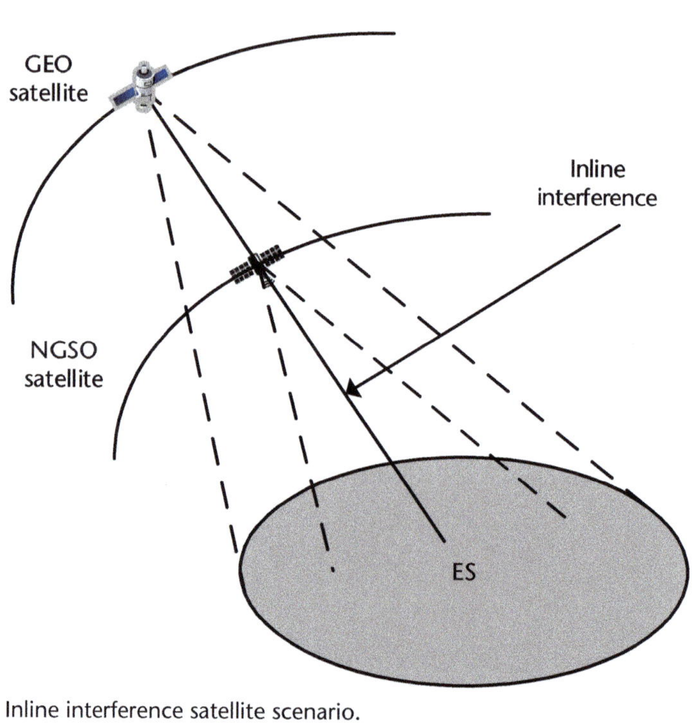

Figure 10.12 Inline interference satellite scenario.

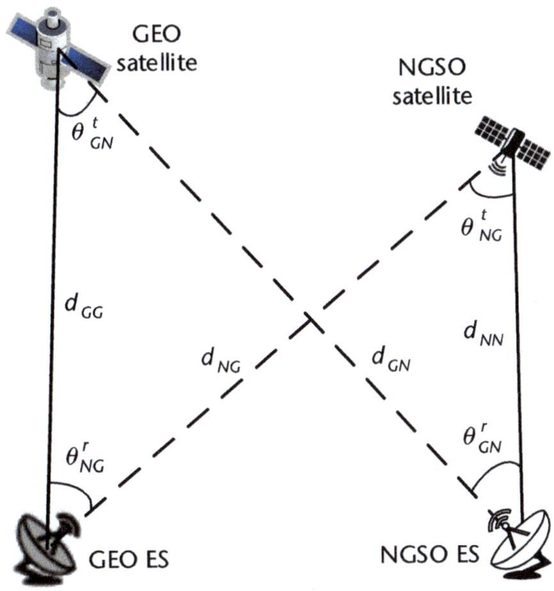

Figure 10.13 Satellite link inline scenario.

interference power received by GEO ES from NGSO satellites is

$$P^{\mathrm{r}}_{\mathrm{NG}} = P^{\mathrm{t}}_{\mathrm{NGSO}}\,(d_{\mathrm{NG}})\,G^{\mathrm{t}}_{\mathrm{N}}(\theta^{\mathrm{t}}_{\mathrm{NG}})G^{\mathrm{r}}_{\mathrm{GE}}(\theta^{\mathrm{r}}_{\mathrm{NG}})\left(\frac{\lambda}{4\pi d_{\mathrm{NG}}}\right)^2 \tag{10.21}$$

where $P^{\mathrm{t}}_{\mathrm{NGSO}}$ represents the NGSO satellite transmission power and $G^{\mathrm{t}}_{\mathrm{N}}$ and $G^{\mathrm{r}}_{\mathrm{GE}}$ represent the transmission and receiving antenna gains of the NGSO satellite and GEO ES, respectively; λ represents the wavelength. Similarly, the power received by the GEO ES from the GEO satellite is

$$P^{\mathrm{r}}_{\mathrm{GG}} = P^{\mathrm{t}}_{\mathrm{GEO}}\,(d_{\mathrm{GG}})\,G^{\mathrm{t}}_{\mathrm{G}}(0)G^{\mathrm{r}}_{\mathrm{GE}}(0)\left(\frac{\lambda}{4\pi d_{\mathrm{GG}}}\right)^2 \tag{10.22}$$

where $P^{\mathrm{t}}_{\mathrm{GEO}}$ and $G^{\mathrm{t}}_{\mathrm{G}}$ represent the transmission power and antenna gain of the GEO satellite, respectively.

Therefore, the expressions of carrier-to-noise ratio and interference-to-noise ratio at GEO ES are

$$(C/N)_{\mathrm{GEO}} = \frac{P^{\mathrm{r}}_{\mathrm{GG}}}{KT^{\mathrm{r}}_{\mathrm{GE}}B} = \frac{P^{\mathrm{t}}_{\mathrm{GEO}}\,(d_{\mathrm{GG}})\,G^{\mathrm{t}}_{\mathrm{G}}(0)G^{\mathrm{r}}_{\mathrm{GE}}(0)\left(\frac{\lambda}{4\pi d_{\mathrm{GG}}}\right)^2}{KT^{\mathrm{r}}_{\mathrm{GE}}B} \tag{10.23}$$

$$(I/N)_{\mathrm{GEO}} = \frac{P^{\mathrm{r}}_{\mathrm{NG}}}{KT^{\mathrm{r}}_{\mathrm{GE}}B} = \frac{P^{\mathrm{t}}_{\mathrm{NGSO}}\,(d_{\mathrm{NG}})\,G^{\mathrm{t}}_{\mathrm{N}}(\theta^{\mathrm{t}}_{\mathrm{NG}})G^{\mathrm{r}}_{\mathrm{GE}}(\theta^{\mathrm{r}}_{\mathrm{NG}})\left(\frac{\lambda}{4\pi d_{\mathrm{NG}}}\right)^2}{KT^{\mathrm{r}}_{\mathrm{GE}}B} \tag{10.24}$$

where $K = 1.38 \times 10^{-23}\mathrm{W/(Hz \cdot K)}$ represents the Boltzmann constant, $T^{\mathrm{r}}_{\mathrm{GE}}$ represents the noise temperature at the receiving end GEO ES, and B represents the bandwidth.

Similarly, the NGSO ES will receive signals from the NGSO satellite and interference signals from the GEO satellite. Similarly, the carrier-to-noise ratio and interference-to-noise ratio expressions at the NGSO ES can be expressed as

$$(C/N)_{\text{NGSO}} = \frac{P_{\text{NN}}^{\text{r}}}{KT_{\text{NE}}^{\text{r}}B} = \frac{P_{\text{NGSO}}^{\text{t}}(d_{\text{NN}})\,G_{\text{N}}^{\text{t}}(0)G_{\text{NE}}^{\text{r}}(0)\left(\frac{\lambda}{4\pi d_{\text{NN}}}\right)^2}{KT_{\text{NE}}^{\text{r}}B} \tag{10.25}$$

$$(I/N)_{\text{NGSO}} = \frac{P_{\text{GN}}^{\text{r}}}{KT_{\text{NE}}^{\text{r}}B} = \frac{P_{\text{GEO}}^{\text{t}}(d_{\text{GN}})\,G_{\text{N}}^{\text{t}}(\theta_{\text{GN}}^{\text{t}})G_{\text{NE}}^{\text{r}}(\theta_{\text{GN}}^{\text{r}})\left(\frac{\lambda}{4\pi d_{\text{GN}}}\right)^2}{KT_{\text{NE}}^{\text{r}}B} \tag{10.26}$$

To alleviate the transmission interference caused by the inline transmission of satellites in different orbits during operation, taking GEO ES as an example, a corresponding optimization problem is formulated to ensure the normal transmission of NGSO satellite-NGSO ES link while minimizing the interference power of the GEO satellite-GEO ES link:

$$\begin{aligned} &\min P_{\text{NG}}^{\text{r}} \\ &\text{s.t.} \frac{P_{\text{GEO}}^{\text{t}}(d_{\text{GG}})G_{\text{G}}^{\text{t}}(0)G_{\text{GE}}^{\text{r}}(0)\left(\frac{\lambda}{4\pi d_{\text{GG}}}\right)^2}{KT_{\text{GE}}^{\text{r}}B} \geq (C/N)_{\text{th}} \\ &\frac{P_{\text{NGSO}}^{\text{t}}(d_{\text{NG}})G_{\text{N}}^{\text{t}}(\theta_{\text{NG}}^{\text{t}})G_{\text{GE}}^{\text{r}}(\theta_{\text{NG}}^{\text{r}})\left(\frac{\lambda}{4\pi d_{\text{NG}}}\right)^2}{KT_{\text{GE}}^{\text{r}}B} \leq (I/N)_{\text{th}} \end{aligned} \tag{10.27}$$

That is, the optimization goal of the problem is to minimize the interference power, where $(C/N)_{\text{th}}$ and $(I/N)_{\text{th}}$ are the tolerable thresholds of the carrier-to-noise ratio and the interference-to-noise ratio, respectively.

10.4.2.3 Cross-Polarization Interference

Polarization refers to the orientation of the electric field vector in the plane of the radiated waveform. In most cases, the polarization of an antenna can be determined through detection. For instance, a vertical whip antenna generates and receives vertically polarized waves. Similarly, a horizontal antenna element produces horizontally polarized waves. Both vertical and horizontal polarizations are classified as linear polarizations. Another type of polarization is circular or elliptical polarization. Unlike linear polarization, the polarization vector in circular or elliptical polarization rotates either clockwise or counterclockwise. This rotation results in right-hand circular polarization (RHCP) or left-hand circular polarization (LHCP), respectively. Circular polarization is a special case of elliptical polarization where the vertical and horizontal components of the polarization vector have equal magnitudes. Aperture antennas are generally capable of supporting vertical, horizontal, or elliptical polarizations, making them versatile for various communication needs.

According to the relationship between the amplitude and phase difference of the orthogonal electric field components, the polarization of electromagnetic waves can be divided into three categories:

1. *Completely polarized wave*: The electric field intensity of the electromagnetic wave is decomposed into orthogonal vectors. If the amplitude of these two components and their phase difference are constant, the polarization

state can be called constant. At this time, the projection of the electric field vector trajectory is a continuous ellipse. As the propagation direction of the electromagnetic wave changes, its propagation trajectory is a straight line, a circle, and an ellipse, corresponding to linear polarization, circular polarization, and elliptical polarization (as shown in Figure 10.14), and the elliptical polarization wave can be regarded as the superposition of two orthogonal circular polarization waves or linear polarization waves, the linear polarization wave is the superposition of two orthogonal circular polarization waves, and the circular polarization wave is the superposition of two orthogonal linear polarization waves.

2. *Partially polarized wave*: This has a specific bandwidth and can be regarded as the sum of a wholly polarized and nonpolarized wave. It exists frequently in life, and the projection of its electric field vector trajectory is a time-varying ellipse.
3. *Completely unpolarized wave*: There is no determined polarization parameter, and the electric field vector trajectory projection is an irregularly changing graph, such as sunlight.

In recent years, polarization reuse technology has become a research hotspot for improving satellite remote-sensing transmission systems [37]. The remote sensing satellite uses polarization reuse technology transmission data to increase the information transmission rate on the star [38]. Polarization reuse technology uses the same star to receive antennas with the same terrestrial and, at the same time, the transmitting frequency of the transmission and different polarization

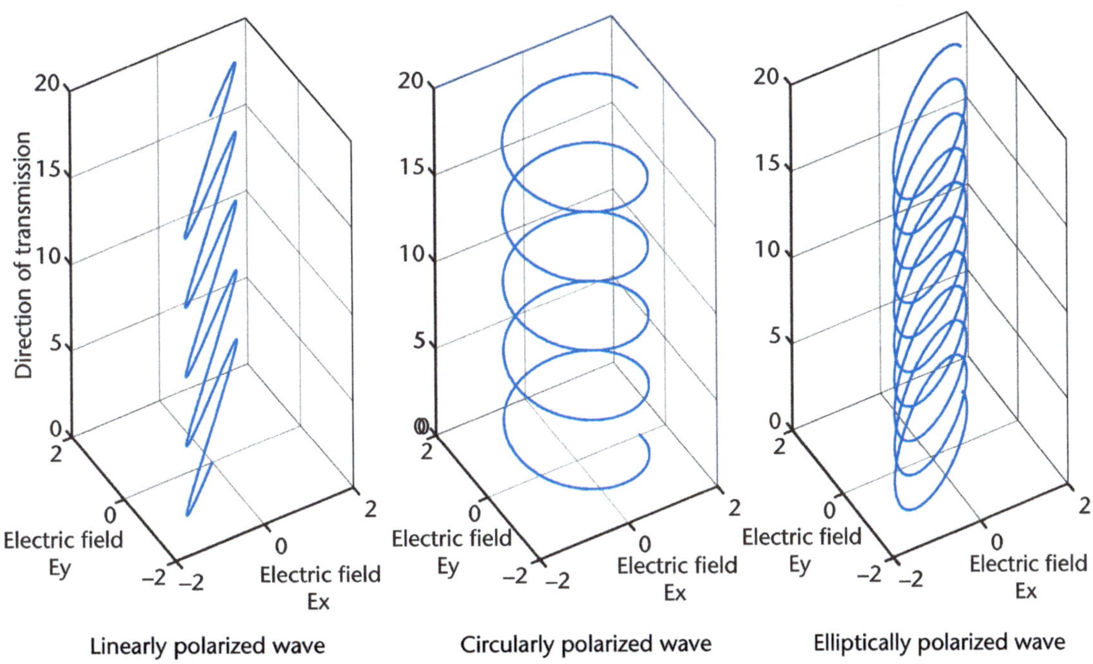

Figure 10.14 Schematic diagram of electric field vector propagation trajectory.

directions. When using the circular polarization carrier, the left rotary round polarization wave and the suitable rotary round polarization wave transmit information simultaneously [39].

Due to the design and transmission of the antenna itself, polarization has always been an interference between polarization and reusing signals. To transmit star data by polarization reuse, the problem of polarization interference must be solved. Regardless of weather reasons, polarization interference is mainly affected by three aspects [40]:

1. Star antenna polarization isolation;
2. Terrestrial receiving antenna polarization isolation;
3. Satellite polarization partial angle.

Theoretically, the two orthodontic polarization waves are entirely isolated. One antenna can be equipped with two receiving or sending ports. Each port only matches one polarized wave and is orthogonal with the other. In the satellite communication system, the characteristic of polarization wave orthogonal is used as an additional isolation when sending and receiving in the adjacent channel. However, due to the actual sending and receiving equipment error and the rainwater exfoliating during the radio wave transmission, the digitization direction of the receiving end has a mistake, resulting in the following two results: (1) the helpful signal transmission of the positive polarization mode will leak in the cross-polarization direction, and cross-polarization interference is formed, and (2) the valuable signals received in the direction of positive polarization will be weakened due to leakage and interference. The polarization isolation index does not meet the standard, which not only causes the transmission signal to polarize attenuation in the direction of the positive polarization but also causes the leakage signal interference frequency band business in the reflux direction, affecting the communication quality. The polarized corner angle of the star floor refers to the difference between the position of the terrestrial station and the dialogue between the base station and the satellite, as well as an antenna feed telephoto ports caused by the earth's curvature. Therefore, the perfect corner enables the receiving antenna to match satellite polarization better and efficiently receive weak satellite signals.

According to the Jones vector expression, the transmission signal can be expressed as

$$E(t) = \begin{bmatrix} E_{\mathrm{h}}(t) \\ E_{\mathrm{v}}(t) \end{bmatrix} = \begin{bmatrix} \cos\varepsilon \\ \sin\varepsilon e^{-j\theta} \end{bmatrix} E\mathrm{e}^{-j\omega t} \tag{10.28}$$

where ω represents the carrier frequency, ε and θ represents the signal polarization state, specifically:

1. When $\theta = 0$ or $\theta = \pi$, the signal is a linearly polarized wave;
2. When $\varepsilon = \pi/4$, the signal is LHCP wave $((\theta = \pi/2))$ or RHCP wave$((\theta = -\pi/2))$;
3. When $\varepsilon, \theta \in (0, 90^\circ)$, the signal is a left-handed elliptically polarized wave, and when $\varepsilon, \theta \in (0, -90^\circ)$, the signal is a right-handed elliptically polarized wave.

Next, we analyze the downlink satellite MIMO scenario in detail, as shown in Figure 10.15. Assume that the satellite transmits polarized signals to two earth stations through the downlink satellite-to-terrestrial link, where the satellite transmits sound signals to ES 1 through LHCP and to ES 2 through RHCP. The transceiver is configured with M and N dual-polarized antennas to match the polarized signal transmission scenario. Therefore, the channel includes spatial fading and polarization fading and can be modeled as follows:

$$H = \begin{bmatrix} H_{\mathrm{LL}} & H_{\mathrm{LR}} \\ H_{\mathrm{RL}} & H_{\mathrm{RR}} \end{bmatrix} = \underbrace{\left(D \otimes \mathbf{1}_{N\times M}\right)}_{\text{polarizationfading}} \odot \underbrace{\begin{bmatrix} \bar{H}_{\mathrm{LL}} & \bar{H}_{\mathrm{LR}} \\ \bar{H}_{\mathrm{RL}} & \bar{H}_{\mathrm{RR}} \end{bmatrix}}_{\text{spatialfading}} \tag{10.29}$$

where $\otimes$ represents the Kronecker product, $\odot$ represents the inner product, H_{LL} is the useful signal channel matrix between the satellite and ES 1 through LHCP, H_{LR} is the interference signal channel matrix between the satellite and ES 2, and the matrix D represents the polarization leakage ratio between polarization components, and has

$$D = \begin{bmatrix} \sqrt{1-r} & \sqrt{r} \\ \sqrt{r} & \sqrt{1-r} \end{bmatrix} \tag{10.30}$$

where r represents the polarization leakage ratio, and $r = 0$ represents that there is no leakage between polarization components and vice versa, which means cross-polarization interference exists. Cross-polarization rate (CPD) can be used to evaluate the degree of cross polarization. Next, since the signal includes two different polarization components x_{L} and x_{R}, correspondingly, considering the allocation of different transmission powers P_{L} and P_{R} to different polarization components,

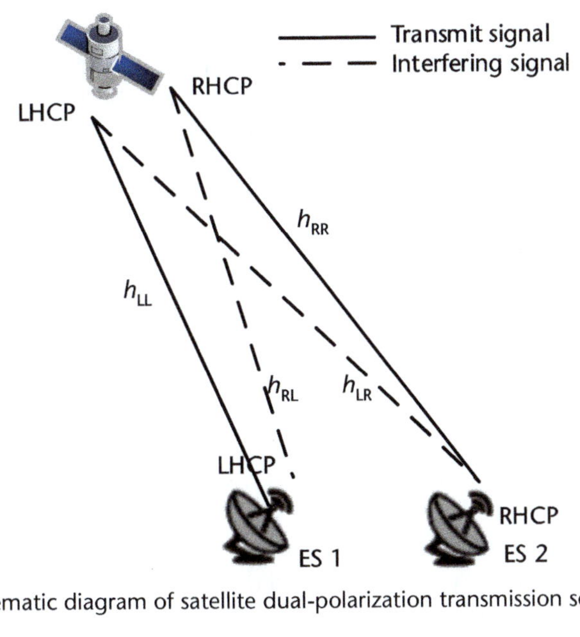

Figure 10.15 Schematic diagram of satellite dual-polarization transmission scenario.

the satellite transmission signal expression can be expressed as

$$x = \begin{bmatrix} p_L x_L \\ p_R x_R \end{bmatrix} = \begin{bmatrix} p_L \cos \varepsilon \\ p_R \sin \varepsilon e^{-j\theta} \end{bmatrix} s \cdot e^{-j\omega t} \tag{10.31}$$

therefore, the receiving end signal is

$$\begin{aligned} & y = Hx + n \\ & \Rightarrow \begin{bmatrix} y_L \\ y_R \end{bmatrix} = \begin{bmatrix} H_{LL} & H_{LR} \\ H_{RL} & H_{RR} \end{bmatrix} \begin{bmatrix} p_L x_L \\ p_R x_R \end{bmatrix} + \begin{bmatrix} n_L \\ n_R \end{bmatrix} \end{aligned} \tag{10.32}$$

where n_L and n_R represent the noise vectors corresponding to different channels, and the transmission rates of different polarization components are, respectively,

$$R_L = \log(1 + \text{SINR}_L) = \log\left(1 + \frac{P_L |H_{LL}|^2}{P_R |H_{LR}|^2 + \sigma_L^2}\right) \tag{10.33}$$

$$R_R = \log(1 + \text{SINR}_R) = \log\left(1 + \frac{P_R |H_{RR}|^2}{P_L |H_{RL}|^2 + \sigma_R^2}\right) \tag{10.34}$$

where $P_R + P_L \leq P_{\max}$; P_{total} represents the maximum transmission power of the satellite; σ_L^2 and σ_R^2 represents the noise power in different polarization directions.

10.5 Interference Management Technology

10.5.1 Adaptive Antenna Anti-Interference Technology

Satellites are widely distributed and subject to various interferences. Therefore, antenna anti-interference technology must be applied to expand coverage and ensure that signals can be received promptly. At the same time, interference signals can be weakened and eliminated to improve the quality of satellite communications. Among them, adaptive zeroing measures are widely used to close radio beams promptly, reduce interference, and achieve deep zeroing. With the continuous development of contemporary technology, the application of intelligent wires is becoming more and more common. The directional pattern can be adaptively changed according to specific environmental conditions to ensure the regular operation of the antenna, effectively suppress interference signals, and improve the signal-to-interference ratio. The intelligent antenna architecture mainly includes channels, antenna arrays, and adaptive signal processing units. The processing unit ensures the operation of the antenna adaptive function, accelerates positioning speed, improves frequency measurement efficiency, strengthens the confidentiality of satellite communications, and has excellent electromagnetic compatibility so that it can meet the needs of various devices.

Adaptive antenna anti-interference technology is a spatial domain signal processing method. Its principle is based on the specific signal transmission environment. It uses an adaptive anti-interference algorithm to automatically adjust the weight coefficients of each array element, optimize the maximum gain direction of the antenna pattern, align with the main lobe direction of the transmitted signal, and form a null in the direction of useless signals (interference, etc.). It improves the reception effect of valuable signals in the spatial domain, reduces transmission interference, and optimizes the antenna's anti-interference level.

As shown in Figure 10.16, the adaptive antenna system consists of an antenna array, a radio frequency chain, an analog/digital (A/D) conversion module, and an adaptive signal processing module. The adaptive signal processing module is the key to the antenna system. Its working process is as follows: the mixed (including proper signals and useless signals) received signal is processed by the radio frequency chain, converted into a digital intermediate frequency signal by a high-speed A/D conversion module, and then weighted by the digital signal processing module to reduce useless signal interference adaptively.

Adaptive antennas can be divided into adaptive nulling antennas and equal sidelobe needle beam antennas based on different principles. Both are based on adaptive antenna technology. The adaptive nulling antenna forms the antenna pattern through an adaptive algorithm, and the weighting coefficient of the array element in the antenna is related to the signal transmission environment; the equal sidelobe needle beam antenna pattern forms an equal sidelobe pattern based on the weights calculated in advance. However, due to the difficulty in predicting the direction of arrival of satellite signals, the minimal power reaching the navigation receiver, and the easy masking by noise, the equal sidelobe needle beam antenna technology is not suitable for satellite navigation anti-interference receivers [41].

10.5.2 On-Satellite Processing Technology

On-satellite processing technology adjusts the uplink and downlink transmission processes through decoupling processing, alleviates transmission interference, and improves the link transmission environment. Many types of existing on-satellite processing technologies exist, such as decoding/encoding technology, multibeam switching technology, despreading/respreading technology, and intelligent automatic gain control technology. Fast frequency modulation and TDMA can be applied to the downlink, and full-band frequency modulation and FDMA can be used to the uplink, effectively increasing power while controlling the antenna size,

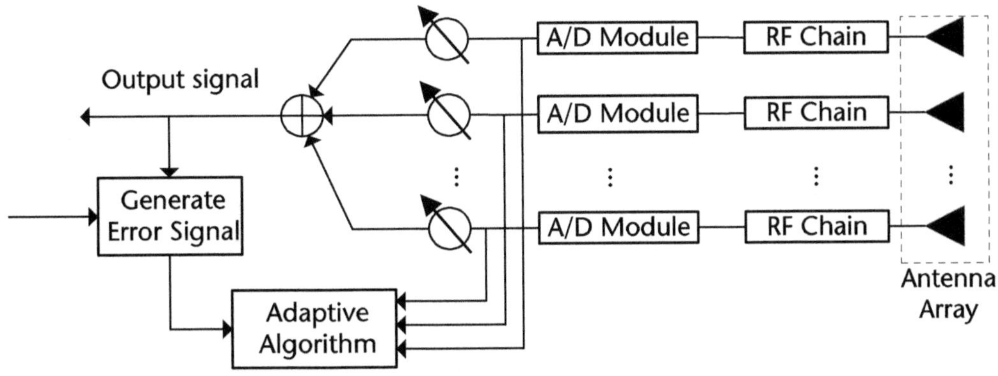

Figure 10.16 Adaptive antenna system block diagram.

thereby reducing the performance requirements of terrestrial equipment and playing an excellent anti-interference role.

On-satellite processing can be divided into two types: regenerative processing and nonregenerative processing. Specifically, the working principle of regenerative on-satellite processing is: after the satellite receives the signal, it demodulates and decodes it in the baseband, performs on-satellite processing, and then encodes and modulates the signal to the carrier frequency; nonregenerative on-satellite processing is different from the former and does not need to demodulate and decode the received signal, but directly performs signal processing. In early satellite communication systems, satellites forwarded signals without any processing, which severely limited the performance of satellite systems in terms of spectrum utilization, link performance, communication capacity, and networking flexibility. Satellite systems with on-satellite processing and on-satellite switching capabilities usually have the following advantages [42]:

1. *Flexible networking*: Users can use single-hop transmission within the satellite coverage area to achieve point-to-point or even point-to-multipoint communication between users.
2. *High spectrum utilization*: Information transmission is completed through single-hop transmission through satellite nodes, greatly reducing system transmission delay and significantly improving spectrum utilization.
3. *On-satellite regenerative processing is adopted to design specific signal modulation and coding methods*: According to the actual channel environment of the uplink and downlink, overcome the cumulative impact of uplink and downlink noise, and minimize the link bit error rate.
4. *It has the potential to achieve satellite-terrestrial integration*: It can realize satellite support for IP services and promote the development of satellite internet in the future.

10.5.3 Spread Spectrum Technology

Spread spectrum technology uses specific codes to broaden the signal spectrum so the system has good concealment and strong antinoise and anti-interference capabilities. Therefore, it is widely used in military and civilian communications, especially in satellite remote control and telemetry.

Spread spectrum technology uses a spread spectrum signal at the transmitting end to broaden the spectrum of the signal to be transmitted into a signal with a broader bandwidth and then transmits it to the channel for transmission; at the receiving end, the original data information is restored from the wideband received signal according to the corresponding rules. The robustness against external narrowband interference is enhanced by transmitting data over a larger bandwidth. The wider the bandwidth of any transmitted signal, the smaller the relative impact of interference on a small part of the bandwidth. Spread spectrum communication systems overcome the defects of traditional communication transmission technology with this unique signal transmission method. Therefore, they occupy an essential position in modern wireless communications. Their characteristics are as follows:

1. *Strong antinarrowband interference capability*: The spread spectrum communication systems have strong anti-interference capabilities because spread

spectrum signals are unpredictable. During the transmission process, the original information is spread, and its spectrum is expanded; that is, the bandwidth of the signal increases after the spread, which means that the bandwidth ratio of the signal before and after the spread increases, the system processing gain increases, and the anti-interference ability is correspondingly intense. The direct sequence spread spectrum system is good at resisting continuous time narrowband interference, while the frequency hopping spread spectrum system can resist pulse interference. Table 10.1 compares the advantages and disadvantages of the two systems.

2. *Good concealment and low interception rate*: Spread spectrum communication has good concealment and a low interception rate. The reason is that the signal to be transmitted is spread spectrum modulated, and its spectrum is broadened to a wide band. The signal power per unit bandwidth and the signal power spectrum density are reduced. Therefore, the signal will be integrated into the noise environment for transmission and will not be easily detected by the enemy. In addition, in the direct sequence spread spectrum system, the spread spectrum principle is that the transmitter multiplies the original data information by a pseudorandom code to achieve a spread spectrum and then modulates and sends it. The receiving end must use the pseudorandom code the same as the transmitting end information to despread the received signal to restore the original data information. However, it is difficult for the enemy or the interference party to obtain the complete spread spectrum code sequence, so it is almost impossible to crack the spread spectrum received signal, and it is impossible to intercept it, ensuring the high reliability of spread spectrum communication.
3. *Good antimultipath interference performance*: Multipath interference refers to the phenomenon in which the signal encounters various reflectors, such as mountains, ionosphere, and buildings, during the transmission process, generating scattering or reflection. At the receiving end, these scattered or reflected signals interfere with the direct signal, thus forming multipath interference. The spread spectrum communication system uses spread spectrum modulation to transmit information at the transmitting end, and there is a corresponding despreading process at the receiving end. The

Table 10.1 Comparison of the Advantages and Disadvantages of Direct Sequence Spread Spectrum and Frequency Hopping Spread Spectrum

System	*Advantages*	*Shortcomings*
Direct sequence spread spectrum	• Low power spectrum density, signal concealment • Resistant to multipath interference • Resistant to selective fading • With ranging capability • With multiaddress capability	• Severe near-far effect • Long capture time • Limited processing gain
Frequency hopping spread spectrum	• Faster capture • No near-far effect • Antimultipath interference • Antifrequency selective fading	• Limited ability to resist multifrequency interference • Inconvenient to use coherent demodulation technology • Limited fast frequency hopper • Poor concealment

received signal is correlated with the local spread spectrum code to find the valuable signal with the best correlation from the multipath signal to achieve antimultipath interference.

4. *Easy to implement frequency division multiplexing and code division multiple access [43].*

Based on the characteristics of solid autocorrelation of pseudonoise (PN) codes and the close-to-zero correlation of different PN codes, different PN codes are assigned to different users. All users only need to use their PN codes to communicate simultaneously on the same frequency band without interfering with each other, realizing frequency division multiplexing and significantly improving spectrum utilization. In addition, the sender and receiver can also perform multiaddress communication. The sender uses different PN codes to spread different data and send them to different receivers. The receivers use their PN codes to spread the received signals and recover the data from different senders. Different senders and receivers will not interfere with each other, realizing CDMA.

10.5.4 Adaptive Modulation and Coding Technology

Adaptive modulation and coding (AMC) technology is based on channel estimation, sends state information through the return channel, and adjusts and optimizes the specific AMC mode according to the real-time signal-to-noise ratio information. It is one of the most commonly used anti-interference technologies in satellite communications. However, the performance of AMC technology is also limited by many reasons, such as the link state estimation algorithm and the delay of the adaptive loop, which may be detrimental to the system's normal operation. The system adopted non-AMC mode in the past, and the power gain improvement was limited. However, AMC technology can improve the power gain by about 20 dB, increase the frequency band utilization rate, and increase the functional efficiency. Therefore, it is widely used in specific practice to optimize system performance. Commonly used AMC technologies in current practice include turbo code adaptive modulation and coding technology, adaptive grid modulation and coding technology, and adaptive bit interleaving modulation and coding technology. Compared with fixed coding modulation, AMC technology significantly improves spectrum efficiency, reduces the reserved link margin, and then selects a high-order optimal coding modulation mode (modulation and coding (MODCOD)), which can further improve the link spectrum efficiency.

The basic framework of AMC is shown in Figure 10.17. Information is sent from the transmitter and reaches the receiver via a satellite relay. According to the signal-to-noise ratio during the transmission process, the optimal modulation and coding scheme are matched, and the scheme signal is sent back to the gateway via the return channel. Finally, the gateway implements a specific scheme for the next information moment based on the received signal and transmits it. The advantage of AMC technology is that it allows the satellite communication system to match the optimal modulation method and coding rate according to the channel status of a certain period, reduce the link design reserve, make full use of limited

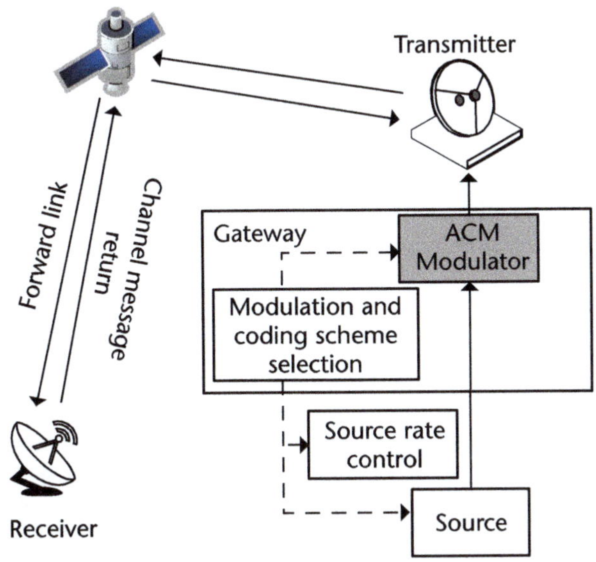

Figure 10.17 ACM basic framework diagram.

resources, and effectively improve the performance of the satellite communication system.

10.5.5 Digital Predistortion Technology

As one of the components with the highest power consumption in the RF link, the power amplifier (PA) seriously affects the energy efficiency performance of the transmitter, and the PA will cause amplitude and phase distortion to the system; that is, after passing through the PA, the phase offset between the input and output signals is not a simple linear relationship. Still, it is affected by the input signal power. As shown in Figure 10.18, the input and output power of a simple linear communication system changes linearly. However, general communication systems all contain nonlinear devices such as PA. As the input power continues to increase to the saturation region, affected by the gain compression effect, the input and output

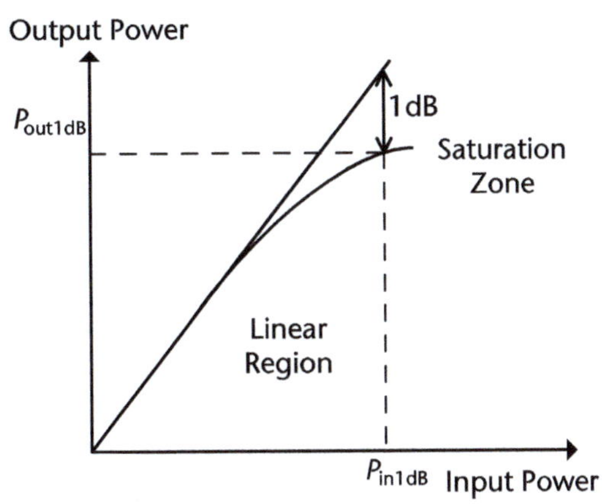

Figure 10.18 PA input and output power characteristics.

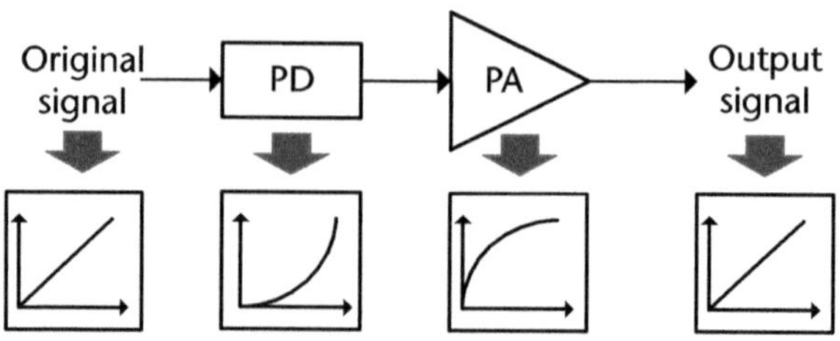

Figure 10.19 Digital predistortion schematic.

power of the saturation region PA is nonlinear, causing nonlinear distortion of the output signal, and the degree of distortion continues to increase with the input power, resulting in out-of-band spectrum leakage, causing interference in adjacent frequency bands, and reducing the system transmission performance.

The predistortion characteristics of digital predistortion (DPD) technology can be entirely complementary to the PA distortion under ideal conditions, so it is widely used in the front end of the system transmitter, and the key to the DPD system is to solve the predistortion function of the PA and do compensation work in advance. In theory, the DPD predistortion characteristic (i.e., distortion compensation) complements the PA distortion, as shown in Figure 10.19. The final theoretical output signal will be a linearly varying signal.

References

[1] Radio Administration Bureau of the Ministry of Information Industry, Radio Frequency Allocation Regulations of the People's Republic of China, Beijing: People's Posts and Telecommunications Press, 2002.

[2] Yi, K-c., Y. Li, C. Sun, and C.-g. Nan, "Recent Development and Future Prospects of Satellite Communications," *Journal of Communications*, 2015, Vol. 36, No. 06, pp. 161–176.

[3] Xia, S., Q. Jiang, C. Zou, and G. Li, "Beam Coverage Comparison of LEO Satellite Systems Based on User Diversification," *IEEE Access*, 2019, Vol. 7, pp. 181656–181667.

[4] Qi, X., B. Zhang, and Z. Qiu, "A Distributed Survivable Routing Algorithm for Mega-Constellations With Inclined Orbits," *IEEE Access*, 2020, Vol. 8, pp. 219199–219213.

[5] Del Portillo, I., B. G. Cameron, E. F. Crawley, "A Technical Comparison of Three Low Earth Orbit Satellite Constellation Systems to Provide Global Broadband," *Acta Astronautica*, 2019, Vol. 159, pp. 123–135.

[6] Tang, J., D. Bian, G. Li, J. Hu, and J. Cheng, "Resource Allocation for LEO Beam-Hopping Satellites in a Spectrum Sharing Scenario," *IEEE Access*, 2021, Vol. 9, pp. 56468–56477.

[7] Sharma, S. K., S. Chatzinotas, and B. Ottersten, "Transmit Beamforming for Spectral Coexistence of Satellite and Terrestrial Networks," *8th International Conference on Cognitive Radio Oriented Wireless Networks, IEEE*, 2013, pp. 275–281.

[8] Yao, L., "Research on Health Prediction Method of Satellite Lithium-Ion Batteries," Changsha: National University of Defense Technology, PhD Dissertation, 2017.

[9] Colavolpe, G., A. Modenini, A. Piemontese, and A. Ugolini, "On the Application of Multiuser Detection in Multibeam Satellite Systems," *2015 IEEE International Conference on Communications (ICC), IEEE*, 2015, pp. 898–902.

[10] Aravanis, A. I., B. S. Mysore, P. D. Arapoglou, and G. Danoy,"Power Allocation in Multibeam Satellite Systems: A Two-Stage Multi-Objective Optimization," *IEEE Transactions on Wireless Communications*, 2015, Vol. 14, No. 6, pp. 3171–3182.

[11] Durand, F. R., and T. Abrão, "Power Allocation in Multibeam Satellites Based on Particle Swarm Optimization," *AEU-International Journal of Electronics and Communications*, 2017, Vol, 78, pp. 124–133.

[12] Wang, A., L. Lei, E. Lagunas, A. I. Pérez-Neira, S. Chatzinotas, and B. Otterson, "NOMA-Enabled Multi-Beam Satellite Systems: Joint Optimization to Overcome Offered-Requested Data Mismatches," *IEEE Transactions on Vehicular Technology*, 2020, Vol. 70, No. 1, pp. 900–913.

[13] Tang, J., D. Bian, G. Li, J. Hu, and J. Cheng, "Optimization Method of Dynamic Beam Position for LEO Beam-Hopping Satellite Communication Systems," *IEEE Access*, 2021, Vol. 9, pp. 57578–57587.

[14] Wang, Y., D. Bian, J. Hu, J. Tang, and C. Wang, "A Flexible Resource Allocation Algorithm in Full Bandwidth Beam Hopping Satellite Systems," *2019 IEEE 3rd Advanced Information Management, Communicates, Electronic and Automation Control Conference (IMCEC), IEEE*, 2019, pp. 920–927.

[15] Caire, G., et al., "Perspectives of Adopting Interference Mitigation Techniques in the Context of Broadband Multimedia Satellite Systems," *in the Proceedings of the International Conference on Satellite Communication (ICSC)*, 2005.

[16] Lei, H., L. Zhang, X. Zhang, and D. Yang, "A Novel Multi-Cell OFDMA System Structure Using Fractional Frequency Reuse," *2007 IEEE 18th International Symposium on Personal, Indoor and Mobile Radio Communications, IEEE*, 2007, pp. 1–5.

[17] Riazul Islam, S. M., N. Avazov, O. A. Dobre, and K.-s. Kwak, "Power-Domain Non-Orthogonal Multiple Access (NOMA) in 5G systems: Potentials and Challenges," *IEEE Communications Surveys & Tutorials*, 2016, Vol. 19, No. 2, pp. 721–742.

[18] Liang, C., R. Duan, S. Ma, L. Tang, and Q. Chen, "Joint Beam-Hopping Scheduling and Power Allocation Algorithm for Low-Orbit Satellites for Energy Efficiency," *Journal of Electronics & Information Technology*, 2023, Vol. 45, No. 02, pp. 436–445.

[19] Anzalchi, J., A. Couchman, P. Gabellini, G. Gallinaro, L. D'Agristina, and N. Algaha, "Beam Hopping in Multi-Beam Broadband Satellite Systems: System Simulation and Performance Comparison With Non-Hopped Systems," *2010 5th Advanced Satellite Multimedia Systems Conference and the 11th Signal Processing for Space Communications Workshop, IEEE*, 2010, pp. 248–255.

[20] Wang, A., L. Lei, E. Lagunas, S. Chatzinotas, A. I. Pérez Neira, and B. Ottersten, "Joint Beam-Hopping Scheduling and Power Allocation in NOMA-Assisted Satellite Systems," *2021 IEEE Wireless Communications and Networking Conference (WCNC), IEEE*, 2021, pp. 1–6.

[21] Lin, Z., Z. Ni, L. Kuang, C. Jian, and Z. Huang, "Dynamic Beam Pattern and Bandwidth Allocation Based on Multi-Agent Deep Reinforcement Learning for Beam Hopping Satellite Systems," *IEEE Transactions on Vehicular Technology*, 2022, Vol. 71, No. 4, pp. 3917–3930.

[22] Zhang, H., Q. Li, Y. Zhang, and X. Li, "Game Theory Based Power Allocation Method for Inter-Satellite Links in LEO/MEO Two-Layered Satellite Networks," *2021 IEEE/CIC International Conference on Communications in China (ICCC)*, Xiamen, China, 2021, pp. 398–403.

[23] Zeng, G.,Y. Zhan, H. Xie, and C. Jiang, "Resource Allocation for Networked Telemetry System of Mega LEO Satellite Constellations," *IEEE Transactions On Communications*, Vol. 70, No. 12, December 2022, pp. 8215–8228.

[24] Zheng, Z., N. Hua, Z. Zhong, J. Li, Y. Li, and X. Zheng,"Time-Sliced Flexible Resource Allocation for Optical Low Earth Orbit Satellite Networks," *IEEE Access*, 2019, Vol. 7, pp. 56753–56759.

[25] He, G., S. Cui, Y. Dai, and T. Jiang, "Learning Task-Oriented Channel Allocation for Multi-Agent Communication," *IEEE Transactions on Vehicular Technology*, 2022, Vol. 71, No. 11, pp. 12016–12029.

[26] Jiang, C., and X. Zhu, "Reinforcement Learning Based Capacity Management in Multi-Layer Satellite Networks," *IEEE Transactions on Wireless Communications*, 2020, Vol. 19, No. 7, pp. 4685–4699.

[27] Kaushal, H., and G. Kaddoum, "Optical Communication in Space: Challenges and Mitigation Techniques," *IEEE Communications Surveys and Tutorials*, 2017, Vol. 19, No. 1, pp. 57–96.

[28] Li, F., B. Li, and J. Lu, "Analysis and Prediction of Solar Eclipse Interference in Satellite Communications," *China New Communications*, 2018, Vol. 20, No. 17, p. 22.

[29] Yuan, W., K. Rong, T. Wei, and W. Nan,"Research on Adaptability Evaluation of Satellite Communication in Complex Electromagnetic Environment," *2021 4th International Conference on Advanced Electronic Materials, Computers and Software Engineering (AEMCSE), IEEE*, 2021, pp. 66–67.

[30] Clegg, A., and A. Weisshaar, "Future Radio Spectrum Access [Scanning the Issue]," *Proceedings of the IEEE*, 2014, Vol. 102, No. 3, pp. 239–241.

[31] Hexian, C., "Research on Frequency Sharing and Interference Suppression of Multi-Satellite Systems." Chengdu: University of Electronic Science and Technology of China, PhD dissertation, 2021.

[32] Leyva-Mayorga, I., et al., "NGSO Constellation Design for Global Connectivity," *arXiv preprint arXiv:2203.16597*, 2022.

[33] Fortes, J. M. P., R. Sampaio-Neto, and J. E. Maldonado, "An Analytical Method for Assessing Interference in an Environment Involving NGSO Satellite Networks," *International Journal of Satellite Communications & Networking*, 1999, Vol. 17, No. 6, pp. 5–7.

[34] Pourmoghadas, A., S. K. Sharma, S. Chatzinotas, and B. Ottersten, "Cognitive Interference Management Techniques for the Spectral Co-Existence of GSO and NGSO Satellites," in *Wireless and Satellite Systems*, I. Otung, P. Pillai, G. Eleftherakis, and G. Giambene (eds.), Cham, Switzerland: Springer, 2016.

[35] Kan, X., and X. Xu, "Energy-and Spectral-Efficient Power Allocation in Multi-Beam Satellites System with Co-Channel Interference," *2015 International Conference on Wireless Communications & Signal Processing (WCSP), IEEE*, 2015, pp. 1–6.

[36] Sharma, S. K., S. Chatzinotas, and B. Ottersten, "In-Line Interference Mitigation Techniques for Spectral Coexistence of GEO and NGSO Satellites," *International Journal of Satellite Communications and Networking*, 2016, Vol. 34, No. 1, pp. 11–37.

[37] Hao, Z., and Z. Zheng, "High Symbol-Rate Polarization Interference Cancelation for Satellite-to-Terrestrial Remote Sensing Data Transmission System," *EURASIP Journal on Wireless Communications and Networking*, 2018, pp. 1–10.

[38] Luo, Z., H. Wang, and K. Zhou, "Polarization filtering Based Physical-Layer Secure Transmission Scheme for Dual-Polarized Satellite Communication," *IEEE Access*, 2017, Vol. 5, pp. 24706–24715.

[39] Popovskyy, V., and A.-W. S. A. Iskandar, "Polarization Multiplexing Modulation in fiber-Optic Communication Lines," *2016 Third International Scientific-Practical Conference Problems of Infocommunications Science and Technology (PIC S&T), IEEE*, 2016, pp. 214–216.

[40] Stapor, D. P., "Optimal Receive Antenna Polarization in the Presence of Interference and Noise," *IEEE Transactions on Antennas and Propagation*, 1995, Vol. 43, No. 5, pp. 473–477.

[41] Xiao, H., "Research on Anti-Interference Technology of Satellite Antenna Based on Multi-Domain Fusion Processing," Harbin: Harbin Engineering University, PhD dissertation, 2013.

[42] Yang, Y., "Research on Broadband Satellite Onboard Switching Technology Based on CWTDM," Xi'an: Xi'an University of Electronic Science and Technology, PhD dissertation, 2014.

[43] Peng, X., K.-B. Png, Z. Lei, F. Chin, and C. Chung Ko, "Two-Layer Spreading CDMA: An Improved Method for Broadband Uplink Transmission," *IEEE Transactions on Vehicular Technology*, 2008, Vol. 57, No. 6, pp. 3563–3577.

CHAPTER 11

Mobility Management for Satellite-Terrestrial Integrated Communication

Wireless communication networks are currently experiencing an unprecedented transformation to address the global deployment demands of the Internet of Everything. The future network is expected to deliver not only communication and computing services but also security for a vast array of devices in a ubiquitous manner. This evolution has necessitated the provision of broadband internet connectivity across all regions of the planet. Despite significant advancements in terrestrial communication networks, coverage remains incomplete, particularly in rural and hard-to-reach areas such as oceans, deserts, polar regions, and high-altitude locations. The extensive reach of satellite networks can substantially enhance communication capabilities for users situated in remote areas. Furthermore, satellite networks play a crucial role in delivering essential and emergency services during and after natural disasters.

In recent years, various industrial groups and standardization organizations including the 3GPP have proposed integrating satellite networks with 5G technology and beyond to facilitate seamless broadband coverage [1]. Satellite networks have witnessed rapid development over the past decade; this is especially true for LEO satellite systems like SpaceX's Starlink constellation. Such satellite networks can be regarded as either an extension of terrestrial IP networks or as independent entities within their own right.

LEO satellites are positioned in low-altitude orbits, typically ranging from 160 to 2,000 km above the earth's surface. In comparison to MEO and GEO satellites, LEO satellites exhibit lower propagation delays and reduced signal attenuation. These characteristics enable LEO satellites to facilitate low-latency communication while also minimizing energy consumption requirements. Additionally, LEO satellites contribute significantly to lowering production costs.

However, the high velocity associated with low Earth orbit results in frequent handovers. For instance, a satellite operating at an altitude of 500 km travels at a speed of approximately 7.6 km per second, completing an orbit around the earth in about 95 minutes and necessitating handover roughly every 5 minutes. Consequently, there is an urgent need for effective mobile management strategies.

Efficient handover management constitutes a critical component of satellite network systems. Mobility management encompasses both handover management and location management.

11.1 Overview of Mobility Management

Handover management refers to the process by which a mobile node (MN) transitions from one access point (AP) to another while maintaining an active connection.

In traditional terrestrial cellular networks, user mobility (i.e., MN movement) results in continuous fluctuations in received signal strength levels. When the received signal strength at any given location falls below an acceptable threshold, the handover process is initiated, facilitating a transition from the current service area (i.e., AP) to a target area.

Regarding location management, most research on cellular networks emphasizes tracking and paging, as illustrated in Figure 11.1. Tracking within cellular networks involves identifying the specific cell where the MN is located based on signal strength received by base stations. Paging serves to indicate a user's location within a cellular network for establishing connections with other users. The tracking area (or location area) utilized in 4G and 5G typically encompasses multiple cells. The objective of location management is to strike a balance between partitioning tracking areas and minimizing costs associated with location updates and paging. To facilitate communication with other devices within the mobile network, MNs must establish end-to-end user plane paths through mobile operators. Idle MNs perform position updates upon crossing boundaries of their tracking areas; these updates are recorded in databases that can be queried to ascertain their locations. To determine the current position of idle MNs, cells within the designated location area are contacted via paging mechanisms. With impending densification in 5G networks, it is anticipated that signaling costs related to location management will rise significantly due to increased frequency of location updates.

In IP-based networks, location management is conducted in a somewhat different manner; the TCP/IP connection of the MN must be maintained while transforming from one access router to another. The introduction of the IP protocol

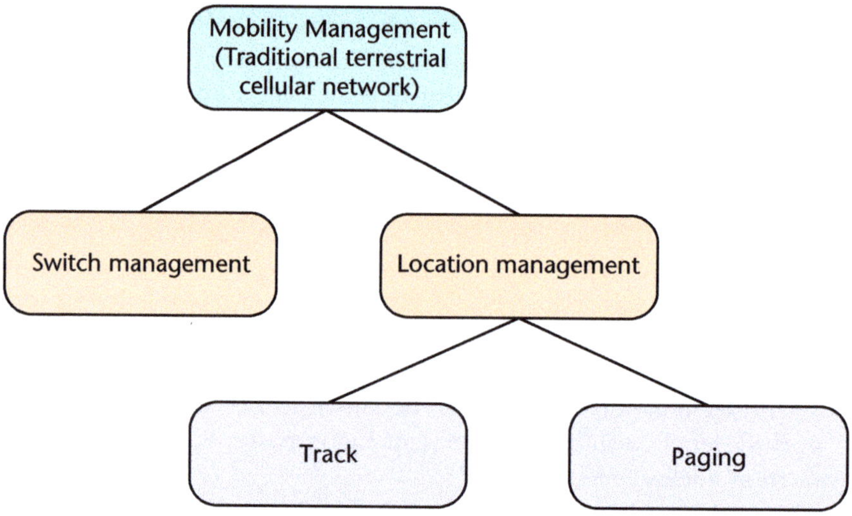

Figure 11.1 Mobility management of traditional terrestrial cellular networks.

in the 1970s marked a significant advancement as an Internet Protocol for transmitting data packets within wired networks, utilizing IP addresses for identification and packet routing. With the evolution of mobile wireless communication devices, it has become imperative to support mobility within IP-based networks. Consequently, the Internet Engineering Task Force (IETF) developed protocols such as MIPv4, followed by MIPv6, PMIPv6, FMIPv6, and HMIPv6. Among these protocols, location management primarily encompasses two key functions: location updates and data transmission.

Location update refers to the process of identifying and updating the logical position of MN within a network. Data transmission specifically routing involves forwarding packets directed towards MN to their new locations. Figure 11.2 illustrates mobility management in IP-based networks.

The network protocol stack can be categorized into five layers: the physical layer, link layer, network layer, transport layer, and application layer, arranged from bottom to top. With the exception of the physical layer, all other protocol layers incorporate mobility management technologies. In LEO satellite networks, research on mobility management primarily concentrates on the link layer, network layer, and transport layer. Each protocol layer employs distinct handover and location management techniques, as illustrated in Figure 11.3.

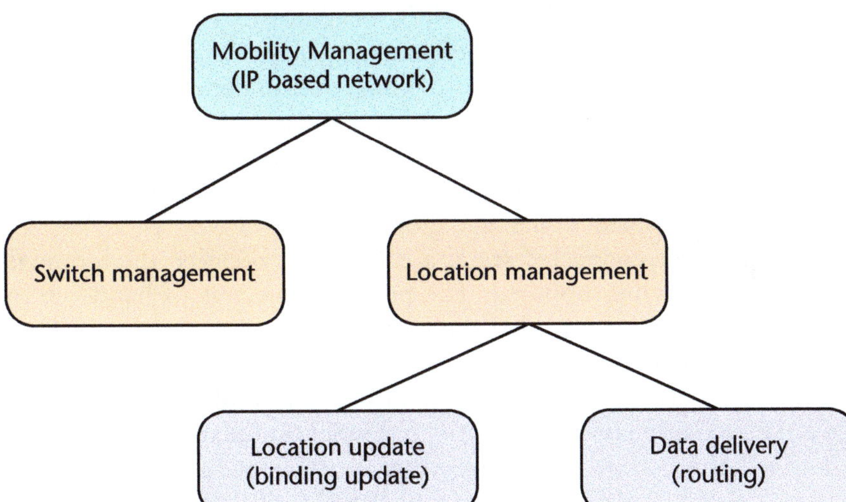

Figure 11.2 Mobility management of IP-based networks.

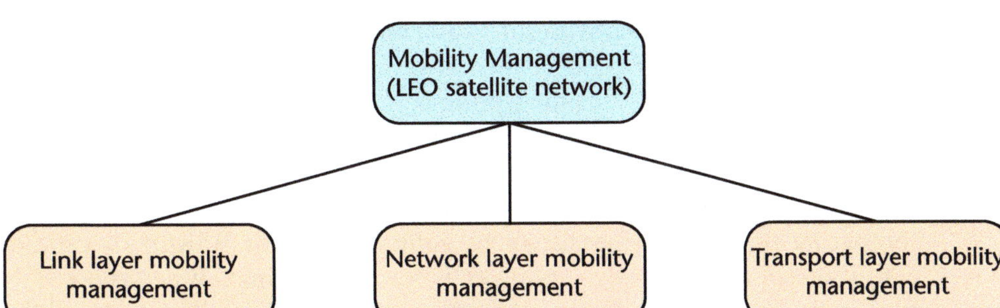

Figure 11.3 Mobility management of LEO satellite networks.

In terms of the role played by satellites, they can act as routers or MNs. When a satellite serves as an MN, the onboard equipment generates and transmits data to the ground gateway, and the satellite is a network user. When satellites act as mobile routers, they can forward data and provide services to other terminals. Satellites are network access points.

This chapter first introduces the link layer management technology and then separately introduces the mobility management technology of the network layer and the transport layer. Finally, the potential mobility technology of satellites will be introduced.

11.2 Link Layer Management Technology

The management of handover processes in link-layer mobility for 5G technology has reached a relatively advanced stage. In satellite networks, handover management primarily encompasses intrasatellite and intersatellite handovers. The terminal establishes a communication link with the nearest satellite node. When there are changes in the coverage area of the satellite hosting the user's terminal device or alterations in beam patterns, effective handover management becomes necessary.

11.2.1 5G Handover Management

In 5G networks, distinct mobility management principles are employed based on the states of user terminals, namely idle, inactive, or connected. In the connected state, handover management is implemented; here, the terminal measures and reports the signal quality of neighboring candidate cells while the network determines when to execute a handover to another cell. Conversely, for inactive and idle states, cell reselection is utilized to manage mobility.

During the connected state, a connection between the terminal and network has already been established. The primary objective of handover management in this context is to ensure that the terminal maintains connectivity while traversing through the network without experiencing communication interruptions or significant degradation in communication quality. Initially, the mobile terminal employs a cell search mechanism to identify potential candidate cells. Upon locating a candidate cell, it proceeds to measure either the reference signal reception power (RSRP) or reference signal reception quality (RSRQ) of that cell. Typically, filtering techniques are applied during these measurements; for instance, averaging values over several hundred milliseconds may be used. Without such average filtering processes in place, measurement report results can exhibit considerable fluctuations that may lead to erroneous handover decisions or result in repetitive transitions between two cells phenomenon known as ping-pong handover.

Measurement events refer to the conditions that terminals must satisfy prior to reporting measurement results to the network. In 5G, six distinct triggering conditions or events can be configured, including A3 events. The A3 event involves comparing the filtered measurement values (RSRP or RSRQ) of candidate cells with those of the currently serving cells for the terminal. When there is a discrepancy in transmission power between the current service cell and a candidate cell, a cell-specific offset may be established to influence the triggering conditions of the measurement event. Furthermore, it is essential to configure a threshold to prevent unnecessary triggering of measurement events when differences in measurement results between two cells are minimal.

If the measurement value from a candidate cell significantly exceeds that of the current serving cell by more than the sum of both the cell-specific offset and threshold, an A3 measurement event will be triggered, prompting a corresponding report to be sent to the network. Additional measurement events can also be configured for terminals; however, both type selection and parameter configuration depend on mobility implementation strategies tailored for specific network deployments.

The utilization of configurable measurement events aims to minimize the transmission of unnecessary measurement reports to the network. An alternative approach for event-driven reporting is periodic reporting; however, in most instances, the costs associated with periodic reports are considerably higher. This is due to the necessity for frequent measurements that accurately reflect terminal movements within the network. From a cost perspective, infrequent periodic reports may be more favorable; nevertheless, they also heighten the risk of overlooking critical handover opportunities. By configuring measurement events, it becomes feasible to transmit measurement reports solely when situational changes occur, which represents a clearly superior option.

Upon receiving a measurement report, the network can determine whether to execute a handover. Handover decisions may also incorporate additional information beyond just measurement reports much as assessing whether there is adequate capacity in the prospective target cell to facilitate handover operations. Even in scenarios where no measurement report has been received, the network retains the capability to initiate a handover for reasons such as load balancing considerations. The signaling interaction is illustrated in Figure 11.4; GNB refers to a base station within 5G networks and oversees all wireless-related functions across one or multiple cells.

The source cell initiates a handover request to the target cell. If both the source and target cells are part of the same gNB, this message is unnecessary, as the gNB is already aware of the status of the target cell. Should the target gNB accept the handover request-keeping in mind that it may refuse due to excessive load the source gNB will instruct the terminal to proceed with handing over to the target cell. This process involves sending RRC reconfiguration messages to the terminal, which contain essential information for accessing the target cell.

To establish a connection with this new area, synchronization is required from the terminal. Consequently, terminals are typically directed to initiate random access procedures toward the target cell. Once synchronization has been successfully achieved, a handover completion message is sent by the terminal to indicate its successful connection with the new cell. Simultaneously, all data cached by the terminal within the source gNB is transferred to its counterpart in the target gNB; furthermore, any new downlink data will be rerouted accordingly to designate it as now serving cell.

11.2.2 Beam Handover

Beam handover, also referred to as intrasatellite handover, occurs within multibeam satellites, as illustrated in Figure 11.5. In this context, the position of user terminal equipment transitions from one point beam of a satellite node to another. The

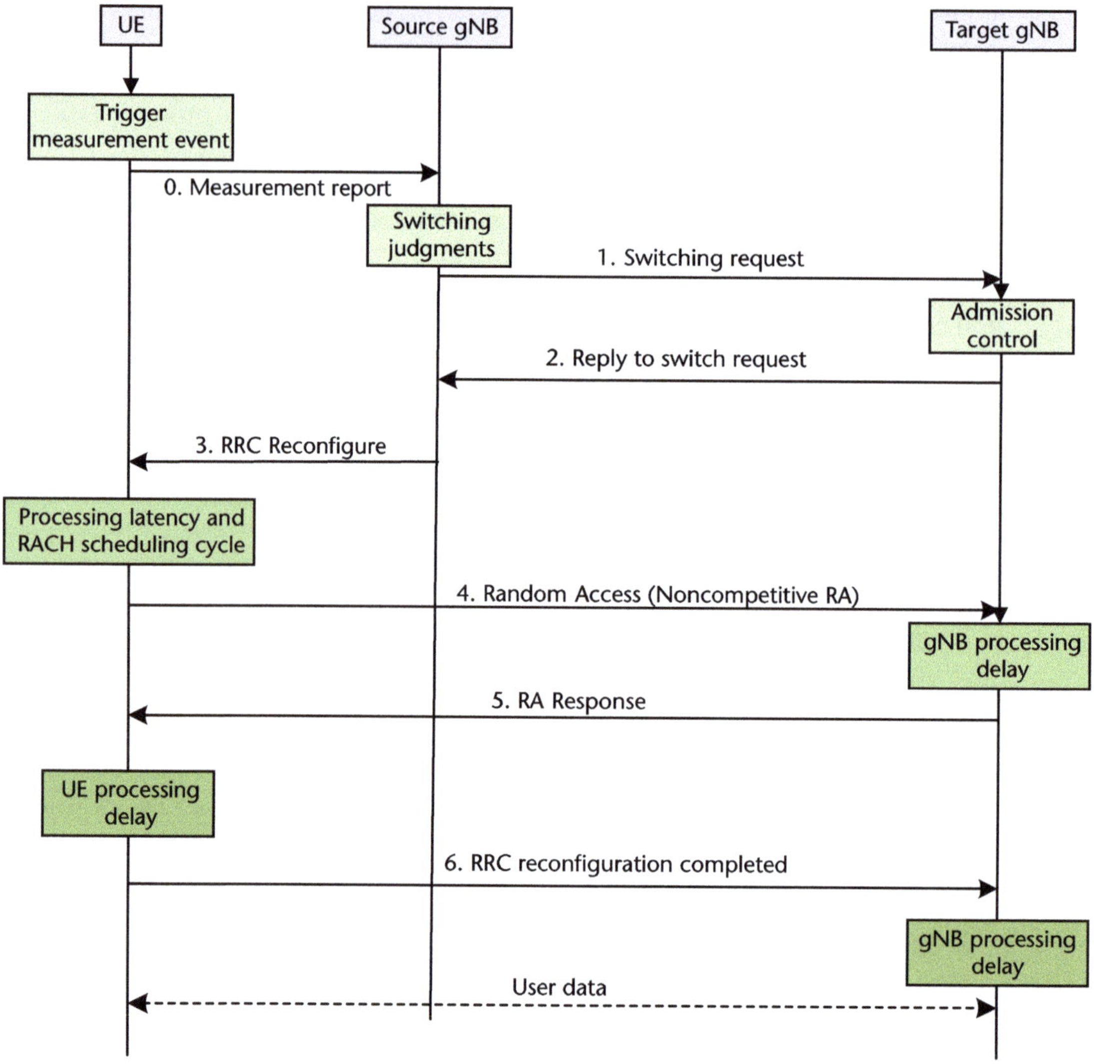

Figure 11.4 Handover process.

ground trajectory movement speed of LEO satellites significantly exceeds that of the terminal's movement speed. Consequently, the visible time for a satellite can extend up to 10 minutes; however, the user's residence time within a single beam may be limited to just 1 minute. This results in a high frequency of handovers between beams.

Research on beam handover primarily concentrates on channel allocation strategies, where each point beam functions as a small cell. When users transition from one beamforming cell to another, they must relinquish previously occupied channel resources and secure new ones. If the subsequent beam cell is unable to provide adequate available channel resources, the handover process will be impeded and user service transmission may be disrupted. Given the diversity of services and varying requirements for QoS, devising effective channel allocation strategies for different QoS demands presents a complex challenge. Existing research

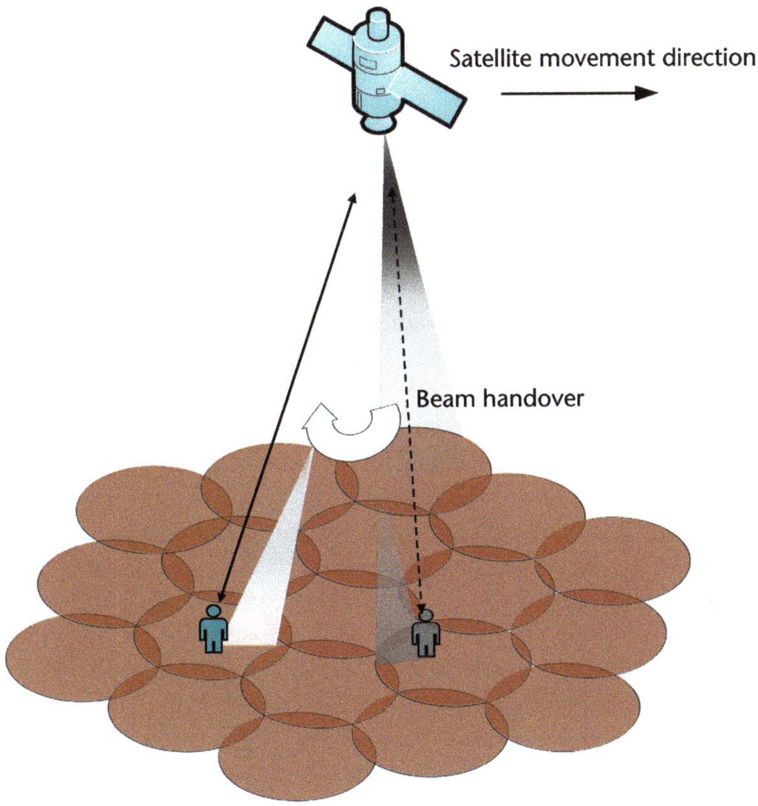

Figure 11.5 Beam handover (intrasatellite handover).

predominantly categorizes into three main schemes: fixed channel allocation (FCA), dynamic channel allocation (DCA), and hybrid channel allocation (HCA) [2].

11.2.2.1 Beam Handover Based on FCA

FCA temporarily allocates all available channels to any requesting cell and can modulate the number of allocated channels in response to fluctuations in business load. The earliest form of FCA was guarantee handover (GH); this strategy [3] pre-allocates the channel resources necessary for handover when initiating new calls, thereby ensuring successful handover. However, this approach results in a significant amount of channel resources being locked prematurely, which may prevent new calls from accessing these resources later on. While it guarantees a 100% success rate for handovers, it is associated with an extremely high call-blocking rate and low utilization of channel resources.

Another typical FCA implementation involves beam handover, which introduces the concept of channel protection. In this framework, each protected channel within the cell is specifically designated to accommodate handover requests; furthermore, the number of such channels can be adaptively and dynamically adjusted based on forecasts regarding future handovers. When a new call request arrives within the community, the algorithm leverages predictable topological information about the satellite constellation to estimate the user's residence time in that area. Simultaneously, it assesses potential handover requests that may arise during this dwell time as well as their expected demand for channels denoted by m. A request will only be accepted if the number of available channels exceeds m. For handover requests, provided there is at least one available channel in the target cell, successful completion of the call transfer is assured.

11.2.2.2 Beam Handover Based on DCA

In a typical DCA beam handover method, when receiving a handover request, if the next cell has no available channel, it will queue and wait in the queue. Due to the higher cost of handover failure compared to blocking new calls, handover requests in the queue have higher priority. Newly arrived call requests are only accepted if there are available channels in both the current beam and the next beam. Otherwise, they will be rejected. This method can ensure a lower call drop rate and simultaneously increase the call loss rate.

Another channel state-based reservation strategy [4] divides the traffic provided by LEO into real-time multimedia traffic and non-real-time data. According to the state of the beam cell, the reservation time for different users is adaptively set, the probability of successful handover P is introduced, and channel resources for users are reserved after successful handover based on the set reservation time. This strategy can reduce the probability of dropped calls and new call blocking, but the adaptive method is not friendly to certain specific services, which, to some extent, reduces the user service experience. To address this issue, an improved adaptive handover strategy has been developed [5]. This strategy takes into account the constantly changing characteristics of wireless channels. In addition, by maintaining the probability of connection blocking and connection disconnection at an acceptable level, a higher QoS is provided to users.

11.2.2.3 Beam Handover Based on HCA

Both fixed channel allocation and dynamic channel allocation have their own shortcomings. With the changing business load of LEO satellites, it is difficult to achieve dynamic channel allocation. In practical satellite communication systems, channel allocation schemes are usually designed by combining the advantages of fixed channel allocation strategies and dynamic channel allocation strategies. Reference [6] proposes a new channel allocation strategy for low Earth orbit mobile satellite systems aimed at reducing the average handover time of calls. This strategy is based on a queuing strategy with a limited queue size, and an algorithm is designed to handle the computational complexity of QoS metrics. This strategy assumes that new calls are evenly distributed in the mobile business area, but in reality, new calls are unevenly distributed in the mobile satellite system service area, and the accuracy is not high enough. Another strategy [7] considers both fixed channel allocation techniques and dynamic channel allocation techniques. By comparing the first in, first out queuing rule, maximum remaining processing time queuing rule, and a measurement-based priority scheme, handover requests between beams that have not been immediately delivered are queued to reduce the handover failure rate.

11.2.3 Interstellar Handover

Interstellar handover refers to the process in which a terminal handovers between two or more satellites. As shown in the Figure 11.6, User 1 and User 2 initially

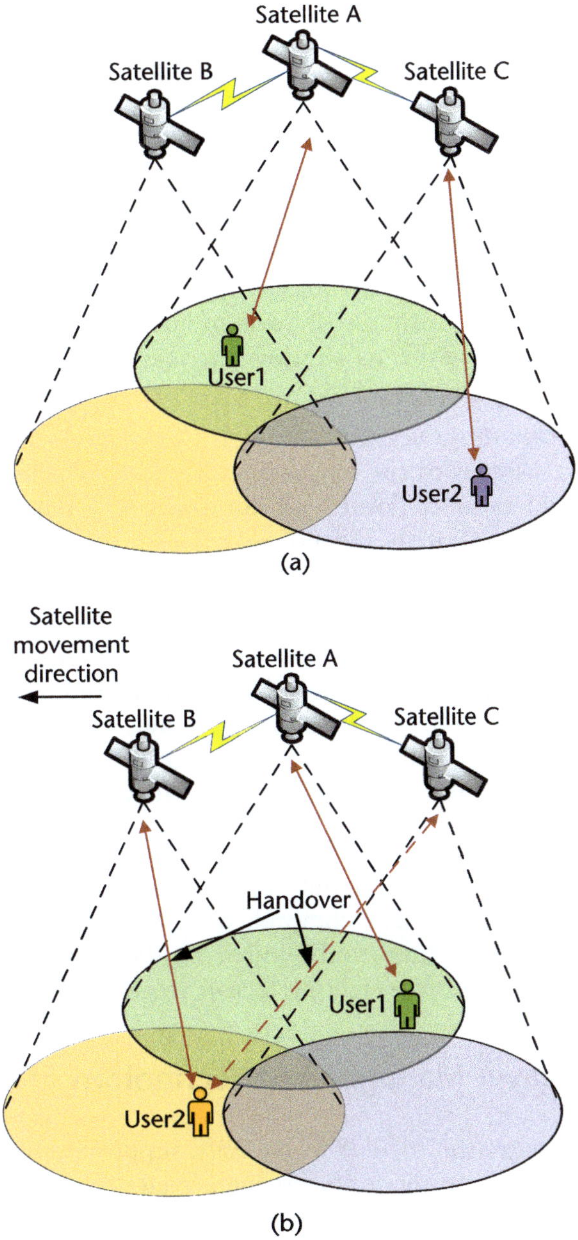

Figure 11.6 Intersatellite handover. (a) User 1 is served by Satellite A, while User 2 is served by Satellite C. (b) User 2 performs handover from Satellite C to Satellite B.

communicate through satellites A and B. After User 2 is handed over to satellite C, they communicate through satellite A B. C communicates. Although intersatellite handover does not occur as frequently as beam handover, it is crucial for satellite diversity systems. Satellite diversity refers to the ability of satellite terminals to establish communication links with multiple satellites at any time, which includes two types: exchange diversity and merge diversity. When the terminal selects a satellite and establishes a communication link with it, this satellite diversity method is exchange diversity. The way in which the ground terminal communicates with multiple satellites simultaneously is called merging diversity. The use of satellite diversity technology can combat signal fading caused by ground buildings or terrain factors, improve communication quality, and provide redundant backup for satellites.

Common intersatellite handover strategies include maximum idle channel quantity strategy, maximum service time strategy, minimum distance strategy, service time and idle channel weighting strategy, and comprehensive weighting strategy:

1. *Maximum number of idle channels strategy*: Priority is given to selecting satellites with the maximum number of available channels, which can achieve a uniform distribution of business loads within the satellite constellation. However, its handover request arrival rate and service quality are relatively low.
2. *Maximum service time strategy*: Prioritize selecting satellites that can provide users with the longest service time, which can reduce the number of handover times during call duration and reduce signaling overhead. However, this strategy did not take into account the fact of uneven business distribution, which can easily lead to satellite overload and increase the handover failure rate.
3. *Minimum distance (maximum elevation) strategy*: Prioritize selecting the closest satellite (i.e., the satellite visible at maximum elevation), which can improve communication quality. However, this strategy does not take into account channel effects and does not fully utilize satellite prior knowledge, so it is generally not used in practical systems

In addition, the weighted strategy of maximum service time and maximum number of idle channels can minimize the number of handovers and play a certain role in load balancing. The comprehensive weighting strategy is to weigh the elevation angle, service time, and number of idle channels, but there is no scientific definition of how to set weighting factors to achieve better handover performance.

11.3 Network Layer Management Technology

With the development of all IP technology, supporting IP mobility management will become a common feature of future integrated networks between sky and earth [8]. The user terminal has a unique IP address in the network. IP address is a means of network identification for user terminals, as well as a means of terminal localization and routing. The most typical network layer management technology is Mobile IPv4 (MIPv4) [9] and Mobile IPv6 (MIPv6) [10], which were designed by the IETF.

11.3.1 MIPv6 Technology

The Mobile Internet Protocol (MIP) comprises three fundamental components: the Home Agent (HA), the Foreign Agent (FA), and the Mobility Node (MN). The HA is responsible for tunneling datagrams and maintaining up-to-date location information for the MN. The FA acts as a router that enables the MN to access the network, providing routing services when the MN is away from its home network. It facilitates communication by sending datagrams to the MN through established tunnels. Additionally, for datagrams transmitted by the MN, the FA can function as the default router for registered mobility nodes.

MIPv6 effectively solves the triangle routing problem [11]. The MIPv6 mechanism is outlined as follows. The MN acquires a permanent address known as the hometown address, which is registered with the HA within the home network for identification and routing purposes. Figure 11.7 illustrates the network architecture and message exchange that occur during the MIPv6 mobility management process.

As a host-based mobility management protocol, MIPv6 enables MN to detect its movement from home networks (previously connected networks) to out-of-town networks through the IPv6 neighbor discovery mechanism. An out-of-town network refers to any network accessible by MN after it has exited its local coverage area. According to the MIPv6 location management process, when MN transitions from its home network into an out-of-town network, it undertakes several steps:

Step 1: Utilize either IPv6 neighbor discovery or automatic address configuration mechanisms to obtain temporary IP addresses from foreign networks, referred to as care-of addresses (CoA).

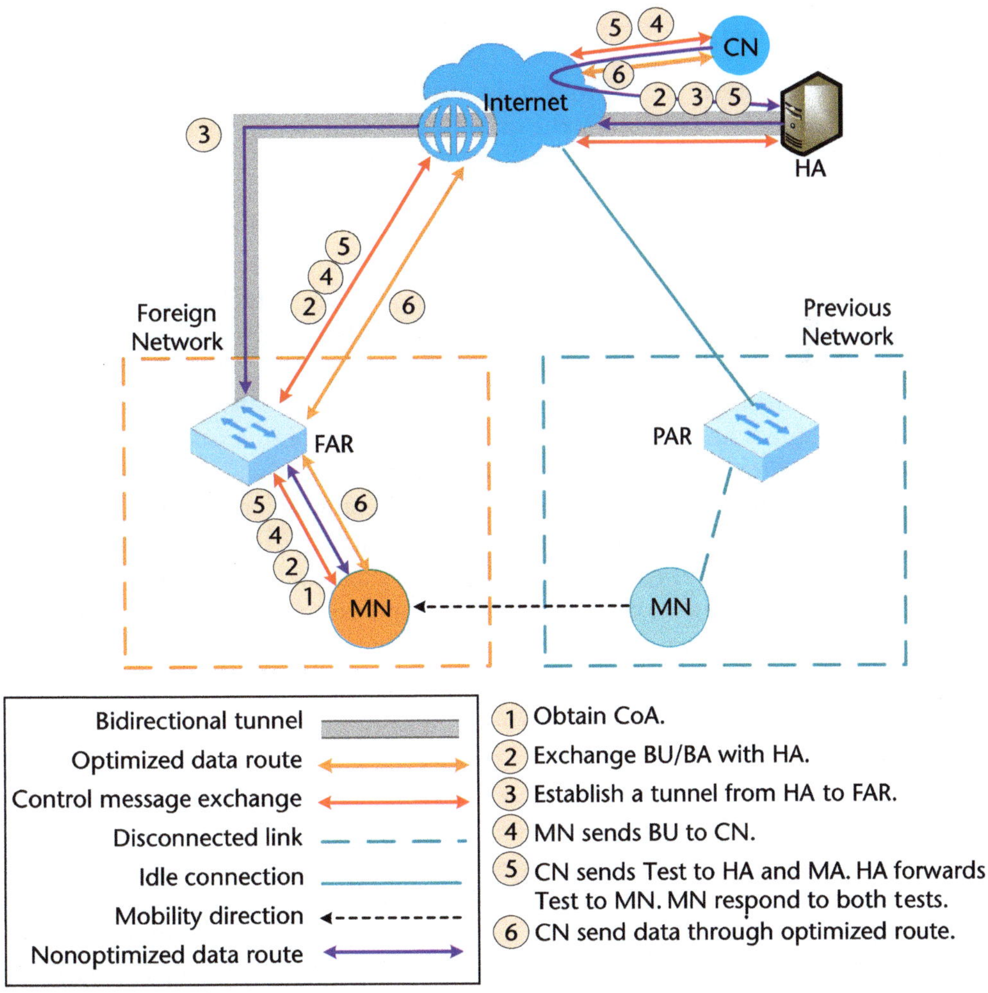

Figure 11.7 MIPV6 management process.

Step 2: The MN informs HA of its current location by dispatching a Binding Update (BU) message; in response, HA sends a Binding Acknowledgment (BA) back to MN.

Step 3: Following completion of this binding update with HA, both HA and Foreign Access Router (FAR) establish a bidirectional tunnel for transferring data packets between Correspondent Node (CN) and MN. In this scenario, packets must traverse through HA, which may not represent the most optimal route.

Step 4: To enhance data forwarding efficiency, MN can optimize routing by sending BU messages directly to CN. Until CN begins utilizing MN's CoA, packets will continue being routed through HA.

Steps 5 and 6: Prior to employing MN's CoA, CN transmits test messages directed at both HA and MN. These messages are forwarded by HA to MN; subsequently, MN must respond via two distinct paths, one through HA and another directly toward CN. Upon receiving responses from both routes, CN can commence communication using NM's CoA; thereafter communication between CN and NM can proceed via FAR without necessitating passage through HA.

When implementing MIPv6 in satellite networks, it is posited that the satellite functions as the MN, while the ground station operates as a straightforward router, and the control center serves as the CN. Each time a satellite establishes contact with a ground station, it acquires a CoA and sets up a virtual tunnel between itself and its HA, which may be another ground station. When the control center transmits a datagram to the satellite, standard routing procedures are employed to direct the packet toward the HA. Upon recognizing an active tunnel leading to the satellite, the HA forwards the data packet through this established tunnel directly to the satellite.

In contrast, when data packets are transmitted from satellites to control centers, reverse tunneling can be employed. Given that bidirectional packets consistently traverse through the HA, this mode of communication necessitates additional network resources. In comparison to direct communication between satellites and control centers, this approach introduces latency. To mitigate such delays, satellites can implement routing optimization by dispatching BU messages to the control center. Subsequently, the control center updates its binding cache and directs data packets straight to the satellite's CoA, bypassing the use of the home address entirely. The ground station functions solely as a default router for the satellites.

11.3.2 PMIPv6 Technology

IETF introduces the PMIPv6 mobile management solution based on MIPV6. PMIPv6 reduces the requirements for terminals on the network side [12, 13].

PMIPv6 introduces two new network entities: the Mobile Access Gateway (MAG) and the Local Mobility Anchor (LMA). The LMA is connected to multiple MAGs, allowing several LMAs to manage the mobility of different MNs within a PMIPv6 domain. When an MN moves within this domain, the MAG facilitates signaling interactions between the MN and LMA to ensure session continuity.

Upon joining the network for the first time, an MN sends a routing request to its first reachable MAG, which subsequently issues a proxy binding update (PBU) to its associated LMA. The LMA responds with Proxy Binding Acknowledgments (PBAs). Additionally, it creates a Binding Cache Entry (BCE) and establishes a bidirectional tunnel with the MAG.

Figure 11.8 illustrates the PMIPv6 management process. When an MN transitions from one MAG's coverage area to another within the same PMIPv6 domain, only local location updates are necessary. Data flow can be adjusted directly in the LMA following these steps:

Step 1: Send a PBU to the LMA, if the MAG1 detects that an MN is moving away from its coverage area.

Step 2: Send a PBA back to MAG1 for LMA to respond.

Step 3: Meanwhile, MAG2 detects that an MN is approaching and sends its own PBU to the LMA.

Step 4: LMA sends a PBA back to MAG2.

Step 5: Throughout this process, it is important that each MN retains its IP address when transforming between different MAGs within the same PMIPv6 domain.

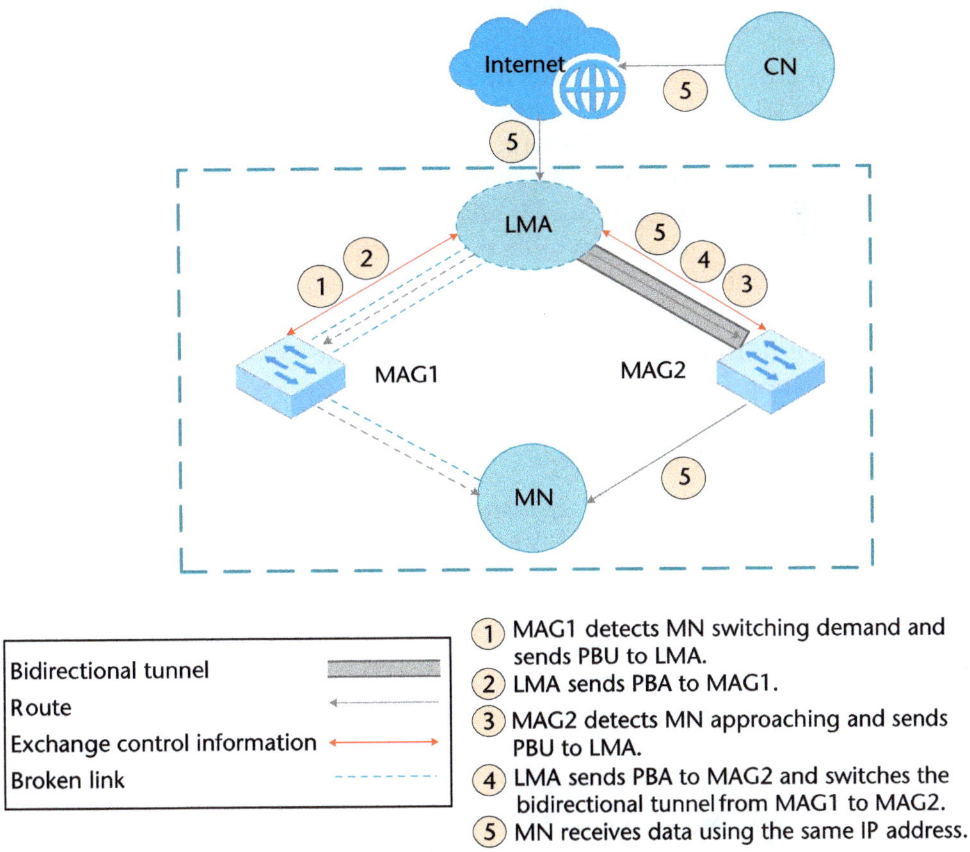

Figure 11.8 PMIPV6 management process.

The example presented in Figure 11.8 demonstrates how PMIPv6 operates. If an MN moves outside of this domain, it must engage in MIPv6's location management processes; herein lies where home network LMAs assume their role as HAs. This framework enables networks to represent mobile nodes effectively during mobility management processes while minimizing signaling interactions between mobile nodes and access routers.

PMIPv6 demonstrates superior performance compared to MIPv6 in terms of handover latency. The network-based mobile services facilitated by the LMA and MAG contribute to a reduction in both handover blocking probability and packet loss, thereby streamlining control signals and handover procedures. Consequently, network-based mobility management protocols can enhance the efficacy of mobility management within low Earth orbit satellite networks.

11.3.3 HiMIPv6 Technology

HiMIPv6 is a hierarchical location management scheme for MIPv6 based on Satellite Moving Anchors (MAP) [14, 15]. By introducing MAP, when a MN roams within the local domain, it sends binding updates to the local MAP instead of its HA and CN, resulting in a decrease in the semaphore outside the local domain. HiMIPv6 defines the satellite handover threshold N for MN to distinguish between global and local updates.

Define N as the threshold for the number of times MN handovers. When MN handovers to a new satellite in the same orbit and the number of times MN handovers does not exceed N, MN updates MAP with an IP address containing the new satellite ID. When a satellite receives a data packet with a target satellite ID equal to its own ID, check if the target MN is in service. If not, then the satellite is the MAP of the target MN. MAP changes the destination satellite ID in the message to the current MN's access satellite ID and forwards the message.

When MN handovers more than N times, or MN handovers to a new satellite in a different orbit, MN in the new access satellite creates a new MAP on NAS and updates its position on HA and CN. Subsequently, MN sends a MAP deletion message to the old MAP satellite, notifying the old satellite MAP to delete MN information. Then, CN redirects the message to MN's new MAP.

11.3.4 VMIPv6 Technology

VMIPv6 is a virtual mobility management solution [16], which is an enhanced version of the MIPv6 protocol. VMIPv6 proposes virtual agent cluster (VAC), virtual agent domain (VAD), and mobile agent anchor (MAA). VAC is a node-set that includes all LEO satellites at the top of a specific LA. The entire coverage area of all nodes in VAC is called VAD. MAA is an onboard router in each LEO satellite that can provide routing services for registered MNs. MAA is similar to the access router in IPv6, but it is mobile. VMIPv6 divides the global ground area into multiple LAs, which can be covered and managed by multiple VACs. All MAAs in VAC can share mobility information of a specific MN to jointly manage that node.

As the network topology changes, satellite nodes in a specific LA may undergo changes. Once the network topology changes, VAC can be reconstructed through two operations: adding new satellites that slide into LA and deleting satellites that slide out of LA. For a given LA, the new satellite and the departing satellite will quickly obtain and forget the binding information of that LA, respectively. The update cost of the new VAC is very low, which means that MAA's information update can be achieved through minimal signal interaction.

To mitigate the challenges associated with mobile management, VMIPv6 delineates two distinct types of MAAs, specifically home MAA and local MAA. The home MAA is responsible for maintaining the connection between the VAC and the HA, while the MN registers its subnet IP address with the home MAA on the HA. Conversely, the local MAA oversees the connection link between MN and VAC; in this context, MN binds its IP address to each relevant local MAA associated with VAC.

11.4 Transport Layer Management Technology

11.4.1 SIGMA Technology

MIP protocols, such as MIPv6, exhibit several performance challenges, including prolonged handover delays, elevated packet loss rates, and reduced throughput. Researchers developed an end-to-end mobility management solution based on the transport layer as an alternative to MIP. This solution is referred to as the Transport Layer Seamless Handoff Scheme for Space Networks (TraSH-SN), also known as SIGMA [17].

As an end-to-end mobility management framework, signaling for internet mobility architecture (SIGMA) does not necessitate any modifications to the existing internet infrastructure [18]. It is applicable to both terrestrial and satellite networks, thereby streamlining the mobility management process in satellite-ground convergence scenarios.

The fundamental concept of SIGMA is to decouple location management from data transmission, thereby achieving seamless handover by utilizing IP diversity. This allows for the retention of the old path during the establishment of a new path throughout the handover process. SIGMA can be integrated with other transport layer protocols that support multihoming and is compatible with both IPv4 and IPv6 infrastructures without requiring mobile IP support.

SIGMA serves as an end-to-end mobility management solution, eliminating the need for HAs or FAs. When a MN communicates data with a CN while nearing the overlapping coverage area of two access routers, it acquires a new IP address from the new access router. During this acquisition process, the old IP address functions as the primary IP address to maintain valid data communication with the CN. Once signals received from the old access router fall below a specified threshold, the MN designates its new IP address as its primary one. The old IP address remains in use to keep connections open while transitioning to this new address.

SIGMA exhibits an exceptionally low packet loss rate during handovers and minimal latency within satellite-ground converged communication networks. For effective location management, SIGMA employs a location manager (LM), which maintains a database correlating MN identities with their current primary IP addresses. Whenever an MN obtains a new address and updates it as its primary one, this change is automatically reflected in the LM's records.

When a CN seeks to initiate communication with an MN, it queries the LM using identifiers such as home addresses, domain names, or public keys associated with that MN. In response, the LM provides information regarding the MN's current primary IP address retrieved from its database. Subsequently, communication between CN and MN proceeds using this updated address.

As previously noted, SIGMA can effectively manage mobility within satellite networks; consider two illustrative scenarios.

Scenario 1: Satellite as a router. A satellite equipped with an onboard IP routing device can function as a router within a satellite network. When a MN connected to one satellite is handed over to another, the high mobility of LEO satellites necessitates frequent handovers for the MN. During these transitions between satellites, it is essential for the MN to maintain a continuous transport layer connection with its CN. In this context, the MN utilizes the satellite as a router to sustain this uninterrupted transport layer connection.

Different satellites and even distinct spot beams within a single satellite may be assigned unique IP subnet addresses. Figure 11.9 illustrates the application of SIGMA in managing mobility within satellite networks when the satellite operates as a router. In this scenario, the MN initially acquires an IP address from Satellite A and communicates through it. As both Satellites A and B move, there comes a point where the MN falls under their overlapping coverage and subsequently obtains an IP address from Satellite B.

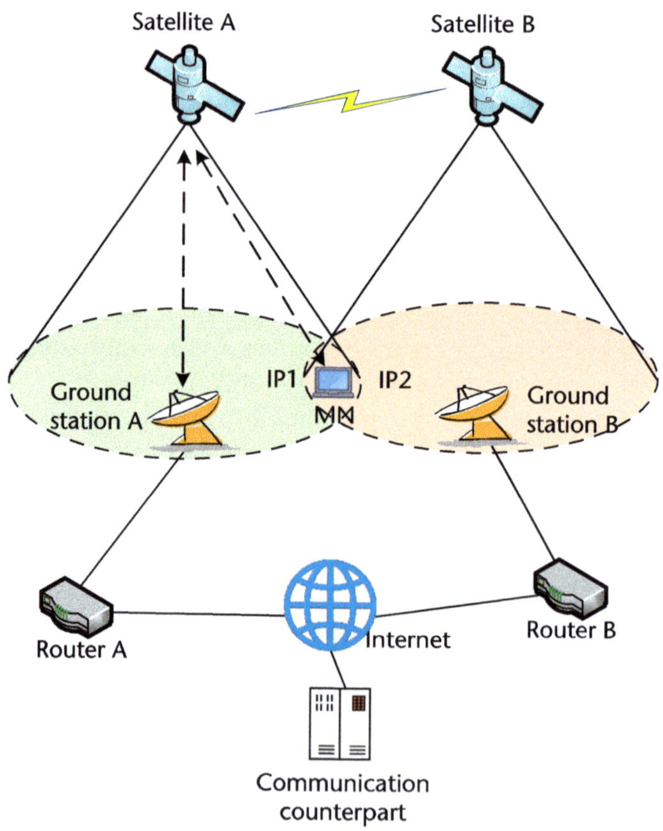

Figure 11.9 SIGMA when a satellite acts as a router.

The MN can accurately predict the trajectories of Satellites A and B; such prior knowledge informs decisions regarding when to configure its primary address with the new IP address while simultaneously removing its old one. This process proves significantly more straightforward than in wireless cellular networks, where user mobility tends to be less predictable.

Scenario 2: In this scenario, the satellite functions as a primary MN. When there are IP devices onboard the satellite capable of generating and transmitting data to earth, such as earth and space observation equipment, or when the satellite receives control signals from terrestrial sources, the communication node on the satellite operates as an MN. As illustrated in Figure 11.10, a CN on earth sends control signals to the MN located on the satellite; subsequently, upon receiving these signals, the MN transmits data back to the CN. Given that ground stations belong to different IP subnets, nodes aboard the satellite must change their IP addresses during handovers between ground stations. This necessitates effective mobility management to ensure uninterrupted connectivity between the satellite and ground nodes.

In this context, both the satellite and access router A correspond respectively to an MN and an access router. Notably, when implementing SIGMA, there are no specific requirements imposed on access routers; thus, modifications to existing internet infrastructure are not necessary, facilitating a more straightforward deployment of SIGMA.

11.4.2 Predictive SIGMA Technology

In order to further reduce the packet loss rate and latency of SIGMA, a prediction mechanism can be introduced [19]; the next access satellite is predicted based on the known satellite orbit information and MN location data.

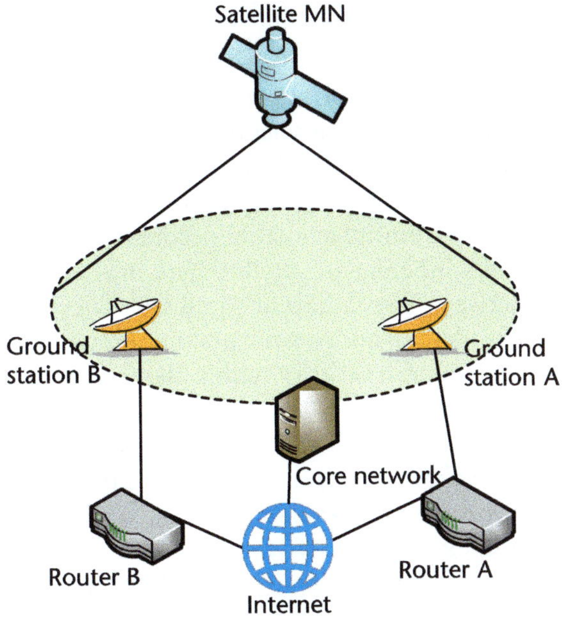

Figure 11.10 SIGMA when satellite acts as a mobile node.

Utilizing the SCTP protocol, the handover process of the prediction-based enhanced SIGMA can be delineated in four steps:

Step 1: Acquire a new IP address and ephemeris information from the new satellite. When the MN enters the overlapping area between two satellites, the handover preparation process commences. Upon receiving an announcement from the new satellite (Satellite B), it should initiate obtaining a new IP address. Concurrently, ephemeris information for this new satellite designated as MN's next router is transmitted via signaling. The FMIPv6 protocol may also be employed here. Given that automatic address configuration ensures uniqueness for each new IP address, rapid address binding can occur between MN and its newly assigned satellite during this step.

Step 2: Calculate handover time t and incorporate the new IP address into the association. After acquiring both a new IP address and ephemeris details from Satellite B, precise link handover time can be computed. Subsequently, CN is informed about both availability of this new IP address and link forwarding time t through SCTPs dynamic reconfiguration option for addresses. At time t, both MN and CN will set their primary addresses to MN's newly assigned IP address; additionally, CN may redirect data traffic toward this updated IP to enhance successful delivery probabilities to MN. The clock synchronization functionality provided by GPS modules within terminals will ensure that both MN and CN execute link-forwarding processes simultaneously with precision.

Step 3: Update location information. Following the delivery of the link, the MN updates its new IP address with the location manager to ensure that subsequent traffic data packets are accurately directed to this new address. Similar to the SIGMA scheme, this approach decouples location management from data traffic forwarding functions, thereby expediting the handover process and minimizing handover delays.

Step 4: Remove obsolete IP address from CN's available destination list. When the MN exits the coverage area of its previous satellite, no further traffic data packets should be transmitted to the old address. The MN informs the CN that it will no longer use this old IP address for data transmission; consequently, CN removes it from its available destination IP list. Given the high mobility characteristics inherent in satellite networks, it is challenging for an MN to oscillate between two adjacent satellites. Therefore, there is no necessity to retain an obsolete IP address for potential reuse by deactivating it. This action also enhances overall security within the framework.

The prediction-based SIGMA scheme consistently allows for packet reception while signaling exchanges occur prior to deleting any old IP addresses on CNs end. In comparison with standard SIGMA schemes, this predictive mechanism reduces signaling exchanges between MN and CN, thus shortening overall handover duration an essential factor in satellite networks that demand rapid and frequent transitions.

11.5 Potential Mobility Management Technology

11.5.1 SDN-Based Mobility Management

The number of satellites in satellite networks continues to grow, and more efficient mobility management technologies need to be developed to reduce system management overhead. The emergence of new network technologies, such as software-defined networking (SDN) and network function virtualization (NFV), has brought new solutions to the mobility management problem of such multi-source heterogeneous networks.

Introducing the concept of SDN into satellite communication networks can separate the data plane and control plane of satellite communication networks. The control plane performs complex routing calculations, resource management, and other operations, and the remaining satellite nodes only need to perform simple hardware configuration and datagram forwarding, reducing the cost of satellite construction and the requirements for onboard computing and processing capabilities. Due to the centralized management mode of the control plane, it is possible to achieve real-time updates of network status, deployment of fine-grained management strategies, flexible network deployment, and resource scheduling. At the same time, the control plane can coordinate network resources across domains, thereby shielding the differences in the underlying heterogeneous networks [20].

Li et al. [21] proposed a satellite architecture based on SDN and NFV, which includes three planes: data layer, control layer, and management layer, to achieve flexible satellite communication network traffic engineering and fine-grained QoS guarantee. The management layer deploys the satellite network management center (SNMC) on the ground, which is responsible for collecting resources and issuing policies for the entire network. The control layer is deployed on GEO satellites, satellite gateways, and ground networks and is centrally managed by SNMC. The SDN architecture can manage the entire network. Combining SDN technology with the mobility of complex satellite networks and multilayer satellite communication networks is one of the potential research directions.

Service function chains (SFC) is an emerging technology under the SDN/NFV architecture that combines multiple virtual network functions (VNFs) to manage the network. The combination of VNFs needs to refer to the needs of business logic and can be dynamically reconstructed as the business logic changes. For example, when the ground network is congested, the communication link between two ground terminals can automatically handover from the ground network to the satellite network to achieve communication, thereby significantly improving network performance. How to consider and achieve efficient mobility management in this dynamic orchestration of SFC is a direction worthy of future research.

11.5.2 Mobility Management Based on O-RAN

5G wireless systems have begun to see widespread deployment across the globe [22]. Both academia and industry are concentrating on the development and standardization of the next generation of mobile networks, specifically beyond 5G and

6G networks. Beyond 5G is poised to fundamentally transform communication systems by providing seamless connectivity in both time and space, thereby creating a unique ecosystem that integrates digital, physical, and human domains. In this context, nonterrestrial networks comprising satellites, drones, and other entities will be fully integrated into terrestrial networks. This integration aims to deliver services to users that are accessible anytime and anywhere.

Satellite networks play a crucial role in 5G as well as subsequent network generations. However, integrating nonterrestrial networks such as satellites with terrestrial infrastructures presents several challenges. Achieving interoperability between nodes is essential for successful integration; however, satellite networks operate as closed systems with strictly defined architectures. The nodes within these satellite networks may be supplied by various operators, which results in limited flexibility regarding optimization and interface management factors that impede effective interoperability.

Open Access Network (O-RAN) is predicated on decomposed and virtualization components interconnected by open interfaces, facilitating interoperability among different vendors. O-RAN encompasses four fundamental concepts: decomposition, virtualization, RAN Intelligent Controller (RIC), and open interfaces [23].

The principle of decomposition extends the functional separation proposed by 3GPP for 5G base stations, effectively partitioning the base station into distinct functional units. The virtualization principle asserts that all components of the O-RAN architecture can be deployed on a cloud computing platform. To manage the RAN, O-RAN introduces the RIC, which collects data from users and base stations via the E2 interface. By processing this data through AI and ML algorithms, the RIC can optimize and implement control policies for the RAN.

O-RAN delineates two types of RICs: non-real-time RICs and near-real-time RICs, each differing in their respective roles and time scales within the context of the RAN. Lastly, O-RAN specifies technical standards for open interfaces that connect various components; these interfaces are essential for enabling the RIC to gather network data and apply control policies effectively across the RAN.

Mobility management technology based on O-RAN represents a promising avenue for future research. Satellite networks are inherently dynamic systems due to the rapid movement of satellites along their orbits, which leads to temporal variations in user channel parameters. Consequently, it is imperative that mobility management be executed swiftly to prevent decisions from being made based on outdated information. In this context, AI and ML emerge as the most effective tools for expedited and data-driven decision-making processes.

However, current AI/ML-based methodologies face intrinsic limitations as they have been tailored to existing satellite architectures that were not originally designed with AI/ML considerations in mind [24]. The implementation of an O-RAN-based satellite network architecture will facilitate dynamic mobility management: essential near-real-time data will be gathered from all network nodes particularly the E2 nodes through open interfaces; subsequently, the AI/ML model will be trained using this collected data. Ultimately, the trained AI/ML model will be deployed within the RIC, utilizing KPI data as input to execute fine-grained control over both centralized unit (CU) and distributed unit (DU) nodes within the RAN for optimized mobility management.

Specifically, the near-real-time RIC must aggregate information regarding traffic demand and user locations in high-density areas alongside satellite ephemeris data to ascertain optimal mobility management strategies informed by auxiliary information. Given that the RIC is situated on terrestrial infrastructure, this RIC-centric approach to mobility may introduce increased latency in scheduling calculations compared to onboard satellite AI solutions; however, it permits comprehensive resource optimization.

References

[1] Li, B., Z. Fei, C. Zhou, and Y. Zhang, "Physical-Layer Security in Space Information Networks: A Survey," in *IEEE Internet of Things Journal*, Vol. 7, No. 1, January 2020, pp. 33–52.

[2] Thakurta, P. K. G., M. Sen, A. Singh, and Anujendra, "A New Hybrid Channel Allocation Scheme for Mobile Networks: Markov Chain Representation," *2014 2nd International Conference on Business and Information Management (ICBIM)*, Durgapur, India, 2014, pp. 53–57.

[3] Liu Z., X. Zha, X. Ren, and Q. Yao, "Research on Handover Strategy of LEO Satellite Network," *2021 2nd International Conference on Big Data and Informatization Education (ICBDIE), IEEE*, 2021, pp. 188–194.

[4] Maral, G., J. Restrepo, E. del Re, R. Fantacci, and G. Giambene,"Performance Analysis for a Guaranteed Handover Service in an LEO Constellation with a 'Satellite-fixed Cell' System," *IEEE Transactions on Vehicular Technology*, Vol. 47, No. 4, 1998, 1200–1214.

[5] Wang X., and X. Wang, "The Research of Channel Reservation Strategy in LEO Satellite Network," *2013 IEEE 11th International Conference on Dependable, Autonomic and Secure Computing, IEEE*, 2013, pp. 590–594.

[6] Rahman M., T. Walingo, and F. Takawira, "Adaptive Handover Scheme for LEO Satellite Communication System," *AFRICON 2015, IEEE*, 2015, pp. 1–5.

[7] Chen, L., Q. Guo, H. Wang, "A Handover Management Scheme Based on Adaptive Probabilistic Resource Reservation for Multimedia LEO Satellite Networks," *2010 WASE International Conference on Information Engineering, IEEE*, 2010, Vol. 1, pp. 255–259.

[8] Musumpuka, R., T. M. Walingo, and J. M. Smith. "Performance Analysis of Correlated Handover Service in LEO Mobile Satellite Systems," *IEEE Communication Letters*, 2016, Vol. 11, No. 20, pp. 2213–2216.

[9] Wang, X., Q. Wu, M. Guo, Y. Li, and L. Zhu, "Current Status and Future Prospects of Mobility Management Technologies for Satellite Communication System," *Telecommunication Engineering*, Vol. 6, No. 11, 2022, pp. 1704–1714.

[10] Perkins, C. (ed.), Telecommunication Engineering,IP mobility support for IPv4[R], 2022, Vol. 62, No. 11, 2002, pp. 1704–1714.

[11] Johnson D., C. Perkins, and J. Arkko, Mobility support in IPv6[R], June 2004.

[12] Dai W., H. Li, Q. Wu, and X. Wang, "Flexible and Aggregated Mobility Management in Integrated Satellite-Terrestrial Networks," *2020 International Wireless Communications and Mobile Computing (IWCMC), IEEE*, 2020, pp. 982–987.

[13] Xie, P., Q. Wang, J. Chen, "A Survey of Mobility Management for Mobile Networks Supporting LEO Satellite Access," *2022 IEEE/CIC International Conference on Communications in China (ICCC Workshops), IEEE*, 2022, pp. 59–64.

[14] He, D., P. You, and S. Yong, "Comparative Handover Performance Analysis of MIPv6 and PMIPv6 in LEO Satellite Networks," *2016 Sixth International Conference on Instrumentation & Measurement, Computer, Communication and Control (IMCCC), IEEE*, 2016, pp. 93–98.

[15] Jinglin, W., and C. Zhigang, "Research on Hierarchical Location Management Scheme in LEO Satellite Networks," *2010 2nd International Conference on Future Computer and Communication, IEEE*, 2010, Vol. 1, pp. V1-127–V1-131.

[16] Zhang, X., K. Shi, S. Zhang, D. Li, and R. Xia, "Virtual Agent Clustering Based Mobility Management Over the Satellite Networks," *IEEE Access*, 2019, Vol. 7, pp. 89544–89555.

[17] Fu, S., L. Ma, M. Atiquzzaman, and Y.-J. Lee, "Architecture and Performance of SIGMA: A Seamless Mobility Architecture for Data Networks," *IEEE International Conference on Communications, 2005, ICC 2005, IEEE*, 2005, Vol. 5, pp. 3249–3253.

[18] Shahriar, A. Z. M., M. Atiquzzaman, and S. Rahman, "Mobility Management Protocols for Next-Generation All-IP Satellite Networks," *IEEE Wireless Communications*, Vol. 15, No. 2, April 2008, pp. 46–54.

[19] Zhang, Z., Q. Guo, and Z. Gao, "A Prediction Based SCTP Handover Scheme for IP/LEO Satellite Network," *2010 6th International Conference on Wireless Communications Networking and Mobile Computing (WiCOM),IEEE*, 2010, pp. 1–4.

[20] Yang, D., et al., "SDN-Based Satellite Networks:Progress, Opportunities and Challenges," *Space-Integrated-Ground Information Networks*, 2020, Vol. 1, No. 2, pp. 34–41.

[21] Li, T., H. Zhou, H. Luo, and S. Yu, "SERvICE: A Software Defined Framework for Integrated Space-Terrestrial Satellite Communication," *IEEE Transactions on Mobile Computing*, Vol. 17, No. 3, March 1, 2018, pp. 703–716.

[22] Qian, Y., "Beyond 5G Wireless Communication Technologies," *IEEE Wireless Communications*, Vol. 29, No. 1, February 2022, pp. 2–3.

[23] O-RAN Architecture Description v5.0, technical specification, July 2021.

[24] Campana, R., C. Amatetti, and A. Vanelli-Coralli, "O-RAN Based Non-Terrestrial Networks: Trends and Challenges," *2023 Joint European Conference on Networks and Communications & 6G Summit (EuCNC/6G Summit)*, Gothenburg, Sweden, 2023, pp. 264–269.

List of Acronyms and Abbreviations

1G	First Generation Mobile Communication
2G	Second Generation Mobile Communication
3G	Third Generation Mobile Communication
3GPP	3rd Generation Partnership Project
4G	Fourth Generation Mobile Communication
5G	Fifth Generation Mobile Communication
6G	Sixth-Generation Mobile Communication
ACeS	Asia Cellular Satellite
ACM	Adaptive coding and modulation
A/D	Analog/digital
AFDM	Affine frequency division multiplexing
AGC	Automatic gain control
AMC	Adaptive modulation and coding
AMPS	Advanced Mobile Phone System
AoP	Argument of perigee
AP	Access point
ARQ	Automatic repeat-request
ATSC	Advanced Television Systems Committee
BCDMA	Broadband code division multiple access
BCE	Binding cache entry
BDS	BeiDou Navigation Satellite System
BER	Bit error rate
BFN	Beamforming network
BH	Beam hopping
BHTP	Beam-hopping time plan
BP	Belief propagation
BPL	Belief propagation list
BPSK	Binary phase shift keying
BSS	Broadcast Satellite Services
CASC	China Aerospace Science and Technology Corporation
CCSDS	Consultative Committee for Space Data Systems
CDMA	Code division multiple access
CN	Correspondent node
CNAV	Civil navigation
CNN	Convolutional neural networks
COA	Care-of-address
CP	Cyclic prefix
CPD	Cross-polarization discrimination

CP-OFDM	Cyclic prefix OFDM
CSI	Channel state information
CSIT	Channel state information at transmitter
CU	Centralized unit
D2D	Device-to-device
DAFT	Discrete affine Fourier transform
DCA	Dynamic channel allocation
DFnT	Discrete Fresnel transform
DFT	Discrete Fourier transform
DFT-S-OFDM	Discrete Fourier transform spread orthogonal frequency division multiplexing
DPD	Digital predistortion
DPPM	Differential pulse position modulation
DPSK	Differential phase-shift keying
DRX	Discontinuous reception
DSSS	Direct sequence spread spectrum
DU	Distributed unit
DVB	Digital video broadcasting
DVB-RCS	Digital video broadcasting-return channel by satellite
DVB-S	Digital video broadcasting-satellite
DVB-S2	Digital video broadcasting-satellite-second generation
DVB-SH	Digital video broadcasting-satellite services to handheld
ECEF	Earth-centered earth-fixed
ECI	Earth-centered inertial
eMBB	Enhanced mobile broadband
ERS	Empirical roadside shadowing
ES	Earth station
ESA	European Space Agency
ETSI	European Telecommunications Standards Institute
FA	Foreign agent
FAR	Foreign access router
FBMC	Filter bank multicarrier
FC	Fully connected
FCA	Fixed channel allocation
FCC	Federal Communications Commission
FDD	Frequency division duplex
FDM	Frequency division multiplexing
FDMA	Frequency division multiple access
FFO	Fractional frequency offset
FFT	Fast Fourier transform
FH	Frequency hopping
FM	Frequency modulation
FOU	Field of uncertainty
FSM	Fast steering mirror
FSS	Fixed satellite services
GBBF	Ground-based beamforming
GEO	Geostationary orbit

GFDM	Generalized frequency division multiplexing
GH	Guaranteed handover
gNB	5G base stations
GNSS	Global Navigation Satellite System
GMSK	Gaussian minimum frequency-shift keying
GPS	Global Positioning System
GSM	Global System for Mobile Communications
GW	Gateway
HA	Home agent
HARQ	Hybrid Automatic Repeat Request
HAPS	High altitude platform station
HCA	Hybrid channel allocation
HM	Handover management
HPA	High power amplifier
HTS	High throughput satellite
IBI	Interband interference
IDAFT	Inverse discrete affine Fourier transform
IDFT	Inverse discrete Fourier transform
IDFnT	Inverse discrete Fresnel transform
IEEE	Institute of Electrical and Electronics Engineers
IETF	Internet Engineering Task Force
IF	Intermediate frequency
IFDMA	Interleaved FDMA
IFO	Integer frequency offset
IFFT	Inverse fast Fourier transform
INC	Inclination
INMARSAT	International Maritime Satellite Organization
IM	Intermodulation
IMTS	International Mobile Telecom System
IOT	Internet of things
IS-95	Interim Standard-95
ISDB-S	Integrated Services Digital Broadcasting-Satellite
ISL	Intersatellite links
ISS	International Space Station
ISFFT	Inverse symplectic finite Fourier transform
ITU	International Telecommunication Union
ITU-R	ITU-Radio Communications Sector
LA	Location area
LEO	Low Earth orbit
LEOSN	Low Earth orbit satellite network
LFDMA	Localized FDMA
LHCP	Left-hand circular polarization
LLA	Longitude latitude altitude
LLR	Log-likelihood ratio
LM	Location management
LMA	Local mobility anchor
LOS	Line of sight

LSTM	Long short-term memory
LTE-A	Long Term Evolution-Advanced
MA	Multiple access
MAA	Mobile agent anchor
MAC	Medium Access Control
MAG	Mobile access gateway
MAP	Mobility anchor point
MBAs	Multibeam antennas
MEO	Middle Earth orbit
MFB	Multiple feed per beam
MF-TDMA	Multifrequency TDMA
MIMO	Multiple input multiple output
MN	Mobility node
MMSE	Minimum mean square error
mMTC	Massive machine type of communication
MODCOD	Modulation and coding
MPSK	Multiple phase shift keying
MSS	Mobile satellite service
MU-MIMO	Multiple user-MIMO
MUSA	Multiuser shared access
MZM	Mach-Zehnder modulator
NAS	New access satellite
NASA	National Aeronautics and Space Administration
NBC	National Broadcasting Corporation
NCC	Network control center
NGAT	Next-generation access technologies
NG-RAN	Next Generation Radio Access Network
NGSO	Non-geostationary orbit
NLOS	Non-line of sight
NMT	Nordic Mobile Telephony
NOMA	Nonorthogonal multiple access
NR	New Radio
NTN	Nonterrestrial network
O2I	Outdoor-to-indoor
OCDM	Orthogonal chirp division multiplexing
OCTR	Offered capacity to requested traffic ratio
OFDM	Orthogonal frequency division multiplexing
OFDMA	Orthogonal frequency division multiple access
OMA	Orthogonal multiple access
OOK	On-off keying
OPA	Optical phased array
OQPSK	Offset-QPSK
O-RAN	Open-RAN
Orbcomm	Orbit Communications
OTFS	Orthogonal time frequency space
PA	Power amplifier
PAPR	Peak-to-average power ratio

PBA	Proxy Binding Acknowledgement
PBU	Proxy binding update
PCCC	Parallel concatenated convolutional code
PCHIP	Piecewise cubic hermite interpolating polynomial
PDCP	Packet Data Convergence Protocol
PD-NOMA	Power domain NOMA
PDMA	Pattern division multiple access
PN	Pseudonoise
PPM	Pulse position modification
PSK	Phase shift keying
PSO	Particle swarm optimization
QAM	Quadrature amplitude modulation
QoS	Quality of service
QPSK	Quadrature phase shift keying
RAAN	Right ascension of ascending node
RF	Radio frequency
RFHO	Receiver FH only
RHCP	Right-hand circular polarization
RIC	RAN intelligent controller
RLC	Radio Link Control
RNN	Recurrent neural networks
RSRP	Reference signal receiving power
RSRQ	Reference signal receiving quality
RTN	Radial, tangential, normal
RTT	Round-trip time
RU	Resource unit
SBA	Service-based architecture
SBBF	Space-based beamforming technology
SC	Successive cancellation
SC-FDE	Single carrier-frequency domain equalization
SC-FDMA	Single carrier-frequency division multiple access
SCL	Serial cancellation list
SCMA	Sparse code multiple access
SDAP	Service Data Adaptation Protocol
SDN	Software-defined networking
SFB	Single feed per beam
SFC	Service function chains
SFFT	Symplectic finite Fourier transform
SFR	Soft frequency reuse
SIC	Serial interference cancellation
SINR	Signal-to-noise ratio
SISO	Single-input single-output
SNMC	Satellite network management center
SNR	Signal-to-noise ratio
SMA	Semimajor axis
SMS	Short Message Service
ST	Sidereal Time

SU-MIMO	Single user-MIMO
TA	True anomaly
TACS	Total Access Communication System
TAI	International Atomic Time
TBCC	Tail-biting convolutional code
TC8PSK	Trellis-coded eight PSK
TDL	Tapped delay line
TDM	Time division multiplexing
TDMA	Time division multiple access
TDRS	Tracking and Data Relay Satellite
TD-SCDMA	Time division-synchronous code division multiple access
TFHO	Transmitter FH only
TN	Terrestrial networks
TMCC	Transmission and multiplexing configuration control
T/R	Transmit/receive
TRFH	Transmitter-receiver FH
TT	Terrestrial Time
UAV	Unmanned aerial vehicle
UE	User equipment
UFMC	Universal filtered multicarrier
UP	User plane
UPF	User plane functions
URLLC	Ultra-reliable low latency communication
UT	Universal Time
UTC	Coordinated Universal Time
UW	Unique word
VAC	Virtual agent cluster
VAD	Virtual agent domain
VSAT	Very small aperture terminal
VLEO	Very low Earth orbit
WCDMA	Wideband code division multiple access
Wireless MAN-Advanced	Wireless metropolitan area network-advanced
WLAN	Wireless local area network
ZF	Zero forcing

About the Authors

Lixia Xiao is currently a professor in the School of Cyber Science and Engineering, Huazhong University of Science and Technology, Wuhan, P. R. China. She received BE, ME, and PhD degrees from the UESTC in 2010, 2013, and 2017, respectively. From 2016 to 2017, she was a visiting student with the School of Electronics and Computer Science, University of Southampton. From 2018 to 2020, she has been a research fellow with the Department of Electrical Electronic Engineering, University of Surrey. Her research interests mainly focus on physical layer transmission techniques for integrated satellite-terrestrial communications, including waveform design, multiantenna technique design, and multiple access design. She has authored or coauthored more than 100 technical papers and has served as associate editor of some technical journals in communications, including *IEEE Network, IEEE Internet of Things Journal, Digital Communications and Networks*, and *China Communications*.

Pei Xiao is a professor in wireless communications in the Institute for Communication Systems (ICS) at University of Surrey. He is currently the technical manager of 5GIC/6GIC and leads the research team in the new physical layer work area, and coordinates/supervises research activities across all the work areas. His main research interests and expertise span a wide range of areas in communications theory and signal processing for wireless communications. His recent research has extensively explored nonterrestrial networks (NTN) and satellite communications, which are crucial for integrating satellite systems into 5G and future 6G networks. His work has addressed significant challenges in energy-efficient communication, security, and waveform optimization for NTN, contributing to the development of robust and efficient satellite communication systems.

Tao Jiang (M'06, SM'10, F'19) is currently a distinguished professor in the School of Cyber Science and Engineering, Huazhong University of Science and Technology, Wuhan, P. R. China. He received a PhD degree in information and communication engineering from Huazhong University of Science and Technology, Wuhan, P. R. China, in April 2004. From August 2004 to December 2007, he worked in some universities, such as Brunel University and University of Michigan-Dearborn, respectively. He has authored or coauthored more than 500 technical papers in major journals and conferences and 13 books/chapters in the areas of communications and networks. His proposed parity-check-concatenated (PCC) polar codes have become the standard of 5G control channel. He served or is serving as symposium technical program committee membership of some major IEEE conferences, including *INFOCOM*, *GLOBECOM*, and *ICC*. He was invited to serve as TPC symposium chair for the *IEEE GLOBECOM 2013*, *IEEEE WCNC 2013*, and

ICCC2013. He has served or is serving as an associate editor of some technical journals in communications, including *IEEE Network, IEEE Transactions on Signal Processing, IEEE Communications Surveys and Tutorials*, and *IEEE Transactions on Vehicular Technology*, and he is the area editor of *IEEE Internet of Things Journal* and associate editor-in-chief of *China Communications*.

Index

A
Acronyms and abbreviations, this book, 263–68
Active phased array, 156
Adaptive antenna anti-interference technology, 230–31
Adaptive modulation and coding (AMC) technology, 234–35
Affine frequency division multiplexing (AFDM) modulation
  about, 127, 143
  advantages, 145–46
  CPP, 144
  DAFT domain, 145–46
  DAFT parameters, 143
  disadvantages, 145–46
  modern architecture, 143
AI-aided adaptive waveform design, 149
Analog beamforming, 158–59
Antennas
  adaptive, 230–31
  beamforming techniques, 158–62
  beam hopping (BH), 208
  feeder link technology, 162–69
  satellite configuration, 153–57
  spaceborne lens multibeam, 157
  spaceborne phased array multibeam, 155–57
  spaceborne reflector multibeam, 153–54
  user link technology, 162–69
  user terminal, 169
Aquisition pointing and tracking (APT)
  about, 51, 52
  laser subsystem, 60
  system flowchart, 55
Atmospheric absorption, 74
ATSC standard, 126
Automatic gain control (AGC), 168

B
Baseband modulation, 111–15
Beamforming
  analog, 158–59
  full-digital, 158
  ground-based (GBBF), 170
  hybrid, 159–62
  multisatellite, 164–69
  single satellite, 162–64
  space-based (SBBF), 163–64
  techniques, 158–62
  technology, 207
Beamforming network (BFN), 162
Beam handover
  about, 245–46
  based on DCA, 248
  based on FCA, 247–48
  based on HCA, 248
  illustrated, 247
  research on, 246–47
Beam hopping (BH)
  about, 203
  antennas, 208
  system composition, 209
  technology, 208–10
  time plan (BHTP), 209–10
BeiDou-3 network, 100
Belief propagation (BP), 89
Belief propagation list (BPL), 89
Binary phase shift keying (BPSK), 59
Building penetration loss, 74–75

C
Carrier frequency offset, 73–74
C-band, 6, 198
CDMA2000, 92, 93, 121
Cell-splitting technology, 207
Cellular mobile communication. *See* Terrestrial mobile communication
Channel coding
  about, 83–84
  adaptive construction, 104
  for CDMA2000, 93
  challenges, 103–4
  classical, 83–91
  convolutional codes, 89–91
  current research status, 101–2
  hardware resources and, 103–4

high-mobility communication scenarios, 103
integrated satellite-terrestrial, 100–108
linear block codes, 84–89
low-complexity decoder design, 104–5
low SNR and, 103
for LTE-Advanced, 93
physical layer security coding, 107–8
possible schemes, 104
principles, 85
for satellite communication, 95–100
synchronous-free transmission, 105–7
for TD-SCDMA, 93
for terrestrial communication, 91–95
for 2G communication, 93
for WCDMA, 93
Channel coding theorem, 83–84
Channel models
about, 75–76
Corazza, 77–78
ERS, 75–76
Lutz, 78–79
TDL, 76, 79–80
Channels
atmospheric absorption, 74
building penetration loss, 74–75
characteristics, 70–75
C.Loo, 76–77
clutter loss, 72
Doppler effect, 73–74
fading, classification, 70
free-space loss, 71
fundamentals, 69–70
ionospheric scintillation, 71
multipath fading, 72–73
propagation effects, 76
rain fading, 72
shadow fading, 72
standards, evolution of, 80–81
Chirp-periodic prefix (CPP), 144
C.Loo model, 76–77
Clutter loss, 72
Code division multiple access (CDMA)
about, 2, 92, 175
illustrated, 175
signal transmission quality, 176
system characteristics, 176–77
wideband (WCDMA), 92–93, 121
Coherent modulation, 58–59
Coherent optical communication techniques, 56–57
Communication capacity evaluation metrics, 48
Communication quality evaluation metrics, 48
Communication subsystem, laser, 60
Consultative Committee for Space Data Systems (CCSDS), 97–98
Convolutional codes, 89–91
Convolutional neural networks (CNN), 80
Coordinated Universal Time (UTC), 44
Coordinate systems, 43–44
Corazza model, 77–78
Coverage
area, 47, 120, 169, 201–4, 222
band illustration, 41
cellular, 22
gap, 48
global, 33, 36, 38, 40, 41, 199–200
low-altitude, 18
multibeam, 153
multiple layers, 29
performance, 41, 43
performance evaluation metrics, 48
single satellite characteristics, 46–47
Cross-orbit multilayer cooperative transmission, 213–14
Cross-polarization discrimination (CPD), 166
Cross-polarization interference, 226–30
CubeSat Laser Infrared Crosslink (CLICK) system, 63–64
Cyclic codes, 86–87
Cyclic prefix, 116–17, 122, 126, 179
Cyclic Prefix OFDM (CP-OFDM), 122

D

Deep-space communication, 95–98
Delta constellation, 36–37
Design factor analysis, 47–48
DFT-S-OFDM-based SC modulation, 117–19
Differential phase shift keying (DPSK), 59
Differential pulse position modulation (DPPM), 58
Digital predistortion technology, 235–36
Digital video broadcasting (DVB) standard, 98–99
Digital Video Broadcasting-Return Channel by Satellite (DVB-RCS), 186
Digital Video Broadcasting-Satellite (DVB-S) standard, 123
Digital Video Broadcasting-Satellite-Second Generation (DVB-S2) standard, 123–24
Digital Video Broadcasting-Satellite Services to Handheld (DVB-SH) standard, 125–26
Discrete affine Fourier transform (DAFT), 143, 145–46
Discrete Fourier transform (DFT), 115
Doppler effect, 73–74
Dual-polarization transmission, 229

DVB-S2X standard, 124
Dynamic channel allocation (DCA), 247, 248
Dynamic frequency allocation, 29–30
Dynamic time slot allocation, 212–13

E
Earth-Centered Earth-Fixed (ECEF) Coordinate System, 43–44
Earth-Centered Inertial (ECI) Coordinate System, 44
8PSK modulation, 112
Electric field vector propagation, 227
Emergency communication, 19
Empirical roadside (ERS), 75
European Data Relay System (EDRS), 62–63

F
Fast Fourier transform (FFT), 115, 134–36
Feeder link antenna technology, 169–70
Filter bank multicarrier (FBMC) modulation
  about, 127–30
  advantages, 132
  disadvantages, 132–33
  real number symbol, 131, 132
  system block diagram, 130
  time-frequency data blocks, 131, 132
First-generation satellites, 19
5G communication
  about, *xi*, 2–3
  handover management, 244–45
  modulation, 121–22
5G nonterrestrial networks, *xi*
Fixed channel allocation (FCA), 247–48
Flower constellation, 38–39
Folding hybrid structure, laser, 60
4G communication, 2
4G modulation, 121
Fourth-generation satellites, 20
Fractional frequency offset (FFO), 74
Free-space loss, 71
Free-space optical communication
  aquisition pointing and tracking (APT) and, 51, 52
  Asia, 64–65
  challenges, 65–68
  complex routing protocols, 66–67
  current status, 62–65
  Europe, 62–63
  fundamentals, 51–54
  high nodes mobility, 66
  initial pointing, 52–53, 54–55
  key techniques, 54–62
  laser antenna technology, 59–61
  lightweight design, 67
  link construction, 54–56
  microwave antenna technology, 61–62
  networking and security, 67
  scanning and acquisition, 53, 55–56
  signal modulation technique, 56–59
  signal processing, 66
  system components diagram, 52
  system flowchart, 53
  tracking, 53, 56
  transmission environment, 65–66
  United States, 63–64
Free-space transmission loss, 219
Frequency band interference, 221–23
Frequency division multiple access (FDMA), 173–74
Frequency modulation (FM), 119
Frequency reuse technology
  about, 205–6
  multicolor, 206–7
  soft, 207–8
Full-digital beamforming, 158

G
Galileo satellite navigation system, 100
Gaofen series satellites, 201
Gaussian minimum frequency-shift keying (GMSK), 119
Generalized frequency division multiplexing (GFDM) modulation
  about, 128, 136
  advantages, 138
  disadvantages, 139
  fragmented spectrum, 137
  principle, illustrated, 138
  subcarriers, 137
  system block diagram, 137
  time slots, 136–37
Geostationary orbit (GEO) satellites, 6–7, 72
Global Navigation Satellite System (GNSS), 99–100
Global Positioning System (GPS), 99–100
Globalstar, 9, 11, 35
GOES satellites, 201
Golden amplitude modulation (GAM), 127, 128, 146
Ground-based beamforming (GBBF), 170

H
Hamming distance, 84
Handover, 29
High-orbit broadband systems, 8
High-orbit narrowband systems, 7–8
High power amplifiers (HPAs), 173–74
HiMIPv6 technology, 254

Hongyan, 35
Hybrid Automatic Repeat Response (HARQ), 27
Hybrid beamforming
  about, 159
  fully connected architecture, 159–60
  lens antenna architecture, 161–62
  partially connected architecture, 160
  switch network-based architecture, 161
  See also Beamforming
Hybrid channel allocation (HCA), 247, 248
Hybrid TDMA/CDMA, 189–90, 191

I
IMT-2000, 120
Initial pointing, 52–53, 54–55
Inline interference, 224–26
Inmarsat, 7–8
Integer frequency offset (IFO), 74
Integrated coding, 148
Integrated Services Digital Broadcasting-Satellite (ISDB-S) standard, 126–27
Integration models
  about, 19–20
  networking models, 21–23
  service models, 20–21
  terminal development models, 23
Integration security, 30
Intensity modulation/direct detection (IMDD) approach, 56–57
Interface design, 29
Interference
  cross-polarization, 226–30
  free-space transmission loss, 219
  frequency band, 221–23
  inline, 224–26
  natural, 214–20
  pointing loss, 219–20
  rain fade, 214–17
  solar eclipse, 217–19
  space, 220–30
Interference management
  adaptive antenna anti-interference technology, 230–31
  adaptive modulation and coding (AMC) technology, 234–35
  digital predistortion technology, 235–36
  interference types and, 214–30
  natural interference, 214–20
  on-satellite processing technology, 231–32
  spread spectrum technology, 232–34
  technology, 230–36
Interleaved FDMA (IFDMA), 179–80
Interleave division multiple access (IDMA), 193–94
International Atomic Time (TAI), 44
International standards, 23–28
Intersatellite resource management
  about, 211–12
  channel availability in short time slots, 213
  cross-orbit multilayer cooperative transmission, 213–14
  dynamic time slot allocation, 212–13
  limited on-satellite power, 212
  optical phased array (OPA), 214, 215
  performance comparison, 211–12
Interstellar handover, 248–50
Intrasatellite handover. *See* Beam handover
Inverse discrete Fourier transform (IDFT), 115
Inverse fast Fourier transform (IFFT), 115
Ionospheric scintillation, 71
IoT intelligent connectivity, 18
IPSTAR, 8, 11
Iridium, 8–9, 11, 35
Iridium Next Generation (Iridium-NEXT), 9
Irregular baseband modulation, 128
Irregular constellation configuration design, 148

J
Japanese Data Relay Satellite System (JDRS), 64

K
Ka-band, 6, 199–200
Kronecker product, 86, 89, 166, 229
Ku-band, 6, 199

L
Lagrange multiplier, 166–68
Laser antenna technology, 59–61
Laser Communications Relay Demonstration (LCRD), 63
Lattice partition multiple access (LPMA), 194–95
L-band, 6, 7, 197
LDPC codes, 87–88, 98, 102
Left-hand circular polarization (LHCP), 226
Linear block codes
  about, 84
  cyclic code, 86–87
  LDPC code, 87–88, 98, 102
  polar code, 88–89, 104–5, 106–8
  RM code, 84–86
  See also Channel coding
Link layer management technology
  about, 244

beam handover, 245–48
5G handover management, 244–45
interstellar handover, 248–50
See also Mobility management
Localized FDMA (LFDMA), 179–80
Log-likelihood ratio (LLR), 105
Longitude Latitude Altitude (LLA) Coordinate System, 44
Long short-term memory (LSTM), 80
Low-altitude coverage, 18
Low-Earth orbit (LEO) satellites, 6–7, 23–24, 33, 205, 243
Low Earth orbit satellite network (LEOSN), 205
Low-orbit broadband systems, 9–10
Low-orbit narrowband systems, 8–9
LTE-Advanced, 121
Luders polar orbit constellation, 42–43
Lutz model, 78–79

M

Machine learning, 80
Mach-Zehnder modulator (MZM), 58–59
Medium Earth orbit (MEO), 6, 72–73, 211
Microwave antenna technology, 61–62
Minimum mean square error (MMSE), 168
Minislot-/slot aggression-based scheduling, 204
MIPv6 technology, 250–52
Mobile nodes (MNs), 242–43, 255–58
Mobility management
about, 241
based on O-RAN, 259–60
of LEO satellite networks, 243
link layer management technology, 244–50
network layer management technology, 250–55
overview, 242–44
potential mobility management technology, 259–61
SDN-based, 259
of traditional terrestrial cellular networks, 242
transport layer management technology, 255–59
Modulation
AFDM, 127, 143–46
AI-aided adaptive waveform design and, 149
for ATSC communication, 126
baseband, 111–15
classic waveforms, 111–19
design guidelines, 148–50
DFT-S-OFDM-based SC, 117–19
for DVB-S2 communication, 123
for DVB-S2X communication, 124
for DVB-S communication, 123
for DVB-SH communication, 125
FBMC, 128–33
5G communication, 121–22
4G communication, 121
GAM, 127–50
GFDM, 128, 136–39
integrated coding and, 148
irregular baseband, 128
irregular constellation configuration design and, 148
for ISDB-S communication, 126–27
OCDM, 127, 141–42
OFDM-based MC, 115–17
1G communication, 119
OTFS, 139–41
performance analysis, 146–48
potential for integrated communication, 127–50
3G communication, 120
2G communication, 119–20
UFMC, 128, 133–36
versatile carrier waveform design and, 148–49
Modulation standard
for cellular mobile communication, 119–22
for satellite communication, 122–27
Multibeam antennas (MBAs)
about, 153
configurations, 154
lens, 157–58
phased array, 155–57
reflector, 153–54
Multibeam satellites, 202, 205, 208, 223
Multicolor frequency reuse technology, 206–7
Multifrequency TDMA (MF-TDMA)
about, 186–87
advantages and disadvantages, 189
dynamic, 187
principle of, 187
receiver FH Only, 188
static, 187
transmitter FH Only, 188
transmitter-Receiver FH, 188–89
types of, 188–89
Multipath fading, 72–73
Multiple access (MA)
about, 173
CDMA, 175–77
FDMA, 173–74
hybrid TDMA/CDMA, 189–90
for integrated communication, 190–95
interleave division (IDMA), 193–94
lattice partition (LPMA), 194–95

MF-TDMA, 186–89
MUSA, 181–82
nonorthogonal (NOMA) schemes, 173, 180–85
OFDMA, 177–78
orthogonal (OMA) schemes, 173–80
PDMA, 184–85
rate splitting (RSMA), 190–93
for satellite communication, 186–90
SC-FDMA, 178–80
SCMA, 182–84
TDMA, 174–75
for terrestrial cellular communication, 185–86
Multiple feed per beam (MFB), 153–54
Multiple-input multiple-output (MIMO) technology, 3, 158, 229
Multiple phase shift keying (MPSK), 111–13
Multisatellite beamforming
about, 164
collaborative principle, 166–69
illustrated, 164
problem formulation, 165–66
system model, 164
See also Beamforming
Multiuser shared access (MUSA), 180, 181–82

N
Natural interference, 214–20
Near-space communication, 98–100
Network control center (NCC), 209
Network function virtualization (NFV), 259
Networking models, 21–23
Network layer management technology
about, 250
HiMIPv6 technology, 254
MIPv6 technology, 250–52
PMIPv6 technology, 252
VMIPv6 technology, 254–55
See also Mobility management
New Radio (NR) NTN
Release-15, 25
Release-16, 15–16
Release-17, 26–27
Release-18, 27–28
Next Generation Access Technologies (GNAT), 28
Next Generation Radio Access Network (NG-RAN), 26
NLOS-TDL-A/B/C/D, 80–81
Noncoherent modulation, 57–58
Nongeostationary orbit (NGSO), 221–26
Nonorthogonal multiple access (NOMA)
about, 173, 180
MUSA, 181–82
PDMA, 184–85
PD-NOMA, 180–81
power allocation algorithm based on, 203
SCMA, 182–84
See also Multiple access (MA)
Nonslot based scheduling, 204
Nonterrestrial networks (NTNs)
combined provision of, 17
CU-DU split architecture, 26
high-speed relay model of, 20
networking models, 21–22
New Radio, 25–26
propagation delay characteristics, 26

O
OFDM-based MC modulation, 115–17
Offset-QPSK (OQPSK), 120
1G communication, *xi*, 1–2
1G modulation, 119
OneWeb, 10, 35
On-off keying (OOK), 57–58
On-satellite processing technology, 231–32
Open Access Network (O-RAN), 260
Optical mechanical subsystem, laser, 59
Optical phased array (OPA), 214, 215
Orbcomm, 11, 35
Orbital orientation, 45
Orbital size and shape, 45
Organization, this book, *xi–xii*
Orthogonal chirp division multiplexing (OCDM), 127
Orthogonal frequency division multiple access (OFDMA), 177–78
Orthogonal frequency division multiplexing (OFDM)
about, 121, 141
advantages, 142
cyclic prefix, 116–17, 122, 126, 179
DFT-S, 117–19
MC modulation, 115–17
modulation, 141–42
multicarrier modulation, 121
receiving and transmitting process block diagram, 142
subcarrier orthogonality diagram, 117
system block diagram, 116
time-frequency data blocks, 131, 132
Orthogonal multiple access (OMA)
about, 173
CDMA, 175–77
FDMA, 173–74
OFDMA, 177–78
SC-FDMA, 178–80
TDMA, 174–75

See also Multiple access (MA)
Orthogonal time frequency space (OTFS) modulation
about, 139
advantages, 141
description, 140
disadvantages, 142
principle, illustrated, 139
OSIRIS program, 63

P
Parallel concatenated convolutional code (PCCC), 90–91
Particle swarm optimization (PSO), 203
Passive phased array, 155
Pattern division multiple access (PDMA)
about, 180, 184
illustrated, 184
pattern design, 184–85
pattern distribution, 185
power optimization, 185
Personal communication, 17
Phased array antennas, 155–57
Physical layer security coding, 107–8
Piecewise cubic hermite interpolating polynomial (PCHIP), 74
PMIPv6 technology, 252
Pointing loss, 219–20
Polar codes, 88–89, 104–5, 106–8
Polarization reuse technology, 227
Polar orbit constellation, 37–38, 40–42
Potential mobility management technology
based on O-RAN, 259–61
SDN-based, 259
See also Mobility management
Power allocation algorithms, 202–3
Power amplifiers (PAs), 235–36
Power domain NOMA (PD-NOMA), 180–81
Power resources, 200
Precoding technology, 206–7
Predictive SIGMA technology, 257–58
Pseudonoise (PN) codes, 234
Pulse position modification (PPM), 58

Q
Quadrature amplitude modulation (QAM), 111, 113–15, 128
Quadrature phase shift keying (QPSK), 59, 120

R
Radial, Tangential, Normal (RTN) Coordinate System, 44
Rain fade interference, 214–17
Rain fading, 72
RAN Intelligent Controller (RIC), 260–61
Rate splitting (RSMA)
about, 190–91
design principles, 193
illustrated, 191
model with two users, 192
user, 193
See also Multiple access (MA)
Recurrent neural network (RNN), 80
Reed-Muller (RM) codes, 84–86
Reflective structure, laser, 60
Release-15 for NR NTN, 25
Release-16 for NR NTN, 25–26
Release-17 for NR NTN, 26–27
Release-18 for NR NTN, 27–28
Reliability evaluation metrics, 48
Resource management
beam hopping technology, 208
frequency reuse technology, 205–8
intersatellite, 211–14
technology, 205–10
Resources
multidimensional, overview, 197–205
power, 200–203
spectrum, 197–200
time slots, 203–5
Restrictive constellation, 39–40
Right ascension of ascending node (RAAN), 38–40, 41–43
Right-hand circular polarization (RHCP), 226
Round-trip time (RTT), 27

S
Sat5G project, 28
Satellite constellation design
classical solution, 40–43
configuration design, 43–46
coordinate and time systems, 43–45
coverage design, 46–47
factor analysis, 47–48
for inclined circular orbit constellation, 42–43
orbital orientation, 45
orbital size and shape, 45
parameters, 45–46
for polar/near-polar orbit constellation, 40–42
purpose of, 43
satellite position in orbit, 46
single satellite coverage characteristics, 46–47
Satellite constellations
about, 33
classification of, 36–43

configuration summary and comparison, 39
defined, 33–34
delta, 36–37
deployment plans, 35
development of, 34–35
flower, 38–39
functions, 36
Globalstar, 35
Hongyan, 35
inclined circular orbit, 42–43
Iridium, 35
OneWeb, 35
Orbcomm, 35
overview, 33–35
parameters, 45–46
polar orbit, 37–38, 40–42
restrictive, 39–40
schematic diagram, 34
star, 37–38
Starlink, 35
Walker, 36–37, 42–43
Satellite mobile communication
about, 4
channel coding for, 95–100
deep-space, 95–98
gateway station, 4–5
high-orbit broadband, 8
high-orbit narrowband, 7–8
illustrated, 4
intersatellite link, 5
low-orbit broadband, 9–10
low-orbit narrowband, 8–9
MA for, 186–90
mobility management, 243–44
modulation standard, 122–27
near-space, 98–100
subscriber station, 5
system illustration, 13
typical systems, 5–10
Satellite network management center (SNMC), 259
Satellite position in orbit, 46
Satellite-terrestrial integrated communication
about, 10–11
application scenarios, 17–19
challenges, 28–30
channel coding, 83–91, 100–108
channel models, 69–81
competition with terrestrial communication, 11
complement to terrestrial communication, 11–12
constellation design for, 33–48
convergence with terrestrial communication, 12
demand for, 15–17
development directions, 19
evolution of, 15–30
international standards evolution, 23–28
mobility management for, 241
models, 19–23
multiantenna technique for, 153–70
multiple-access for, 173–95
network characteristics, 13–14
resource management for, 197–236
shared industrial chain for, 17
signal modulation for, 111–50
vision of, 12–14
S-band, 6, 197–98
Scanning and acquisition, 53, 55–56
SDN-based mobility management, 259
Second-generation satellites, 19
Serial cancellation list (SCL) decoding algorithm, 89
Service function chains (SFCs), 259
Service models, 20–21
Shadow fading, 72, 79
Shared Access Terrestrial Satellite Backhaul Network (SANSA), 28
Short Message Service (SMS), 21
Sidereal Time (ST), 44
Signaling for internet mobility architecture (SIGMA)
about, 255
concept, 255
predictive, 257–58
satellite acting as mobile node, 257
satellite acting as router, 256
Signal level attenuation, 217
Signal modulation. *See* Modulation
Signal modulation technique
about, 56–57
coherent modulation, 58–59
noncoherent modulation, 57–58
Single carrier-frequency division multiple access (SC-FDMA), 178–79
Single feed per beam (SFB), 153–54
Single satellite beamforming, 162–64
6G communication, 3
6G space-terrestrial integrated technology, *xi*
Slot-based scheduling, 204
Small Optical Transponder (SOTA), 64
Soft frequency reuse (SFR) technology, 207–8
Software-defined networking (SDN), 259
Solar eclipse interference, 217–19
Space-based beamforming (SBBF) technology, 163–64
Spaceborne lens multibeam antenna, 157
Spaceborne phased array multibeam antenna, 155–57

Spaceborne reflector multibeam antenna, 153–54
Space interference
about, 220–21
cross-polarization interference, 226–30
frequency band interference, 221–23
inline interference, 224–26
joint spectrum coexistence scenarios, 221
Spaceway-3 satellite, 8
SpaceX, 10, 201
Sparse code multiple access (SCMA)
about, 180, 182
codebook design, 183–84
encoder, 183
illustrated, 182
Spatial division multiple access (SDMA), 192–93
Spread spectrum technology, 232–34
Standardization evolution process, 24
Star constellation, 37–38
Starlink, 9–10, 35, 201
Successive cancellation (SC) decoding algorithm, 89
Successive cancellation list (SCL) decoding, 105, 106
Synchronous-free transmission, 105–7
System cost evaluation metrics, 48

T
Tapped delay line (TDL) models, 76, 79–80
Telecommunications, 17–18
Terminal development models, 23
Terrestrial mobile communication
about, 1–3
cellular, 91–94
channel coding for, 91–95
competition with, 11
complement to, 11–12
convergence with, 12
MA for, 185–86
mobility management, 242
modulation standard, 119–22
WLAN, 94–95
Terrestrial Time (TT), 44–45
3rd Generation Partnership Project (3GPP), *xi*, 15, 24, 25
Third-generation satellites, 20
3G communication, 2
3G modulation, 120
Time division multiple access (TDMA), 174–75, 203–4
Time division synchronous code division multiple access (TD-SCDMA), 2, 93
Time slot resources, 203–5
Time systems, 44–45
Total radiated power, 168
Tracking, 53, 56
Tracking and data acquisition system (TDAS) satellites, 61
Transportation communication, 18
Transport layer management technology
predictive SIGMA technology, 257–58
SIGMA technology, 255
See also Mobility management
Transport Layer Seamless Handoff Scheme for Space Networks (TraSH-SN), 255
Trellis-coded eight PSK (TC8PSK), 127
Triangular QAM (TQAM), 128
Turbo codes, 91
2G communication, 2, 91–92, 93
2G modulation, 119–20

U
Universal filtered multicarrier (UFMC) modulation
about, 128, 133
advantages and disadvantages, 136
FFT and, 134–36
system block diagram, 133
User link antenna technology
about, 162
multisatellite beamforming, 164–69
single satellite beamforming, 162–64
See also Antennas
User plane functions (UPF), 28
User terminal antennas, 169

V
Versatile carrier waveform design, 148–49
Very small aperture terminal (VSAT) equipment, 8, 19
Viasat, 12
VMIPv6 technology, 254–55

W
Walker constellation, 36–37, 42–43
Wideband CDMA (WCDMA), 92–93, 121
Wild Blue, 11–12
Wireless local area networks (WLAN) communication, 94–95

X
X-band, 6, 198

Artech House Mobile Communications Library

William Webb, Series Editor

3G CDMA2000 Wireless System Engineering, Samuel C. Yang

3G Multimedia Network Services, Accounting, and User Profiles, Freddy Ghys, Marcel Mampaey, Michel Smouts, and Arto Vaaraniemi

5G and Satellite RF and Optical Integration, Geoff Varrall

5G-Enabled Industrial IoT Networks, Amitava Ghosh, Rapeepat Ratasuk, Simone Redana, and Peter Rost

5G New Radio: Beyond Mobile Broadband, Amitav Mukherjee

5G Spectrum and Standards, Geoff Varrall

802.11 WLANs and IP Networking: Security, QoS, and Mobility, Anand R. Prasad and Neeli R. Prasad

Achieving Interoperability in Critical IT and Communications Systems, Robert I. Desourdis, Peter J. Rosamilia, Christopher P. Jacobson, James E. Sinclair, and James R. McClure

Advances in 3G Enhanced Technologies for Wireless Communications, Jiangzhou Wang and Tung-Sang Ng, editors

Advances in Mobile Information Systems, John Walker, editor

Advances in Mobile Radio Access Networks, Y. Jay Guo

Artificial Intelligence in Wireless Communications, Thomas W. Rondeau and Charles W. Bostian

Broadband Wireless Access and Local Network: Mobile WiMax and WiFi, Byeong Gi Lee and Sunghyun Choi

CDMA for Wireless Personal Communications, Ramjee Prasad

CDMA RF System Engineering, Samuel C. Yang

CDMA Systems Capacity Engineering, Kiseon Kim and Insoo Koo

Cell Planning for Wireless Communications, Manuel F. Cátedra and Jesús Pérez-Arriaga

Cellular Communications: Worldwide Market Development, Garry A. Garrard

Cellular Mobile Systems Engineering, Saleh Faruque

Cognitive Radio Interoperability through Waveform Reconfiguration, Leszek Lechowicz and Mieczyslaw M. Kokar

Cognitive Radio Techniques: Spectrum Sensing, Interference Mitigation, and Localization, Kandeepan Sithamparanathan and Andrea Giorgetti

The Complete Wireless Communications Professional: A Guide for Engineers and Managers, William Webb

Designing RF Combining Systems for Shared Radio Sites, Ian Graham

EDGE for Mobile Internet, Emmanuel Seurre, Patrick Savelli, and Pierre-Jean Pietri

Emerging Public Safety Wireless Communication Systems, Robert I. Desourdis, Jr., et al.

From LTE to LTE-Advanced Pro and 5G, Moe Rahnema and Marcin Dryjanski

The Future of Wireless Communications, William Webb

Geospatial Computing in Mobile Devices, Ruizhi Chen and Robert Guinness

Gigahertz and Terahertz Technologies for Broadband Communications, Second Edition, Terry Edwards

GPRS for Mobile Internet, Emmanuel Seurre, Patrick Savelli, and Pierre-Jean Pietri

GSM and Personal Communications Handbook, Siegmund M. Redl, Matthias K. Weber, and Malcolm W. Oliphant

GSM Networks: Protocols, Terminology, and Implementation, Gunnar Heine

GSM System Engineering, Asha Mehrotra

Handbook of Land-Mobile Radio System Coverage, Garry C. Hess

Handbook of Mobile Radio Networks, Sami Tabbane

Handbook of Next-Generation Emergency Services, Barbara Kemp and Bart Lovett

High-Speed Wireless ATM and LANs, Benny Bing

Implementing Full Duplexing for 5G, David B. Cruickshank

In-Band Full-Duplex Wireless Systems Handbook, Kenneth E. Kolodziej, editor

Inside Bluetooth Low Energy, Second Edition, Naresh Gupta

Integrated Satellite-Terrestrial Network Fundamentals for Mobile Communications, Lixia Xiao, Pei Xiao, and Tao Jiang

Interference Analysis and Reduction for Wireless Systems, Peter Stavroulakis

Interference and Resource Management in Heterogeneous Wireless Networks, Jiandong Li, Min Sheng, Xijun Wang, and Hongguang Sun

Internet Technologies for Fixed and Mobile Networks, Toni Janevski

Introduction to 3G Mobile Communications, Second Edition, Juha Korhonen

Introduction to 4G Mobile Communications, Juha Korhonen

Introduction to Communication Systems Simulation, Maurice Schiff

Introduction to Digital Professional Mobile Radio, Hans-Peter A. Ketterling

An Introduction to GSM, Siegmund M. Redl, Matthias K. Weber, and Malcolm W. Oliphant

Introduction to Mobile Communications Engineering, José M. Hernando and F. Pérez-Fontán

Introduction to OFDM Receiver Design and Simulation, Y. J. Liu

An Introduction to Optical Wireless Mobile Communications, Harald Haas, Mohamed Sufyan Islim, Cheng Chen, and Hanaa Abumarshoud

Introduction to Radio Propagation for Fixed and Mobile Communications, John Doble

Introduction to Wireless Local Loop, Broadband and Narrowband, Systems, Second Edition, William Webb

IS-136 TDMA Technology, Economics, and Services, Lawrence Harte, Adrian Smith, and Charles A. Jacobs

Location Management and Routing in Mobile Wireless Networks, Amitava Mukherjee, Somprakash Bandyopadhyay, and Debashis Saha

LTE Air Interface Protocols, Mohammad T. Kawser

Metro Ethernet Services for LTE Backhaul, Roman Krzanowski

Mobile Data Communications Systems, Peter Wong and David Britland

Mobile IP Technology for M-Business, Mark Norris

Mobile Satellite Communications, Shingo Ohmori, Hiromitsu Wakana, and Seiichiro Kawase

Mobile Telecommunications Standards: GSM, UMTS, TETRA, and ERMES, Rudi Bekkers

Mobile-to-Mobile Wireless Channels, Alenka Zajić

Mobile Telecommunications: Standards, Regulation, and Applications, Rudi Bekkers and Jan Smits

Multiantenna Digital Radio Transmission, Massimiliano Martone

Multiantenna Wireless Communications Systems, Sergio Barbarossa

Multi-Gigabit Microwave and Millimeter-Wave Wireless Communications, Jonathan Wells

Multiuser Detection in CDMA Mobile Terminals, Piero Castoldi

OFDMA for Broadband Wireless Access, Slawomir Pietrzyk

Practical Wireless Data Modem Design, Jonathon Y. C. Cheah

The Practitioner's Guide to Cellular IoT, Cameron Kelly Coursey

Prime Codes with Applications to CDMA Optical and Wireless Networks, Guu-Chang Yang and Wing C. Kwong

Quantitative Analysis of Cognitive Radio and Network Performance, Preston Marshall

QoS in Integrated 3G Networks, Robert Lloyd-Evans

Radio Resource Management for Wireless Networks, Jens Zander and Seong-Lyun Kim

Radiowave Propagation and Antennas for Personal Communications, Third Edition, Kazimierz Siwiak and Yasaman Bahreini

RDS: The Radio Data System, Dietmar Kopitz and Bev Marks

Resource Allocation in Hierarchical Cellular Systems, Lauro Ortigoza-Guerrero and A. Hamid Aghvami

RF and Baseband Techniques for Software-Defined Radio, Peter B. Kenington

RF and Microwave Circuit Design for Wireless Communications, Lawrence E. Larson, editor

Sample Rate Conversion in Software Configurable Radios, Tim Hentschel

Signal Failure: The Rise and Fall of the Telecoms Industry, John Polden

Signal Processing Applications in CDMA Communications, Hui Liu

Signal Processing for RF Circuit Impairment Mitigation, Xinping Huang, Zhiwen Zhu, and Henry Leung

Smart Antenna Engineering, Ahmed El Zooghby

Software-Defined Radio for Engineers, Travis F. Collins, Robin Getz, Di Pu, and Alexander M. Wyglinski

Software Defined Radio for 3G, Paul Burns

Software Defined Radio: Theory and Practice, John M. Reyland

Spectrum Wars: The Rise of 5G and Beyond, Jennifer A. Manner

Spread Spectrum CDMA Systems for Wireless Communications, Savo G. Glisic and Branka Vucetic

Technical Foundations of the IoT, Boris Adryan, Dominik Obermaier, and Paul Fremantle

Technologies and Systems for Access and Transport Networks, Jan A. Audestad

Third-Generation and Wideband HF Radio Communications, Eric E. Johnson, Eric Koski, William N. Furman, Mark Jorgenson, and John Nieto

Third Generation Wireless Systems, Volume 1: Post-Shannon Signal Architectures, George M. Calhoun

Traffic Analysis and Design of Wireless IP Networks, Toni Janevski

Transmission Systems Design Handbook for Wireless Networks, Harvey Lehpamer

UMTS and Mobile Computing, Alexander Joseph Huber and Josef Franz Huber

Understanding Cellular Radio, William Webb

Understanding Digital PCS: The TDMA Standard, Cameron Kelly Coursey

Understanding WAP: Wireless Applications, Devices, and Services, Marcel van der Heijden and Marcus Taylor, editors

Universal Wireless Personal Communications, Ramjee Prasad

Virtualizing 5G and Beyond 5G Mobile Networks, Larry J. Horner, Kurt Tutschku, Andrea Fumagalli, and ShunmugaPriya Ramanathan

WCDMA: Towards IP Mobility and Mobile Internet, Tero Ojanperä and Ramjee Prasad, editors

Wi-Fi 6: Protocol and Network, Susinder R. Gulasekaran and Sundar G. Sankaran

Wireless Communications in Developing Countries: Cellular and Satellite Systems, Rachael E. Schwartz

Wireless Communications Evolution to 3G and Beyond, Saad Z. Asif

Wireless Intelligent Networking, Gerry Christensen, Paul G. Florack, and Robert Duncan

Wireless LAN Standards and Applications, Asunción Santamaría and Francisco J. López-Hernández, editors

Wireless Sensor and Ad Hoc Networks Under Diversified Network Scenarios, Subir Kumar Sarkar

Wireless Technician's Handbook, Second Edition, Andrew Miceli

For further information on these and other Artech House titles, including previously considered out-of-print books now available through our In-Print-Forever® (IPF®) program, contact:

Artech House
685 Canton Street
Norwood, MA 02062
Phone: 781-769-9750
Fax: 781-769-6334
e-mail: artech@artechhouse.com

Artech House
16 Sussex Street
London SW1V 4RW UK
Phone: +44 (0)20 7596-8750
Fax: +44 (0)20 7630-0166
e-mail: artech-uk@artechhouse.com

Find us on the World Wide Web at: www.artechhouse.com